IP SWITCHING

IP Switching
Protocols and Architectures

Christopher Y. Metz

McGraw-Hill
New York • San Francisco • Washington, D.C. • Auckland • Bogotá
Caracas • Lisbon • London • Madrid • Mexico City • Milan
Montreal • New Delhi • San Juan • Singapore
Sydney • Tokyo • Toronto

Library of Congress Cataloging-in-Publication Data

Metz, Christopher Y.
IP switching : protocols and architectures / Christopher Metz.
p. cm.
Includes bibliographical references and index.
ISBN 0-07-041953-1
1. TCP/IP (Computer network protocol) 2. Telecommunication—Switching systems. 3. Computer network protocols. 4. Computer network architectures. I. Title.
TK5105.585.M48 1998 004.6'2—dc21 98-38791
CIP

McGraw-Hill
A Division of The McGraw-Hill Companies

1 2 3 4 5 6 7 8 9 0 AGM/AGM 9 0 3 2 1 0 9 8

ISBN 0-07-041953-1

The sponsoring editor for this book was Simon Yates, the editing supervisor was Ruth W. Mannino, and the production supervisor was Clare Stanley. It was set in Times Roman by Don Feldman of McGraw-Hill's Professional Book Group composition unit.

Printed and bound by Quebecor/Martinsburg.

This book is printed on recycled, acid-free paper containing a minimum of 50% recycled, de-inked fiber.

This book is dedicated to my brothers—Patrick and Joshua—and my sisters—Leila, Amy, and Rebecca. They exhibit strength of character and love and compassion for those around them. As the oldest brother I will always be filled with pride and envy.

CONTENTS

ACKNOWLEDGMENTS

This book would not have been possible without the love and support of my wife Leonor. She provided gentle prodding and sustained encouragement over the course of this entire effort. Without typing a word, she is co-author in every sense of the word.

A special thanks go to the outstanding professionals at McGraw-Hill. Steven Elliot helped me decide on a topic. But it is Simon Yates and Ruth Mannino who deserve special recognition. They provided encouragement, ideas, guidance, and support throughout this entire effort. And they did it in such a way as to work within the confines of my personal and professional responsibilities. If you are considering authoring a professional computing book, there is no better group of professionals to work with than the first-class folks at McGraw-Hill.

I would also like to acknowledge a long list of individuals whom I have directly and indirectly come across throughout this effort. They have encouraged, facilitated and inspired my exploration of many of these interesting topics. Arun Viswanathan, Vijay Srinivasan, Yakov Rehkter, Brian Carpenter, Bruce Davies, Roch Guerin, John McQuillan, Joel Halpern, Paul Doolan, Hal Sandick, Steve Blake, stand out among those who have and/or continue to contribute to the body of knowledge in this area. At IBM, Nancy Feldman, Rick Boivie, Kishore Jotwani, Fred Allard, Patrick Droz, and Ton Engbersen provided support and assistance. Finally, a big thanks to my former manager at IBM, Bill Zuber, who always encouraged and supported my immersion into these new technology areas.

Part 1

Overview

CHAPTER 1

Introduction

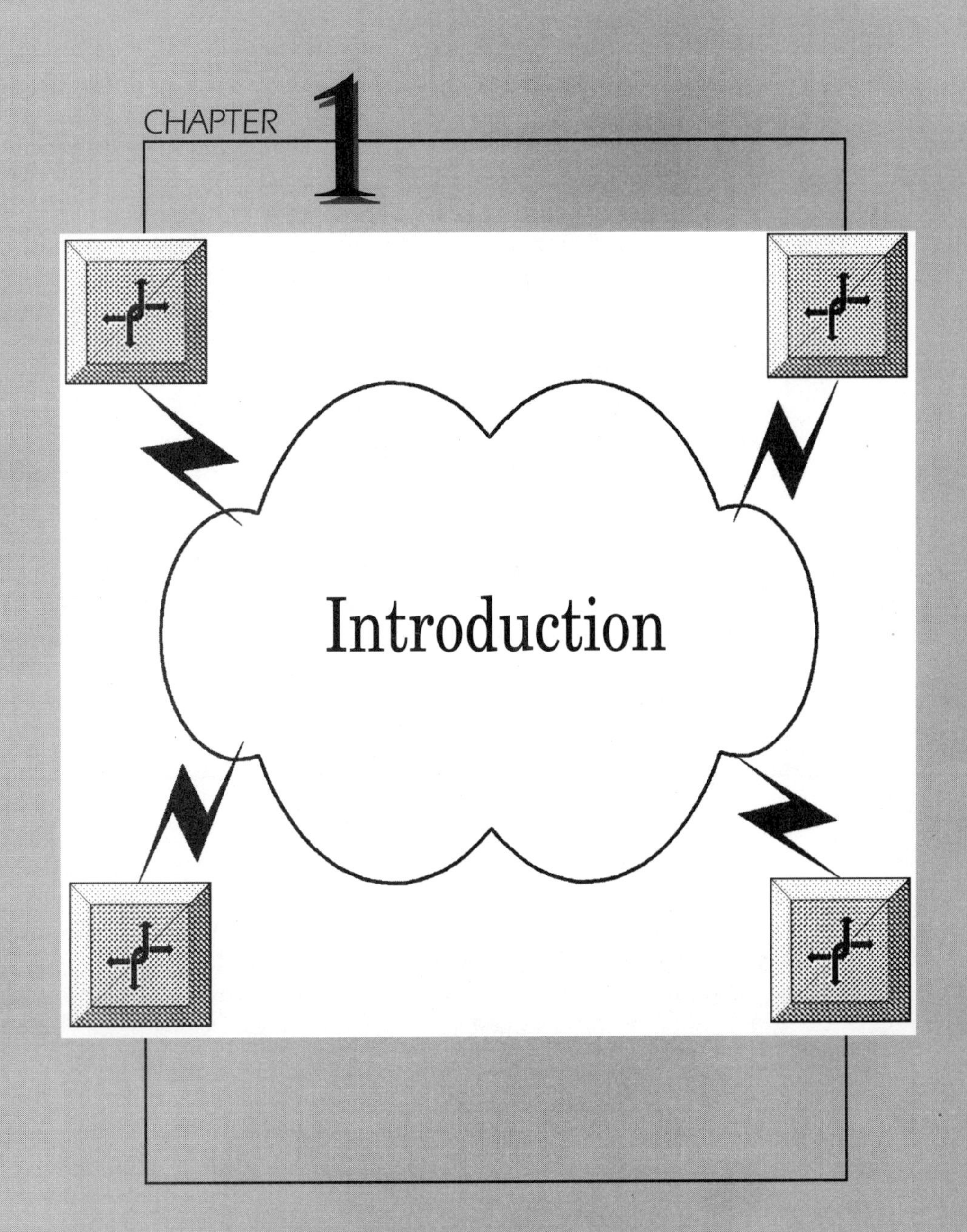

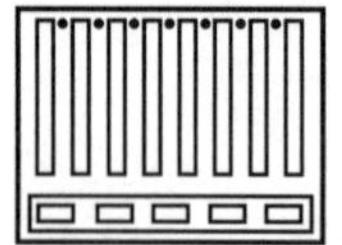

It has become clear to industry participants and observers (and even those folks just back from a multiyear journey into deep space) that the Internet has achieved the status of a phenomenon unlike that of any other technology introduced in this century or the last. Its reach, its scope, the press articles, the venture capitalists, America Online (AOL), AT&T, Microsoft, and many more have all been thrown into the mix. And all have been linked in one way or another with the Internet. It is the only technology that comes to mind that Joe Sixpack and the Fortune 1000 think about and use every day. It is perhaps one of the few entities in the world that treats every user with equal indifference. A CEO who wants to guarantee better service for a company has to build a special network. On the Internet it is fair access to everyone.

This is not to take away from some of the other technology developments of the last 150 years. Sure the telephone was important. It allowed us to communicate with a disembodied voice at the other end of a piece of wire. Yes, the automobile transformed our lives by allowing us to cover an order of magnitude more ground in a single day than what was possible sitting on one of God's creatures. The light bulb, the radio, and the television were all very important additions to improving our daily lot, extending our waking hours, informing us, and entertaining us. For the most part we have conquered that survival problem. Most people have enough food on the table, a warm bed to sleep in, and some place to go to in the morning and return from in the evening. We can sit back in a passive receive mode and take in whatever channel our fingers last pressed on the remote control.

The Internet has the potential to make things even better—or worse. On the upside, it is a tool that enables us to communicate with our human peers around the world. It provides us with access to an infinite amount of information stored on thousands and thousands of networked computers. Powerful search engines make the job of locating meaningful and useful information much quicker and simpler. The different cycles of a business transaction from initial casual interest to full-blown, whip-out-the-credit-I-want-it-now purchases can be performed over the Internet. Indeed, some envision the day when the Internet removes any reason to get out of bed in the morning. All business and some recreation can be conducted by propping oneself up on a couple of pillows and dialing in using a laptop or whatever computers they have in the future.

On the downside, the Internet can make us all electronic zombies, incapable of communicating directly with other human beings. Our only waking hours will be spent in front of a computer screen typing in URLs or pointing to another hyperlink. Most worrisome is the unrecoverable time wasted as a result of the "waiting for reply" effect, or WFRE (pronounced "wee-free"). WFRE occurs when the Internet becomes so congested and so slow that your browser enters an almost permanent state of waiting for reply. Sometimes it is just a matter of seconds; other times it can be several minutes. All of the time spent staring at the computer screen during WFRE might approach some alarming figures on the order of months or even years. The gist of the previous discussion is this:

1. The Internet is experiencing and will continue to experience tremendous growth.
2. It is equally popular with both residential and corporate end users. To the latter it could mean millions in new revenues.
3. Powerful and creative industry forces are at work seeking to exploit the Internet to their and their customers' advantage. This will undoubtedly generate more interest and hype in the Internet for both casual and serious use. Accompanying that will be a proportional growth in usage and traffic.
4. Internet bandwidth and capacity is lacking, resulting in erratic response times, unpredictable performance, and question marks about its ability to support future, high-bandwidth, delay-sensitive applications or even modest increases in residential bandwidth capabilities.

How did we reach such a precarious state? That question has several answers, none really definitive, and probably a few are left out. First, the Internet is really a victim of its own success. New subscribers are signing up every day and more and more people are turning their browsers to the Web looking for the next interesting Web site. Because access is the price of a local phone call plus a modest flat rate, there is no pricing defense to protect Internet resources from chronic browsers. Another deficiency is not necessarily a lack of networking resources but more a lack of server resources. Some servers or server farms cannot handle the load and frequency of connection requests brought on by a "flash" crowd (i.e., a large number of clients attempting to simultaneously access the same Web server). This could be viewed as an interim problem because server vendors will usually swoop in and sell the besieged content host new servers, load balancers, Web caches, etc.[1]

[1]Actually, there are some pretty neat Web server performance-boosting technologies such as proxy servers, Web caches, Transmission Control Protocol (TCP) connection routers, and distributed cache protocols that enable resources to be shared and distributed across the entire network.

Another problem is that some links just may not have enough bandwidth to support the offered traffic load. The partial remedy is to, of course, add more bandwidth, and new technologies such as wave division multiplexing (WDM) seem to promise an unlimited supply. All of these contribute to the extremely erratic performance of the Internet and its well-known clone, the corporate intranet. They are problems that are understood, and although some such as the pricing issue cannot be instituted overnight, there is at least the notion that a solution exists and can be used in the future. But perhaps the most basic problem with Internet performance and more generally Internet Protocol– (IP-) based networking perhaps lies in the fundamental process of IP routing or forwarding and the platforms that perform this function.

1.1 Early Routing

Initially there was routing. Boxes first called *gateways* (as in a gateway from one network to another) and later coined *routers* (not to be confused with the machine tool) connected different physical networks into one larger unified internetwork. Individual packets of information contained a destination network label that the router attempted to match with one of many entries in a table of destination networks that it knew about. Once a match was found, the router could direct the packet out the appropriate interface and wait for the next one to arrive. This relatively simple process is performed for each individual packet that arrives at the router. Even if a large number of packets with a common destination arrive consecutively, the router will still treat each packet in an individual manner.

Consider a first-generation router as illustrated in Fig. 1-1. It consists of a single general-purpose central processor and multiple media interface adapter cards all connected to each other by a shared bus. The processor is responsible for running routing protocols and maintaining a forwarding table of next-hop routers that the packet can be sent to. Packets entering the router traveled over the bus and up into the processor where a standard routing table lookup and next-hop determination were performed. The packet then traveled over the shared bus to the appropriate output interface. The performance of this system was gated by the speed of the bus and the processing capacity of the central

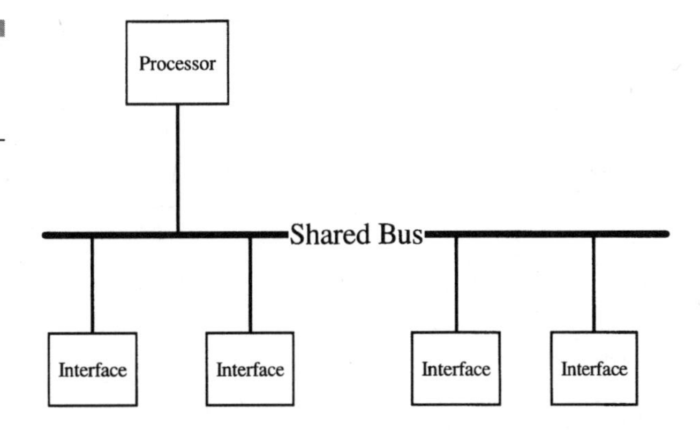

Figure 1-1 First-generation router.

processor. In addition each packet was required to travel over the bus twice. Because early IP networks were software-based and more concerned with connectivity and reachability than performance, this design was satisfactory.

But the Internet grew, first at a modest rate and then more rapidly. More people and more sites resulted in more networks, which led to larger routing tables and slower lookup times. Coupled with a proportional growth in user traffic, this necessitated the introduction and deployment of faster router technologies. One common enhancement involved distributing the forwarding computations out to the media interface adapters. A portion or all of the routing table could be cached in memory on the input adapter. This enabled the input adapter to perform the forwarding computation and direct the packet over the bus to the appropriate output adapter without any central processor intervention. The performance of this model was still limited by the speed of the bus and to some degree by the time required to sort through a large routing table during lookup. Another improvement in router technology replaced the bus with a switch. Thus all input and output ports were connected to each other by a nonblocking switch fabric. This model is illustrated in Fig. 1-2. Despite improvements in internal design and efficiency, the per-packet processing demands placed on the router itself as well as the bursty unpredictable nature of IP traffic serviced through first-in, first-out (FIFO) queues kept the price high, throughput gains minimal, and performance erratic.

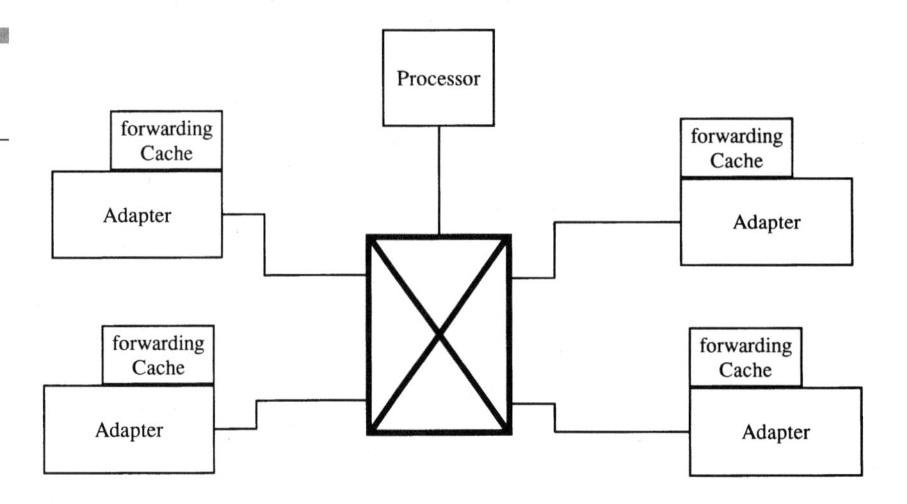

Figure 1-2 Router with switch fabric.

1.2 ATM and IP

Parallel to the growth and emergence of the Internet and faster router technologies was significant interest in the area of switching. Broadband Integrated Services Digital Network (ISDN) was envisaged as a technique to enable real-time communications between communicating endpoints. It uses a simple transport technique called Asynchronous Transfer Mode, or ATM. ATM can take any information stream, data, voice, or video, segment it into small packets called *cells,* and then transport the cells over preestablished paths called *virtual connections.*[2] Because of its ability to support data, voice, and video with quality of service over a number of different high-bandwidth technologies, ATM was viewed as the premier switching technology and certainly one that attracted the most attention.

ATM is different from IP routing in several respects. It is connection oriented, not connectionless. An end-to-end connection must be established between communicating parties before any data can be sent. ATM requires that the connections be manually configured or dynamically established using a signaling protocol (very similar to dialing up a friend on the phone). Another difference is that ATM does not perform a per-

[2]Cells are only 53 bytes in length and are not big enough to carry a destination address. Therefore it is not possible to forward cells based on a destination address as is the case with IP and packets.

packet routing decision at each intermediate node (switch).[3] For ATM the path through the network between two parties is computed prior to any data exchange and is fixed for the duration of the connection. At connection establishment, each intermediate ATM switch assigns an identifier or label that is unique to the switch, the connection, and the input/output port on the switch. This accomplishes two things: Resources along a fixed path of ATM switches are reserved for a specific connection, and the individual ATM switches build essentially a forwarding table that is local and contains only entries about the active connections that will traverse the switch. Contrast this with a router that must maintain a networkwide routing table containing all possible destinations including those that perhaps the router will never forward packets to.

The process of forwarding a cell through an ATM switch is similar to forwarding an IP packet through a router. Both contain information in the header that is used to index a table that lists the interface the packet or cell should be directed to for outbound transmission. ATM is faster, much faster than traditional software-based routers because the labels in a cell header are fixed length, the connection table that is indexed during lookup is typically much smaller than the IP forwarding tables maintained in routers, and the forwarding itself is done in specialized hardware. Thus ATM switching enjoys a distinct and noticeable performance delta over software-based IP routing because hardware can perform a simple task much faster than software can perform a complex task.[4]

Table 1-1 is not an attempt to reopen a useless debate on the merits of one technology over the other but rather points out several fundamental differences between IP and ATM. Once there were heated debates about the pros and cons of IP versus ATM as the next-generation network technology. Those debates are over, and it is clear that IP will be present in all networks. Of course one of the beauties of IP is that it can run on top of just about any data link technology, including ATM. Rather than offering it up as an alternative to IP, ATM should be viewed as a very powerful data link technology that IP can and should exploit.

The first attempts to integrate IP and ATM were tepid at best. The connectionless hop-by-hop forwarding paradigm of IP was unaltered.

---

[3]IP can do this because each packet carries the destination network address in the header.

[4]At some point advances in silicon will place the price and performance of cell switching and packet routing on equal ground.

TABLE 1-1

Comparison of IP and ATM

Attribute	IP	ATM
Orientation	Connection-less	Connection-oriented
Packet size	Variable	Fixed (53 bytes)
QoS	No	Yes
Information	Data	Data, voice, and video
Path determination	Per packet	Connection setup
Forwarding state	All possible networks	Local active transit connections
Forward based	Longest match address prefix	Fixed-length label
Signaling	No	Yes

In fact IP had no idea that it was running over a network capable of reserving bandwidth and bounding delay. From the ATM perspective, a set of separate addressing, routing, and signaling protocols was developed exclusively for ATM that was to operate with no modification. What process could bring these two disparate technologies together? The only real answer to the question was the address resolution process. To successfully communicate with a peer on the same subnet, a source IP host needs to resolve the destination IP address with the corresponding layer-2 address. This enables the source IP host to address packets to the destination IP address and then encapsulate the IP packet in a layer-2 frame with the appropriate layer-2 destination address. With ATM though, the destination ATM address is used only to forward the Signaling Virtual Channel (SVC) setup request prior to connection establishment. Nevertheless the source IP host still needs to resolve the destination IP address with an ATM address so that it can initiate an ATM SVC to the destination. This basic approach is illustrated in Fig. 1-3. The ATM-attached source host queries the address resolution server for the appropriate ATM address of the destination host. The address resolution server returns the ATM address to the source, and the source then sets up an SVC to the destination.

The Internet Engineering Task Force (IETF) first suggested this approach in RFC1577, Classical IP and ARP over ATM. In RFC1577 an ATM Address Resolution Protocol (ATMARP) server was defined that maintained a table of IP and ATM addresses. The ATM Forum later generalized this approach to resolve Media Access Control (MAC) addresses

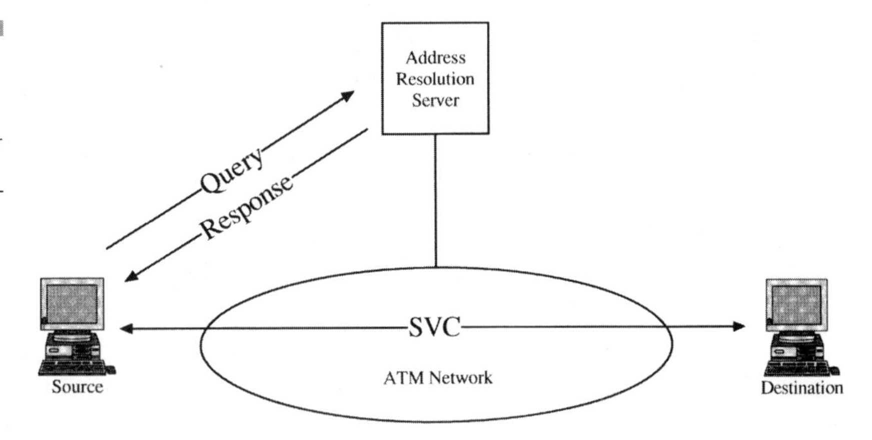

Figure 1-3
Address resolution server approach for data networking over ATM.

with ATM addresses and added support for broadcast as well. The intent was to develop a system by which local area network (LAN) applications could run over an ATM network without any modifications. Their effort was dubbed LAN emulation (LANE) and is quite popular because it enables multiprotocol LAN applications to run transparently over emulated and legacy LANs. Both Classical IP and LANE are similar in that they completely separate the higher-layer IP and LAN functions from the underlying services supported by ATM.

Another area of similarity between the two models is the scope of ATM connectivity. Classical IP restricted ATM communications to only those hosts (or routers) on the same logical IP subnet (LIS). Even if two hosts on different subnets were connected to the same ATM network, they were still required to forward their packets through a router. This introduced the usual overhead and delay associated with hop-by-hop routing. LANE had a similar restriction in that two hosts on the same ATM network but on different logical IP networks could not communicate directly using ATM. They too were required to use a router to exchange information.

The next evolutionary step was to allow two hosts on different networks to communicate with each other through ATM directly. This required a relaxation of the existing concatenated network model that required all intersubnet traffic to pass through routers. The approach was to continue to support traditional hop-by-hop routing but also provide a mechanism by which a source IP device could establish a direct ATM connection to a destination IP device on another network. Therefore the network could provide a path for normal hop-by-hop routing for best-effort, low-volume traffic flows and a second high-performance,

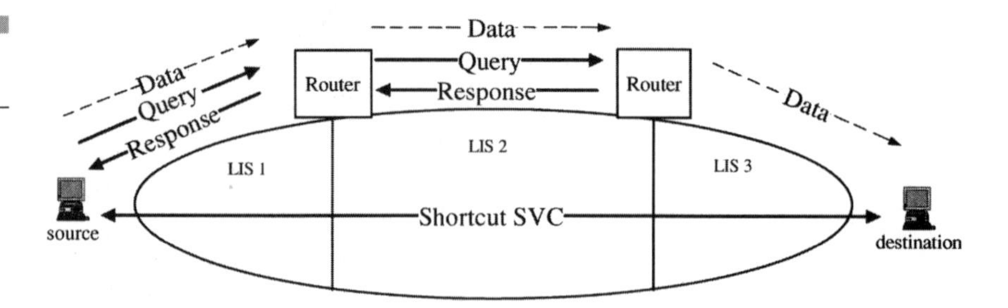

Figure 1-4
Shortcut routing.

high-bandwidth switched path for high-volume, long-duration traffic flows.

This technique was implemented first in the IETF's Next Hop Resolution Protocol (NHRP), which later on was incorporated into the ATM Forum's Multiprotocol over ATM (MPOA). Both extended the address resolution server mechanisms first introduced in classical IP and LANE to resolve addresses beyond the subnet boundaries. Thus a new type of high-performance routing called *cut-through* or *shortcut* routing was born. It augmented traditional IP routing with a separate "pannetwork" address resolution service along with the underlying ATM signaling and routing protocols necessary to manage dynamic SVCs. The concept of shortcut routing is illustrated in Fig. 1-4.

MPOA externalized the distributed forwarding concept that was first implemented in standalone second-generation routers. The central processor is now called a route server and the forwarding function is placed in standalone edge devices connected to the edge of the ATM network. The shared bus in this distributed router model is the ATM network itself. MPOA called it the virtual router, and it served to move packets at gigabit rates between subnets. It seemed like a good idea and perhaps something that might catch on, particularly in those campus networks where a collapsed backbone router was a performance bottleneck.

1.3 IP Switching

Something extraordinary occurred in the spring of 1996. A small startup company from California named Ipsilon introduced a revolutionary idea called *IP switching*. The concept was startling in its simplicity and elegance. Take an IP router processor, strap it onto an ATM switch, remove all of the switch's ATM Forum signaling and routing protocols, and call it an IP switch. Give the IP router processor control of the attached ATM switch and allow the IP switch as a whole to run normal IP routing protocols and perform normal hop-by-hop IP forwarding. When a high-volume, long-duration flow is detected, have the IP router processor tell its upstream neighbor to assign a new virtual path identifier/virtual connection identifier (VPI/VCI) label to the cells of the flow and then update the connection table in the ATM switch accordingly. Once this independent process occurs between every IP switch pair along the routed path, it becomes a simple task for each IP switch to splice together the appropriate upstream and downstream entries in the switch connection table. What started out initially as a hop-by-hop routed traffic flow ends up as an ATM-switched traffic flow. The basic concept is illustrated in Fig. 1-5.

The notion that routed traffic flows could be dynamically switched through an ATM switch fabric without all that messy and complicated ATM Forum signaling, routing, and MPOA stuff turned the industry almost upside down—at least from the hype perspective. Crowds mobbed the Ipsilon booth at the trade shows, and their presentations on the topic at conferences were packed to the rafters. The trade press wrote favorably on this innovative approach. Other vendors jumped on the Ipsilon bandwagon and announced support for IP switching.

And why not? There was no question that Ipsilon was staffed by extremely talented IP and ATM engineers and developers. Tom Lyon, the founder of Ipsilon, developed the ATM Adaptation Layer (AAL5). Peter Newman, a noted engineer and developer of ATM switch technology, joined up. Robert Hinden, influential in the development of IPv6, was a member of the technical staff as were others who came from industry heavyweights such as Cisco. Another reason why IP switching was so attractive was because it was just that—IP switching. It did not pretend to or try to support anything other than regular old IP. In fact an IP switch was just a router running normal routing protocols. It just happened to have an ATM switch that served as the forwarding engine. IP switching also exploited the universal belief that raw ATM cell

Figure 1-5
Ipsilon IP switching.

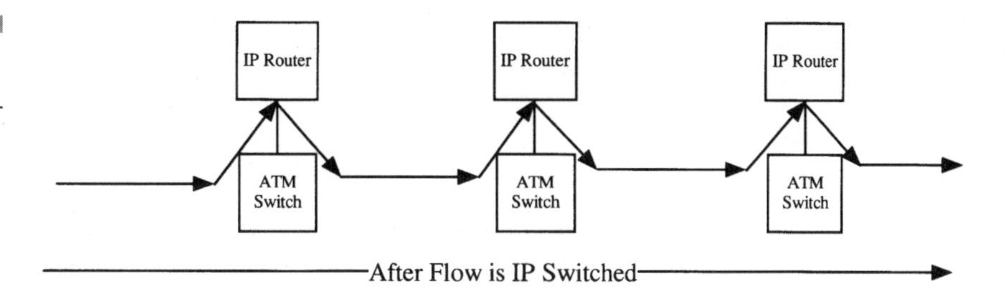

switching was a very fast and efficient means of forwarding large volumes of data—much faster than what any router on the market at the time could do. In this industry when raw speed talks, people listen.

To the delight of the IP purists, Ipsilon went out of its way to disassociate IP switching from anything resembling ATM. First, the company made a point of emphasizing the sheer complexity of ATM Forum protocols such as LANE and Q.2931 signaling versus their lightweight flow setup and switch management protocols (Ipsilon Flow Management Protocol, or IFMP, and General Switch Management Protocol, or GSMP). ATM's powerful traffic management capabilities were summarily dismissed, as were QoS (quality of service) and ATM flow control. As amusing (and prophetically accurate) as some of these pronouncements were, they also exposed a chord of latent concern and reluctance on the part of many about going down the path to ATM. Many in the industry certainly felt that more bandwidth would be needed and that ATM was a viable technology. However, many were equally concerned about the complexity of the ATM Forum suite of protocols and the associated implementation costs, including network management, training, and so on. There were also no real applications at the time that could exploit ATM exclusives such as QoS. It might also be said that the ATM Forum itself did not help the image of ATM technology and opened the door for upstart ideas like IP switching. Recall at the time that the ATM Forum's solution for IP routing over ATM, MPOA, was still under development. In fact it was sort of bogged down in the typical standards-body-abstract-idea-acronym-creation phase that made it difficult to present a clear and simple explanation about how it could work. A look at the draft specification reveals the terms Internet Address Subgroup (IASG), IASG Coordination Functional Group (ICFG), Edge Device Functional Group (EDFG), and several other functional groups. It did not make for easy reading.[5]

1.4 IP Switching for Routers

Given the focus on ATM switch technology and the attention that the Internet and IP was beginning to receive, it is surprising that no one

[5]To the MPOA working group's credit, they vastly simplified the specification later in 1996. The final MPOA specification released in July 1997 is quite accessible and clearly spells out the functions and behaviors of just two components, the MPOA client and the MPOA server.

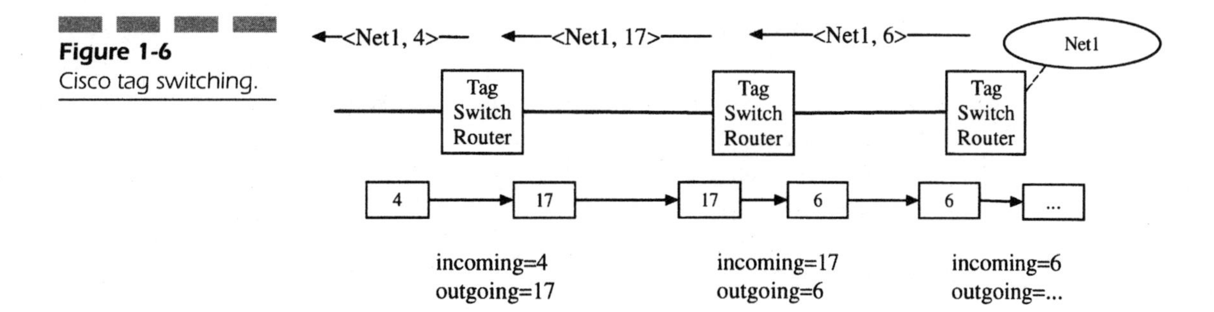

Figure 1-6
Cisco tag switching.

thought of IP switching earlier.[6] Nevertheless that did not prevent the industry heavyweights from developing similar technologies that merged IP routing and switching into a single integrated solution. In the fall of 1996, Cisco Systems announced tag switching. Tag switching generalizes the notion of IP switching to include packets. It was conceived to improve performance in core router networks by enabling routers to use a label-swap forwarding mechanism rather than the traditional longest-match address lookup. The company felt that it was more efficient to label-swap packets and cells based on a tag than it was to extract a per-packet destination address and perform a longest-match address lookup in a routing table. And by separating the network-layer routing process from the actual forwarding mechanism, any number of different layer-3 services could be associated with a simple and fast-forwarding paradigm, label swapping. It works by first distributing a set of tags based on routing table entries to the tag switch routers in the network and then forwarding the traffic through the network by performing a label-swapping operation on the tag contained in each packet. The concept of tag switching is illustrated in Fig. 1-6.

Tag switching differs from IP switching in that it generates and distributes tags based on the presence of control traffic and more specifically by routing protocol updates.[7] Tag switching builds a switched path to each destination that appears in the routing table of a tag switch

[6]Actually Japanese researchers introduced the idea of a Cell Switch Router (CSR) several years earlier that performed connectionless routing and connection-oriented cut-through switching. It was dubbed Conventional IP over ATM and did not receive much attention, probably because mainstream efforts were focused on Classical IP and LANE. The CSR concept will be covered in more detail in Chap. 10.

[7]Other forms of control traffic such as the Resource Reservation Setup Protocol (RSVP) can distribute tags. These techniques will be covered in Chap. 11.

router and *all* data destined for a particular destination network is placed on the same switched path. Contrast this approach with IP switching, which builds a switched path based on the arrival of the actual data flow itself and only for that data flow. Tag switching can be classified as topology- or control-driven, whereas IP switching is flow- or data-driven.

Almost immediately following the announcement of tag switching from Cisco, IBM proposed a solution with the less-pedestrian sounding name of ARIS. ARIS stands for Aggregate Route-Based IP Switching and although it is similar in concept to tag switching, there are several distinct differences. Some are definite enhancements to topology-driven IP switching in general, and others can be viewed as just different methods for accomplishing the same thing. ARIS first introduced the notion of multipoint-to-point switched path. In a routed environment the forwarding path from many ingress sources to a common egress router follows a hop-by-hop tree structure. ARIS maps the layer-3 forwarding tree onto a layer-2 multipoint-to-point tree that is rooted at the egress router and branches out to all ingress routers. All traffic flowing to a destination(s) behind the egress router will ride "up the branches" of the layer-2 switched path toward the egress router. This concept is illustrated in Fig. 1-7.

ARIS enables the establishment of multipoint-to-point switched paths by permitting the process of virtual connection (VC) merging to be performed at each branch point. VC merging is the process in which cells from multiple upstream VCs are merged onto a single downstream VC.

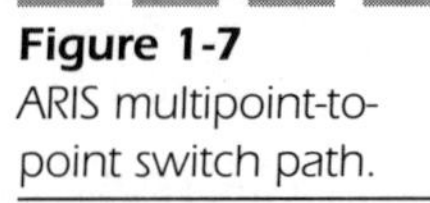

Figure 1-7
ARIS multipoint-to-point switch path.

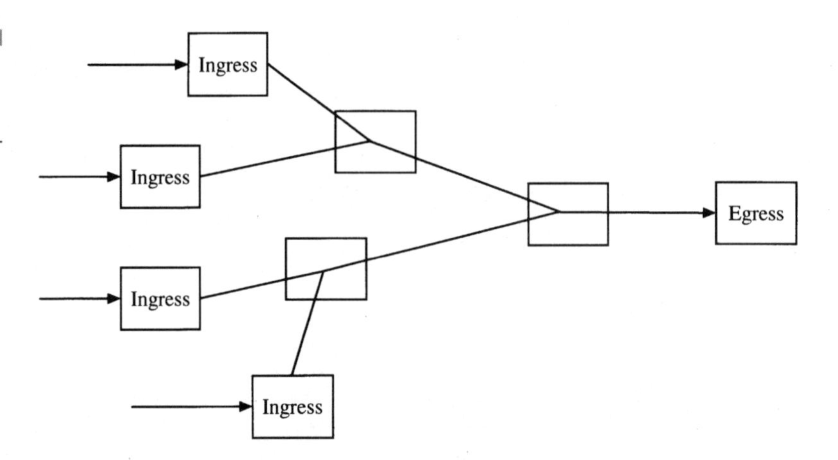

Techniques such as per-VC buffering and queuing are needed at each merge point to maintain frame integrity and prevent cell interleaving on the new downstream VC. A network supporting multipoint-to-point switched paths will consume fewer VC resources while enabling all traffic to be switched. This provides scalability, and VC merging eventually was incorporated into tag switching and the follow-on standards.[8]

Other minor differences between tag switching and ARIS involve loop prevention, local versus egress label distribution, extent of egress aggregation, and the use of TCP or User Datagram Protocol (UDP) to transport signaling messages. But fundamentally both solutions form switched paths based on the existing IP network topology and switch all traffic to particular destinations.

1.5 An IP Switching Standard

Other solutions and techniques based on the use of layer-2 switching to improve IP routing performance emerged. Some came with names cooked up by the marketing departments of the specific vendor: 3Com FastIP, Cascade IP Navigator, Cabletron SecureFast, etc. Others were research efforts or ideas that will probably be best remembered for their clever and exotic sounding acronyms: SITA, PLASMA, and IPSOFACTO. Incorporating both IP support and layer-2 switching in all of these and the aforementioned technologies created a new class of solutions referred to collectively as IP switching. Unfortunately Ipsilon did not trademark that name, so it has come to be associated with not just a single vendor's solution but rather with a whole suite of different technologies that combine IP routing and switching over a single network infrastructure.

By the end of 1996, a working group was formed in the IETF to attempt to standardize an integrated routing and switching solution called Multiprotocol Label Switching (MPLS). The major contributors to this working group are engineers and developers from many of the vendors who were or are in the process of building IP switching solutions.

[8]The ATM Forum is also examining enhancements to User-to-Network Interface (UNI) and PNNI signaling and routing to support multipoint-to-point signaling.

Ipsilon was conspicuously absent from this group, referring to it as "the Cisco Tag Switching Debating Society." Many of the techniques first described in the tag switching and ARIS proposals have been combined with ideas from others to form an initial MPLS frame, architecture, and Label Distribution Protocol (LDP) specification.[9] MPLS is viewed as a solution that will help Internet Service Providers (ISPs) deploy IP routing services in a more scalable and controlled fashion over a frame relay and ATM infrastructure. And with or without a switched infrastructure, the use of labels to forward packets through the network creates many interesting possibilities for directing traffic flows through specific links and nodes. It is perhaps traffic engineering more than performance or scalability that holds the most promise for MPLS.

A number of different techniques and protocols now exist that make it possible to exploit hardware-based switching as a means of improving IP routing performance. The predominant hardware-based switching technology was and continues to be ATM, and so all IP switching solutions that have been introduced to date make use of ATM switching. The initial efforts such as Classical IP and LANE solved the IP over ATM connectivity problem. Recent efforts such as MPOA and MPLS attempt to exploit ATM to provide better performance and reduce network transit time. Additionally the inherent existence of both a default routed path and a separate layer-2 switched path presents a useful set of options to campus and ISP traffic engineers. It is now quite feasible to configure the network to route best-effort traffic and switch high-priority traffic.

What is the price that one must pay for an IP switching system? At the risk of punting on the response, it depends. It may be as simple as providing new software for an installed network of ATM switches. Maybe all that one needs is a route server and a new set of edge devices, or perhaps one may even need to upgrade the entire backbone infrastructure, which could be in the campus or the ISP core. In practice it is not something that will extend out to the desktop because by default that is Ethernet and will remain so. IP switching is really a backbone solution, not an end-to-end solution. One needs to weigh the current backbone costs and performance (could be good, could be bad) against projected future traffic loads and application requirements.

[9]MPLS will be described in detail in Chap. 15.

1.6 Conclusion

Do we even need IP switching? It now seems quite possible to move large volumes of packets between subnets at wire-speed using a new family of hardware devices called layer-3 switches. Layer-3 switches and their ISP-scale cousin, the gigabit router, are simply much faster routers that, like IP switching, have introduced specialized hardware into the forwarding path. In this case it is not an ATM switch but specially designed application specific integrated circuits (ASICS) programmed to do layer-3 routing. It is routing business as usual, and one does not need any ATM Forum or specialized control protocols to cache the forwarding path over an ATM VC. Hardware-based routing could relegate ATM and specific IP switching solutions to niche areas in the Internet and corporate intranets.

But one should keep in mind several points based on past history and future trends. First, bandwidth alone is not the solution. It helps and is the most important and valuable resource to have in any network. But historically the demand for bandwidth has always outpaced the available capacity. That trend may not change for the foreseeable future or ever, and if that is the case, network providers may become selective about which traffic they allow on their networks and how they handle it. One can also imagine network providers becoming more active in providing a better service for those active traffic flows transiting their networks.[10] Assuming an infrastructure with sufficient bandwidth and active queue management, network providers will need either a dynamic routing protocol that factors in offered traffic load, existing network performance, and topology conditions when selecting a path or a mechanism to configure a nondefault route for specific traffic flows. Current dynamic IP routing protocols do not select routes based on traffic or performance metrics, and configuring nondefault paths requires per-router filters and static routes. However, an IP switching solution in the campus may employ ATM in the backbone and thus utilize Private Network-to-Network Interface (PNNI) as a means to compute and then pin routes based on bandwidth, QoS metrics, and reachability. In the ISP space, MPLS offers a simple way to configure and generate nondefault routes for selected traffic flows.

[10]This presumes that the subscriber or customer is paying an extra fee for a predictable level of performance that the network provider has agreed to deliver.

The other point to make here regards future trends. Next-generation IP networks will only be useful if they can support nondefault services such as security, classes-of-service (COS), QoS, multicast, service-level agreements (SLA), etc. IP switching does involve some extra work, but the trade-off is that connectionless IP routed traffic is placed on what is essentially a connection-oriented layer-2 switched path. With this comes the benefits of connection-oriented networking in general such as high bandwidth, traffic policing and shaping, flow separation, QoS, security, nondefault routing, and so on. The point to remember here is that new protocols and developments will allow network providers to offer new IP services. They will offer these services to attract new subscribers and customers or, in the case of an intranet, to make the end-user population more productive. IP switching is one technique that will enable network providers to more efficiently and cost-effectively support many of these essential new services.

The rest of this book will cover in some detail the basics of IP, LAN and wide area network (WAN) switching, and the suite of different IP switching solutions that have developed and been introduced to date. It is curious although not really surprising that the fundamental problems and solutions addressed by IP switching—fast-forwarding lookups, increased bandwidth, lower latency, QoS, nondefault routing, and so on—continue to be worked on within the industry. We may not ever be satisfied, but that is a very positive attitude because it will push us to discover new ways of improving what we already have.

CHAPTER 2

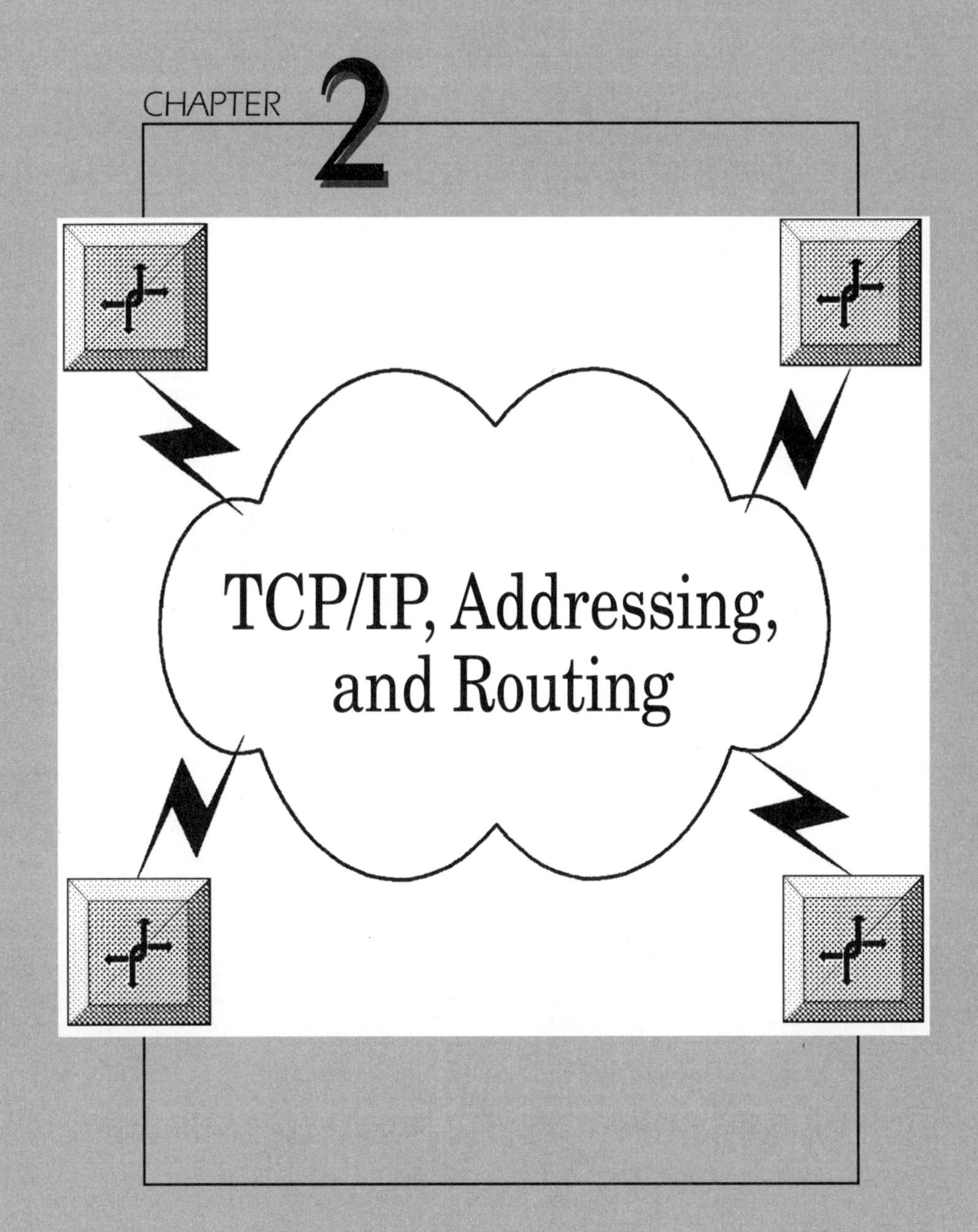

TCP/IP, Addressing, and Routing

Because this book is primarily about IP and switching, it might be helpful to review the basic suite of Transmission Control Protocol/Internet Protocol (TCP/IP) protocols. TCP/IP is in actuality a suite of different protocols that enable computers and end users to communicate with each other over a network of arbitrary size and over different media. This chapter provides a brief overview of the TCP/IP architecture, the addressing structure, and the operation of IP routing.

2.1 TCP/IP History

The history of TCP/IP and more generally the history of internetworking can be traced back to the 1960s. A couple of research papers by Licklider and Clark[1] and Klienrock,[2] respectively, described the ideas behind a "Galactic Network" and packet-switching theory. But it was the Cold War that motivated the military and scientific communities to develop a communications infrastructure that would be resilient to nuclear attack. The notion of using a packet-switched network with destination-addressed packets of data (called datagrams) was described in a paper by Paul Baran in 1964.[3] If some number of communications links or nodes were knocked out during the course of a nuclear exchange, the packet could be rerouted over an alternate and available path and eventually delivered to its destination.

In the late 1960s, the Advanced Research Projects Agency (ARPA) funded research and development in this area, and in 1969 the first ARPANET node called an Interface Message Processor (IMP) was installed and made operational. By the end of 1969, the ARPANET consisted of four nodes.

In addition to the invention of Ethernet in 1973, the 1970s were highlighted by the development of the first internetworking protocols. The first ARPANET host-to-host protocol was the Network Control Program (NCP). This gave way to further research on a suite of internetworking protocols and to what would eventually be called TCP/IP. A first cut, called TCP, which combined the end-to-end transport functions of TCP

---

[1]J. C. R. Licklider and W. Clark, "On-Line Man-Computer Communications," Aug. 1962.

[2]L. Kleinrock, "Information Flow in Large Communications Nets," *RLE Quarterly Progress Report,* July 1961.

[3]P. Baran, "On Distributed Communications Networks," RM-3420-PR, Rand Corp., Aug. 1964.

with the packet forwarding and addressing functions of IP, was developed by Kahn and Cerf in 1974.[4] Later this model was reorganized, and the basic functions of TCP and IP as we know them today were split into separate protocols.

It is interesting to note that the underlying principles of TCP/IP and internetworking that were first introduced so many years ago have for the most part remained unchanged. The basic principles that guided the development of TCP/IP and have allowed it to flourish as it does to this day are as follows:

- ✔ Each network is unique and self-sustaining. It can operate with or without attachment to another network. A collection of one or more networks interconnected by devices called gateways (routers) is called an internetwork.
- ✔ Communication is best-effort. Packets will be retransmitted by the source if they are not acknowledged by the destination.
- ✔ Routers (initially called gateways) interconnect networks. No flow-specific state is maintained in the router.
- ✔ There is no centralized control. The operation of the network is not dependent on any centralized controlling entity.

Given these principles and the flexibility to accommodate different applications over just about any media, the progress in TCP/IP development accelerated through the 1980s and 1990s to where it stands today. Some important dates and milestones (but probably not all) are noted below:

- ✔ *1983.* Hosts migrate from NCP to TCP/IP. UNIX 4.2BSD is released, which contains TCP/IP
- ✔ *1984.* Domain Name Service (DNS) invented.
- ✔ *1986.* NSFNET is formed; it serves along side the ARPANET as a backbone for the Internet
- ✔ *1988.* Famous paper[5] on TCP/IP congestion control and avoidance is published by Van Jacobson after a series of Internet congestion collapses. It describes the fundamentals of TCP slow-start and congestion control.

---

[4] R. E. Kahn and V. G. Cerf, "A Protocol for Packet Network Interconnection," *IEEE Transactions on Communications,* Tech 5, May 1974.

[5] V. Jacobson, "Congestion Avoidance and Control," *ACM Sigcomm Proceedings,* 1988.

- *1989.* First Request for Comments (RFC) on the Open Shortest Path First (OSPF) routing protocol is released.
- *1991.* ATM Forum is formed. First members are Northern Telecom, Sun, Digital Equipment Corporation (DEC), and Sprint.
- *1992.* World Wide Web (WWW) invented by Tim Berners-Lee begins to emerge. IPng (IP Next Generation) working group (WG) is formed to study alternatives for the next generation of IP
- *1993*-1994. Ethernet switching in the LAN and frame relay for WAN begin to take off. First notion of fusing IP routing with ATM switching is introduced by Japanese researchers.
- *1995.* Popularity, promise, and potential of ATM as *the* next-generation network technology reaches its zenith
- *1996.* IP switching is added to the network lexicon when Ipsilon announces a new product and supporting protocols. Cisco and IBM later chime in with more scalable solutions for switching IP packets through a network of hybrid switch-routers. Others follow with similar schemes.
- *1997.* Formation of the MPLS WG in the IETF provides a formal setting for vendors and their adherents to argue the merits of their respective approaches and to address the challenge of integrating network-layer routing with layer-2 switching. A new device, called a *layer-3 switch,* which performs very fast routing in hardware, is introduced by established veterans and brash industry start-ups.

With all due respect to Systems Network Architecture (SNA), Novell Internet Protocol Exchange (IPX), Appletalk, Decnet, etc., and their believers, TCP/IP is the dominant internetworking protocol. Much continues to happen with TCP/IP, and techniques to make it faster, more reliable, scalable, and better will continue to move forward.

2.2 TCP/IP Architecture and Components

Like other networking protocols such as SNA and Open Systems Interconnection (OSI), the architecture and functions of TCP/IP can be most easily explained using a layered model. But rather than use seven dif-

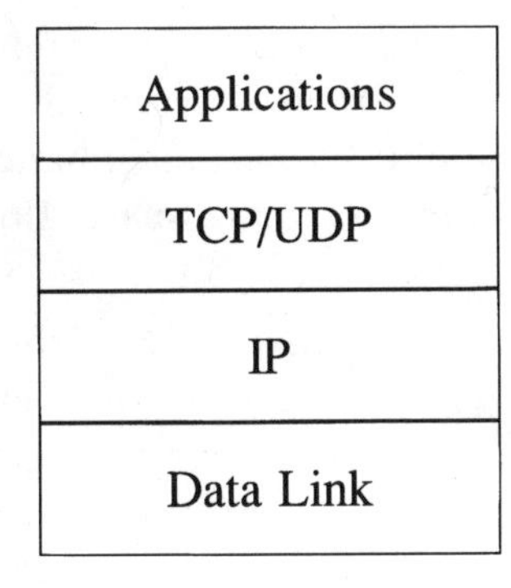

Figure 2-1
TCP/IP architecture.

ferent layers, the TCP/IP protocol architecture only really defines four, as shown in Fig. 2-1 and listed below:

- *Data link.* The data link layer defines the physical connectivity to a specific media and format of the information frames that are sent and received over this media. Examples of data link technologies that are supported by TCP/IP include Ethernet (any speed), ATM, token-ring, Fiber Distributed Data Interface (FDDI), frame relay, etc. The beauty of TCP/IP is that it can flow over just about *any* physical network.
- *IP.* The IP layer forwards packets of data from a source to a destination. Each packet contains the IP address of the destination, and IP uses this information to direct the packet toward the destination.

  IP runs in all hosts (source and destination) as well as in the packet switching/forwarding devices called routers, which are defined below. The IP layer is connectionless, which means that a path (or route) through the network to the destination is not established before any data starts flowing. Indeed, it is theoretically possible for every packet to take a different route. In addition, IP does not guarantee that the packets will arrive in the right sequence or that they will even reach the destination. The designers of IP wanted it this way so that no additional per-connection or per-flow state would need to be maintained in the network. It should also be noted that although there are multiple transport and application functions above the layer and multiple data link technologies below it, the singular component of IP is the point of convergence for TCP/IP and the Internet at large.

There are currently two versions of IP: IPv4 and IPv6. IPv4 is the version that is in use today. IPv6, otherwise known as IPng, is new and contains many simplifications and enhancements, the most significant being the larger address space. IPv6 will be reviewed in the next chapter.

- *TCP/UDP.* The transport layer runs on top of the IP layer and consists of two protocols. TCP provides a connection-oriented and reliable transport service, and UDP provides a connectionless and unreliable transport service between a source and destination. TCP and UDP run on host systems and provide their respective services to different applications.
- *Applications.* Many different applications utilize the underlying TCP/IP services. Telnet for terminal emulation, the File Transport Protocol (FTP) for file transfer, Hypertext Transfer Protocol (HTTP) for Web Browsing, and Simple Mail Transfer Protocol (SMTP) for email are just several well-known examples.

In addition to the protocol stack, the other two elementary components of TCP/IP are the packet and the router. The packet is simple to define and is illustrated in Fig. 2-2. It consists of a data payload followed by a transport header, then an IP header, and finally a data link header. The data link header is used to direct a packet from a source to the destination when that packet is traveling on a specific data link media (e.g., Ethernet). The contents of the data link header will change if the packet travels over different data link technologies that exist in the path between a source and destination. Consider as an example a path between a source and destination consisting of an Ethernet LAN, a frame relay Permanent Virtual Circuit (PVC), and then a token-ring LAN. The transport header may consist of a TCP header, a UDP header, or one may be omitted altogether if the application itself uses its own transport services rather than those provided by either TCP or UDP. The IP header is of course mandatory because it contains the address of the packet's eventual destination.

The router is a device that forwards packets from one network to another. Each interface on a router connects to a separate network.

Figure 2-2
Simple IP packet.

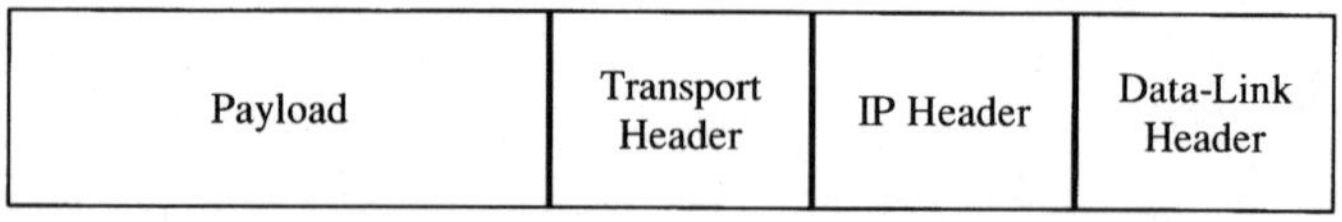

Payload	Transport Header	IP Header	Data-Link Header

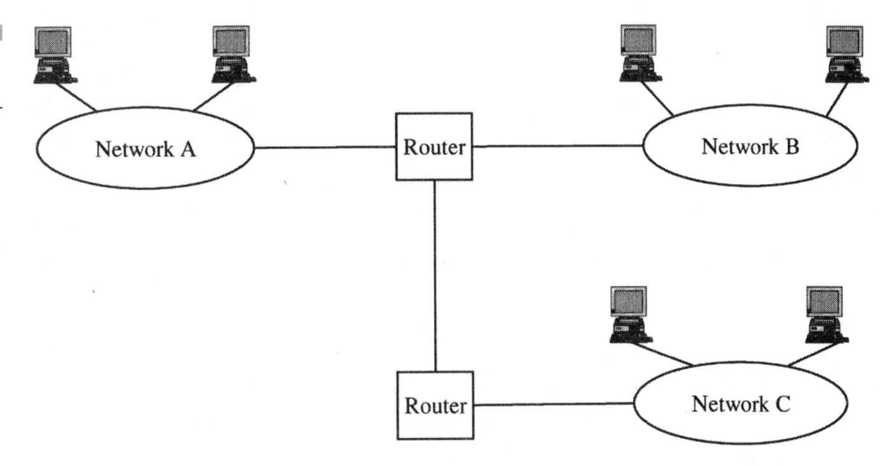

Figure 2-3
Sample internetwork.

Routers exchange information with other routers about the location and reachability of other networks within the entire internetwork. The exchange of information between routers is performed by a routing protocol. The routing protocol is also used by the router to compute the "best-path" (actually the shortest or least-expensive path) that the packet should follow to reach the destination.

Figure 2-3 shows an internetwork consisting of three networks interconnected by routers. Hosts on the same network (those on network A) of course do not need a router to communicate. However, if a source host and destination host are on different networks, the source host must send destination-host-bound packets through a router. The router will look at the destination address contained in packet, compare it with a table of network destinations that the router knows about, and then forward the packet toward the destination. The router is the essential building block of the Internet and more generally of TCP/IP-based networks.

2.3 IPv4 Header Format

The information that is required by the network to direct the packet to its final destination is contained in the header. The IP header is appended to the transport- or application-layer information after it is received by IP from the layer above and before it is dropped down onto the data link layer for transmission over a specific media. The length of an IPv4 header can range from 20 bytes all the way up to 60 bytes if additional

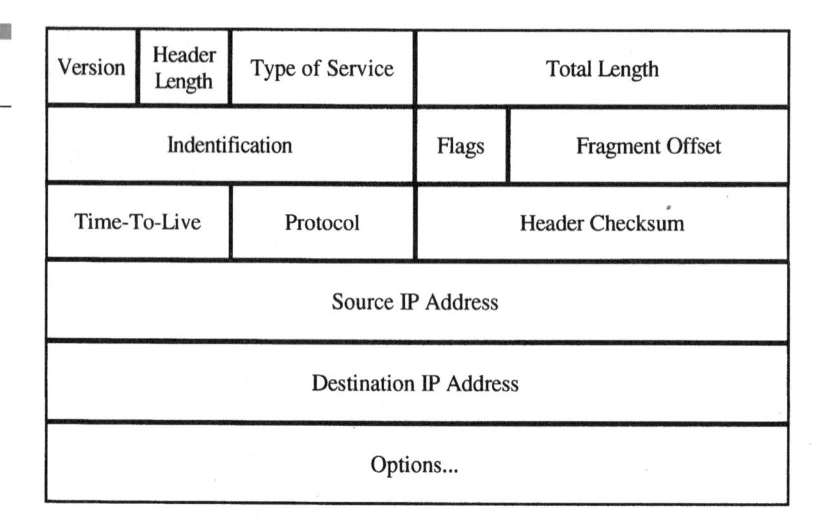

Figure 2-4
IPv4 header format.

options are used. The format of the IPv4 header is shown in Fig. 2-4; each part of it is discussed below:

- *Version.* The current version is 4.
- *Header length (HL).* The length of the IP header in the number of 32-bit words. Thus a typical 20-byte header would consist of five 32-bit words.
- *TOS.* The 8-bit Type-of-Service field consists of two parts: the precedence and type of service (TOS). The precedence field consists of 3 bits and was originally defined as a way to assign a level of priority to a specific packet. The remaining 5 bits define the Type-of-Service that the packet wishes to receive as it flows through the network. The TOS field has never really been used as intended, but a new work effort is under way to redefine the TOS field so that packets are assigned to service classes. This is called Differentiated Services and will be discussed in the next chapter.
- *Total length.* The 16-bit datagram length field is used to define the length of the entire IP packet in bytes. The maximum length of an IP packet is 65,535 bytes.
- *Identification.* This field contains a 16-bit identifier that is used by the destination host to recognize and group fragments of a packet. IP routers will fragment a packet if the maximum transmission unit (MTU) of the data link that the packet came in from is larger than the MTU size of the data link it is going out on. The MTU size

is defined as the largest IP packet size that can be carried in a data link frame. Reassembly is only performed at the destination host.

Fragmentation is not a good thing because it introduces extra work for the routers and the destination host. A technique called Path MTU Discovery[6] enables a sending host to discover the largest MTU that can be sent along a path between a source and destination without any fragmentation occurring.

- *Flags.* The flags field contains 3 bits with the following values:
  - Bit 0. Reserved
  - Bit 1. 0 = may fragment the packet; 1 = don't fragment
  - Bit 2. 0 = last fragment; 1 = more fragments
- *Offset.* The 13-bit fragment offset fields contains the number of 8-octet chunks (called fragment blocks) that the packet fragment contains from the start of the original packet until it is fragmented.
- *TTL.* The time-to-live (TTL) field is used to prevent packets from looping around the network for an indefinite period of time. Each time a packet passes through a router, the TTL value is decremented. When the TTL value reaches zero, the router discards it. Originally this value was in some number of seconds but has since taken on the meaning of a maximum hop count or hop limit.

  The TTL value is also used to scope, or control, how far a multicast packet can travel through the network. The TTL value in a multicast packet must be greater than the configured TTL threshold on an outbound router interface, or the packet will be discarded.
- *Protocol.* The protocol field is used to identify the next-higher-layer protocol that is using the IP service. Examples are 6 = TCP, 17 = UDP, 1 = Internet Control Message Protocol (ICMP), 89 = OSPF, etc.
- *Header checksum.* This 16-bit value is computed on the fields of the IPv4 header. It is updated each time the packet passes through a router because the TTL field changes and values within the options fields may have also changed.
- *Source IP address.* 32-bit IP address of the sending host.
- *Destination IP address.* 32-bit IP address of the destination host.

---

[6]RFC1191, Path MTU Discovery.

The IP header may also contain a series of options that need to be processed by each router along the path. Options that can be appended to an IP packet include source routing and a router alert. Options are not used much anymore because whatever functionality they may add is not worth the extra processing required at each intermediate router.

2.4 IPv Addressing

The IPv4 address space is confined to a fixed length of 32 bits and is usually presented in a dotted decimal format of uu.xx.yy.zz (e.g., 121.17.13.1). Each host or router interface is assigned an IP address that defines the network and host number on the network that the device is attached to. The network portion of the address is called the network prefix and the host portion, the host number.

Classful addressing means that the boundary between the network prefix and the host number falls on one of several octet boundaries. Where that boundary is located in the 32-bit address depends on the class of IP address and the value of the subnet mask. The subnet mask is a bit mask that identifies how many bits in the address comprise the network prefix. There are three classes of IP unicast addresses; they are summarized in Table 2-1. The subnet mask can be represented in the decimal format (e.g., 255.0.0.0) or in "/ # of bits in the mask" notation. For example, a Class B address of 189.23.0.0 with a subnet mask of 255.255.0.0 could be written as 189.23.0.0/16.

TABLE 2- 1 Classful IP Unicast Addresses

Class	Octet 1	Octet 2	Octet 3	Octet 4	Network Range	Total No. of Networks	Subnet Mask
Class A	Network	|-------- Host --------|			1-126	126	255.0.0.0 or /8
Class B	|---- Network ----|		|----- Host ----|		128-191	16,384	255.255.0.0 or /16
Class C	|--------Network --------|			Host	192-223	2,097,152	255.255.255.0 or /24

2.4.1 Subnetting

Subnetting is a technique used to divide a single IP network address into a number of smaller subnetworks. This enables a single classful IP address to be subdivided and shared among many different sites in a single enterprise without requiring a separate classful network address for each location. The process of subnetting is performed by dividing the host portion of a classful IP address into separate subnetwork and host portions. As with traditional classful addressing, the boundary between the network portion (network prefix + subnetwork) and host portions is identified by the subnet mask.

An example is shown in Fig. 2-5. The Class B address of 187.15.0.0 is assigned to the ABC Corp. The network planners at ABC Corp. wish to build a corporate IP network (intranet) that will interconnect over 200 different sites. Because the "187.15" portion of the IP address space is fixed, the network planners have the two remaining octets, or 16 bits, in which to define subnets and hosts on those subnets. They decide to define the third octet as the subnetwork number and the fourth as the host number on a particular subnet. The subnet mask for this internet-

Figure 2-5
ABC Corp. network.

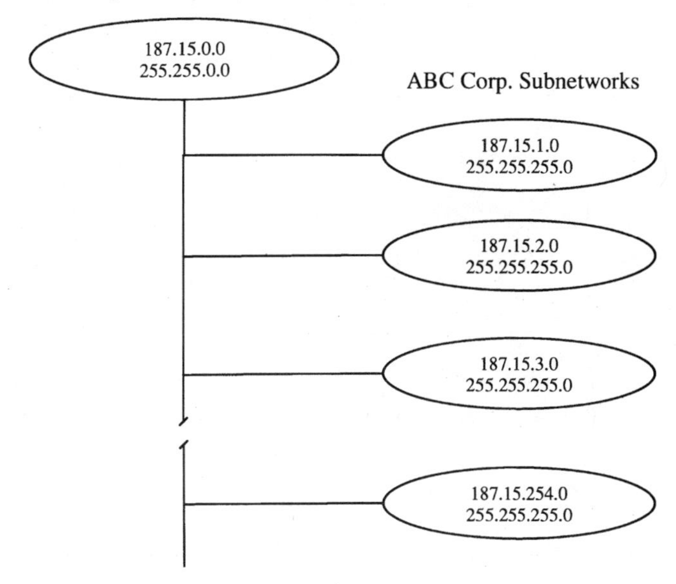

work is now 255.255.255.0. This means that the ABC Corp. intranet can support up to 254 separate subnets and that each supports up to 254 separate hosts. As far as connecting to the Internet is concerned, the ABC Corp. looks like a single network addressed as 187.15.0.0.

This example illustrates the case where a single subnet mask (255.255.255.0) is defined for the entire network. This means that the maximum number of hosts per network is limited to 254. There may be cases where a subnet may need to support more or fewer hosts than is provided in the current fixed-length subnet mask. In that situation a Variable Length Subnet Mask (VLSM) can be used. A VLSM enables a single classful address to be divided into subnetworks of varying size. Suppose for example that the ABC Corp. needed to support a large campus subnet consisting of 500 hosts. Rather than assign two subnets, each supporting 254 hosts, the company planners could assign a single subnet with a mask of 255.255.254.0. that supports up to 512 host addresses. Another site may require only 100 host addresses, in which case a subnet mask of 255.255.255.192.0 (128 host addresses) would be sufficient. Another example is a point-to-point subnet connecting two routers. Because only two possible "hosts" can ever exist on a subnet of this type, a mask of 255.255.255.252 could be used so as not to waste any addresses for a subnet that could never make use of them.

VLSM can more efficiently manage the address space so that the number of allocated host addresses per subnet matches the actual amount. For VLSM to work efficiently in a dynamically routed network, routing protocols that support VLSM must be configured. VLSM-supported routing protocols include the subnet mask along with the address prefix in their advertisements. Current routing protocols such as Routing Information Protocol-2 (RIP-2) and OSPF support VLSM.

2.4.2 Supernetting

Subnetting divides a single IP address into multiple subnetworks. A supernet is the inverse of a subnet and is accomplished by aggregating multiple Class C networks into a single common address prefix. The idea behind the address aggregation operation is twofold:

1. Reduce the size of the routing tables maintained by routers by reducing the number of separate Class C network entries.

2. Make more efficient use of the unused IP address space by only allocating the number of addresses that a network needs. For example, a Class B address contains approximately 65,000 host addresses. A network may only use a small portion of them, thus creating a large chunk of unused addresses that no one else will ever be able to use. By the same token a single Class C network only contains 254 usable addresses. This may be too small.

An example best illustrates this concept. Consider a block of 16 Class C addresses as shown in Fig. 2-6. Rather than represent all 16 Class C network addresses as separate entries, a single supernet of 192.18.0.0/20 is used. Technically speaking a supernet is a network whose subnet mask is less than the natural classful subnet mask. The block of contiguous Class C addresses represented by the "network prefix/" notation is called a Classless Interdomain Routing (CIDR) block. CIDR has been instituted over the last few years as a way of slowing the allocation of IP addresses and reducing the number of entries in the Internet's routers. Organizations that now wish to obtain an Internet address will be assigned a CIDR block rather than the traditional classful addresses described in the previous section.

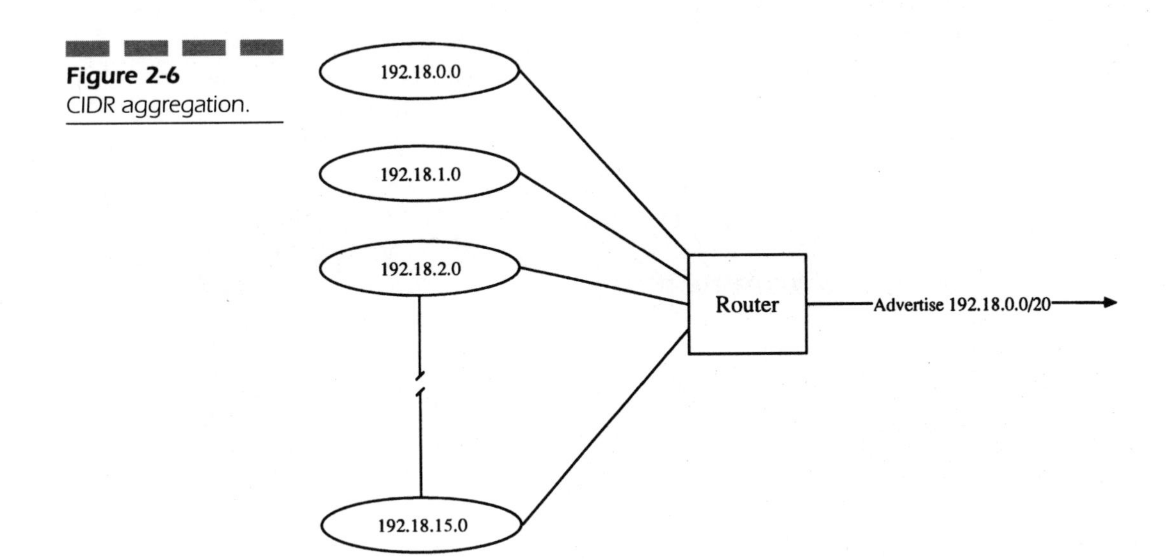

Figure 2-6
CIDR aggregation.

2.5 IP Routing and Forwarding

Routing is the general term that describes the process by which packets from a host on one network are transported through one or more routers to a host on another network. To be more specific, routing consists of two different separate and distinct operations:

- *Routing database management.* Routers maintain a database of destination networks and possibly other topological data that they use to figure out where a packet should be directed based on the destination IP network address contained in the packet header. The maintenance of these databases is performed when routers exchange routing database update messages. The formats of these messages and the information they contain are defined by the use of one or more dynamic routing protocols (e.g., RIP, OSPF). The processes, algorithms, and protocols that routers use to maintain a consistent and accurate database and to build a forwarding (hereafter referred to as a routing table) table[7] is also called IP routing.
- *Packet forwarding.* The actual process of forwarding a packet from one network to another is called packet forwarding or IP forwarding. When a packet arrives at a router, information such as the destination IP address is extracted from the packet header and compared with the entries in the routing table. The routing table that is derived from the contents of the routing database consists of a table of destination networks and the associated next-hop address in the path that the router believes leads to the destination. The header checksum is recomputed, the TTL is decremented, and then the packet is directed to the next-hop address and delivered to the final destination. This is called hop-by-hop routing because the decision of where to forward the packet is made at each router based on the router's own routing table and the destination address contained in the packet header.

Figure 2-7 depicts the basic components of the IP routing process from the perspective of an individual router. Routing database update messages arrive at the router and are passed to the routing protocol entity for processing. Once these messages are processed, a new IP

---

[7]A forwarding table is also called a *forwarding information base* (FIB).

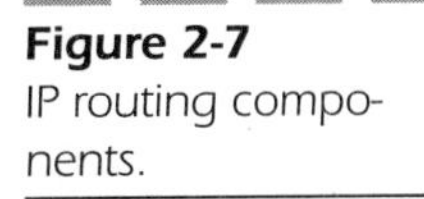

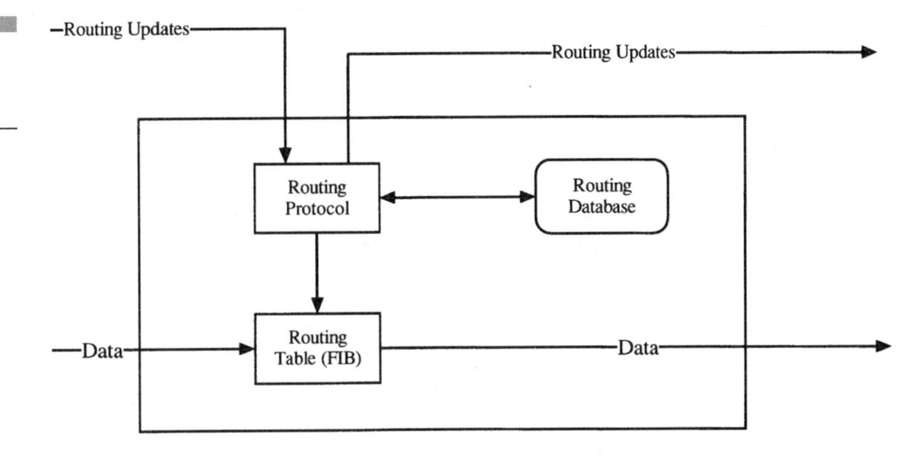

Figure 2-7 IP routing components.

routing table is generated. When a packet arrives at the router, its destination is compared with the entries in the routing table, and the packet is sent on its way. In addition the routing protocol entity may generate new routing database update messages to pass on to other neighboring routers for processing.

Although the process of IP routing and routing protocols in general has become the source of hundreds of papers, standards, debates, and books, it is this simple process of forwarding a packet through the network that is starting to receive much attention. For a variety of different reasons packets can take a long time to reach their destination. Some of these reasons include:

- *Too many router hops in the path.* This is a network design issue that can be solved by replacing some routers with switches, and this is being done.
- *Large routing tables.* The larger the routing table the longer it takes to look up and find an exact match for the packet's destination network address. The size of current routing tables is "somewhat" under control through the use of CIDR and other address aggregation techniques, but it is still expected to grow over time, and that is not even taking into account IPv6, which will probably include at least 64 bits for network addressing.
- *Older router technology.* Legacy equipment will always be with us. In older routers, packets arriving on an inbound interface were required to traverse a bus up to the central processor (where the routing table lived) and then back down and out the outbound

interface, which was not very efficient. However, newer router designs are caching some or all of the entire forwarding table in specialized hardware located on the adapter interface. In addition the shared bus that served as an interadapter transport has been replaced by high-performance switches.

All these of new and yet-to-be-invented techniques will no doubt improve the performance of the IP forwarding process in routers. Another technique to consider involves redirecting or caching the IP forwarding path over or through layer-2 switches and, more specifically, ATM switches. This process is called IP switching and will be covered in later chapters.

2.6 Routing Protocols

Routing protocols enable routers to exchange network reachability and topology information with other routers. The primary objective of any routing protocol is to ensure that all routers in a network have a consistent and accurate database of the network topology. This is important because it is from this database of network topology information that the forwarding table at each router is computed. Correct forwarding tables contribute to a high probability that packets will reach their destination; incorrect or incomplete forwarding tables will mean that packets do not reach their destination and, worse yet, may loop around the network for some period of time, consuming both bandwidth and router resources.

Routing protocols are classified as either interdomain or intradomain. A domain is also called an Autonomous System (AS); an AS is a collection of routers that are controlled and administered by a single entity, and is identified by a single AS number (e.g., AS 3). Intradomain protocols (also called interior gateway protocols, or IGPs) are used between routers in the same AS. Their job is to compute the fastest or least-expensive path between any two networks in the AS, which should result in the best performance. Interdomain protocols (also called exterior gateway protocols, or EGPs) are used between routers in different autonomous systems. Their job is to compute a path through different autonomous systems. Because autonomous systems are controlled by different organizations, the criteria for selecting a path through an AS may depend on certain policies such as cost, security, availability, performance, inter-AS business relationship, etc., rather than just plain old performance. An example of an EGP is the Border Gateway Protocol (BGP), and examples of IGPs are OSPF and RIP. Figure 2-8 shows a network of three different autonomous systems running an EGP between neighboring ASs and an IGP within the AS.

Other criteria that differentiate one routing protocol from another include

- *Scalability.* This is indicated by the routing protocol's ability to support a large number of routers and networks while minimizing the amount of interrouter control traffic (routing database updates) and router resources needed to compute new routing tables.
- *Loop avoidance.* When a routing protocol computes a new routing table, it would be ideal to prevent or avoid paths that cause packets to pass through a router or network more than once. This is difficult to achieve given the time it takes to disseminate topology changes to all routers in the network. Nevertheless loop avoidance is an important property that some protocols such as BGP and EIGRP (Enhanced Interior Gateway Routing Protocol) support.

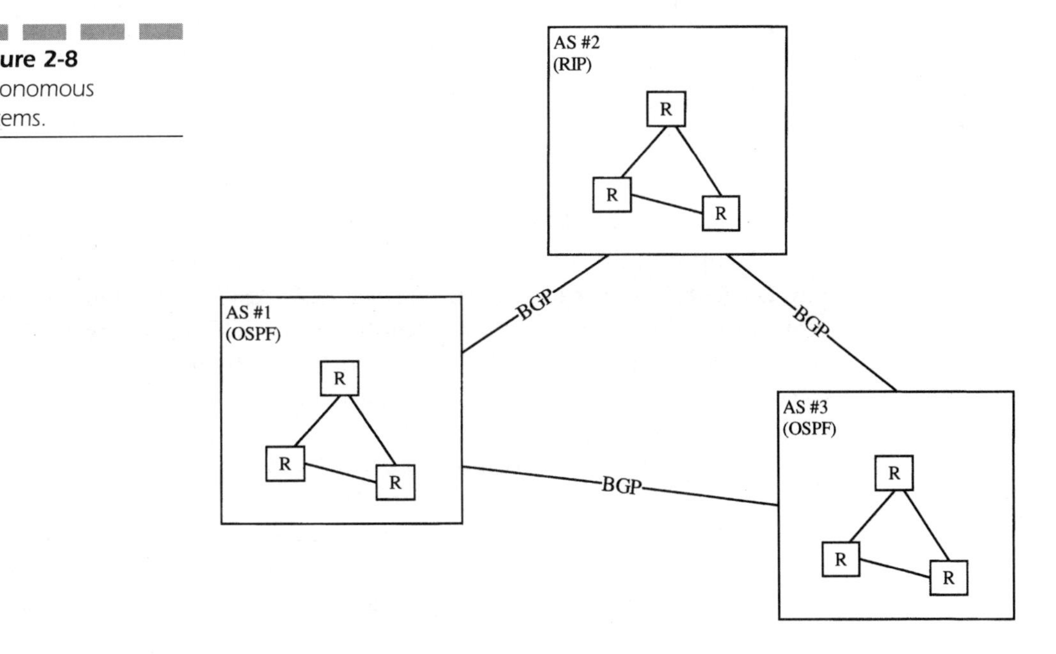

Figure 2-8 Autonomous systems.

- *Convergence.* When the network topology changes (e.g., a link goes down, a new network is added, etc.), routing protocols must distribute this information throughout the entire network of routers and then compute a new routing table at each router that reflects this new information. This process is called convergence. The faster the routers converge on the correct topology, the more quickly the packets will be successfully delivered to the destination.
- *Standards.* Routing protocols that have been developed within the IETF are documented in an RFC. This enables different vendors to implement the routing protocol on their respective platforms and promotes interoperability.
- *Extensibility.* This defines the ability of the routing protocol to incorporate new functions without changing its basic operation and backward compatibility. An excellent example of an extensible routing protocol is OSPF. Examples of new functions added to OSPF are multicast, QoS routing, and link-layer addressing support.
- *Metrics.* These are the parameters or values that are advertised along with the destination network and go into the calculation of the routing table. Examples are number of hops, link cost, bandwidth, and delay.
- *Routing algorithm.* Routing protocols employ one of two fundamental routing algorithms: distance vector or link state.

2.6.1 Distance Vector Routing

The premise of distance vector routing is based on the notion that a router will tell its neighbor routers about all the networks it knows about and the distance to each of those networks. A router running a distance vector routing protocol will advertise to its directly attached neighbor routers one or more distance vectors. A distance vector consists of a tuple {network, cost}, where network is the destination network and cost is a relative value that represents the number of links or routers in the path between the advertising router and the destination network. Thus the routing database consists of a number of distance vectors that indicate the distance or cost to all networks from the router.

When a neighbor router receives a distance vector update from its neighboring router, it adds its own cost value (typically equal to 1) to the cost value received in the update. Then the router compares the newly computed cost to reach a destination network with the information received in earlier distance vector updates. If the cost is less, the router updates the routing database with the new costs, computes a new routing table, and includes the neighbor who advertised the new distance vector information as the next hop.

This elegantly simple process is illustrated in Fig. 2-9. Router C advertises a distance vector {destination network, cost} pointing to net 1, which is directly attached to router C. Router B receives the distance vector, adds its own cost to the cost in the distance vector received from router C, and advertises this to router A. Router A now knows that it can reach net 1 that is two hops away through router B.

Although distance vector routing is quite simple, several well-known problems can occur. For example, if the link between routers B and C breaks, router B may attempt to reroute packets through router A because router A advertised to router B a distance vector of {Net 1, 4}. Router B will receive this and advertise back to router A a distance vector of {Net 1, 5} and so on. This is called the "count-to-infinity" problem, which can result in a longer than necessary time to converge.

The solution to this problem is called *split horizon*. Simply put, split horizon states, "Never advertise reachability to the next hop toward a destination." In other words router A should never advertise a distance vector of {Net 1, 4} to router B because B is the next hop to net 1. In a variation on split horizon called *poison reverse*, reachability to a destination network is advertised to the next hop with an extremely high cost. The intent is to ensure that the original next-hop router (B) never looks to the advertising router (A) for reachability to net 1.

Other problems associated with distance vector routing protocols occur because the entire routing database is sent out to all router interfaces at periodic intervals (typically every 60 s). A large routing data-

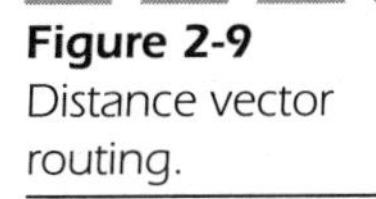

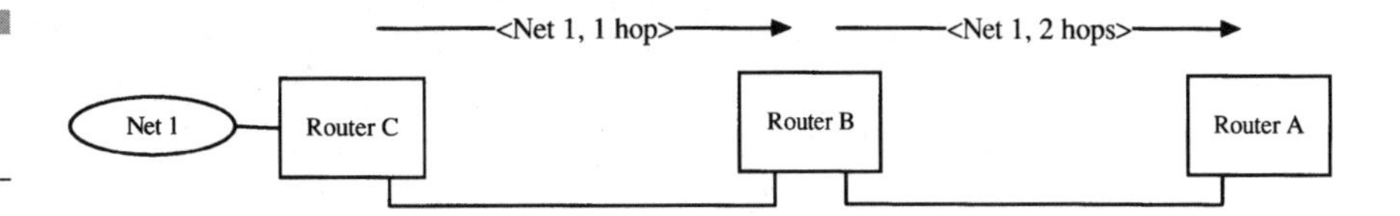

Figure 2-9 Distance vector routing.

base can result in substantial link overhead over a slow link. An implementable fix to this concern is called triggered updates in which the routing database is sent out only after a change in topology.

Distance vector routing is based on the popular Bellman Ford algorithm and has been implemented in a number of routing protocols including RIP and IGRP (Interior Gateway Routing Protocol).

2.6.2 Link State Routing

Link state routing works on the notion that a router can tell every other router in the network the state of the links attached to the router, the cost of the links, and the identity of any neighboring routers attached to those links. Routers running a link state routing protocol will propagate link state packets (LSPs) throughout the network. An LSP generally contains a source identifier, a neighbor identifier, and the cost of the link between the two. LSPs received by all routers are used to construct an identical topology database of the entire network. The routing table is then computed based on the contents of the topology database. In effect all routers in the network maintain a map of the network topology, and from this map they compute a shortest path (least-cost path) from any source to any destination.

Figure 2-10
Link state routing.

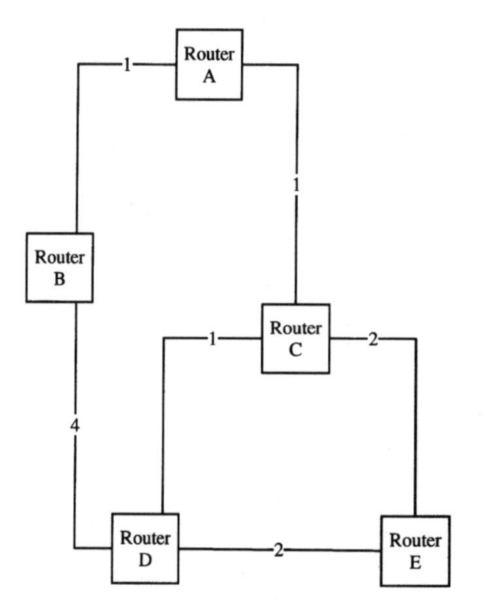

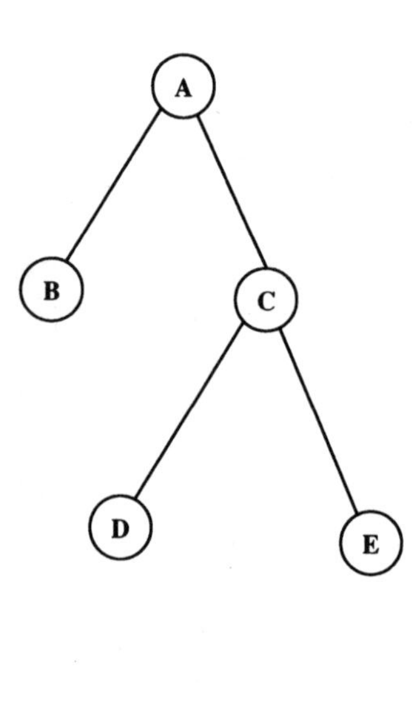

Consider the network topology shown in Fig. 2-10. The value attached to the interrouter links is the cost of the link. The routers propagate LSPs to all the other routers in the network, which are used to build a link state database. Next, each router in the network computes a tree that is rooted at the router and branches to all other routers by taking the shortest or least-cost path. The result for router A is a shortest-path tree as shown in Fig. 2-10, which is then used to compute a routing table. The algorithm to compute the shortest-path tree is called the Dijkstra algorithm.

Link state routing protocols have several advantages over distance vector routing protocols:

- ✓ *Faster convergence.* Link state routing protocols generally converge faster for a couple of reasons. First, LSPs can be quickly flooded throughout the network and used to build an accurate view of the network topology. Second, only the topology change is reflected in the LSP and not the entire routing database. And third, the count-to-infinity problem is not applicable.
- ✓ *Less network overhead.* Link state routing protocols only transmit LSPs that reflect a change in network topology versus transmitting the entire routing database.
- ✓ *Extensibility.* Link state routing protocols can be extended to support and propagate different network metrics, addresses, or other topological information. And because a router maintains a topology database, the new information is available when computing a path to a specific destination.
- ✓ *Scalability.* Link state routing protocols possess better scaling properties because routers in a large network can be divided into multiple groups. Within a group routers exchange LSPs with each other and build an identical topology database of that group. To communicate topology information between groups, a special subset of routers first summarizes the group topology database and then transmits the summarized data in an LSP to other special routers in the adjacent group. This reduces the amount of memory and processing in routers because the topology database is only as big as the number of routers in the group, and only routers in the group where a topology change occurred must compute new shortest-path trees and routing tables—not the routers in other router groups. This notion of hierarchy is illustrated in Fig. 2-11 and is an important concept that has been implemented in link state routing protocols such as OSPF and PNNI.

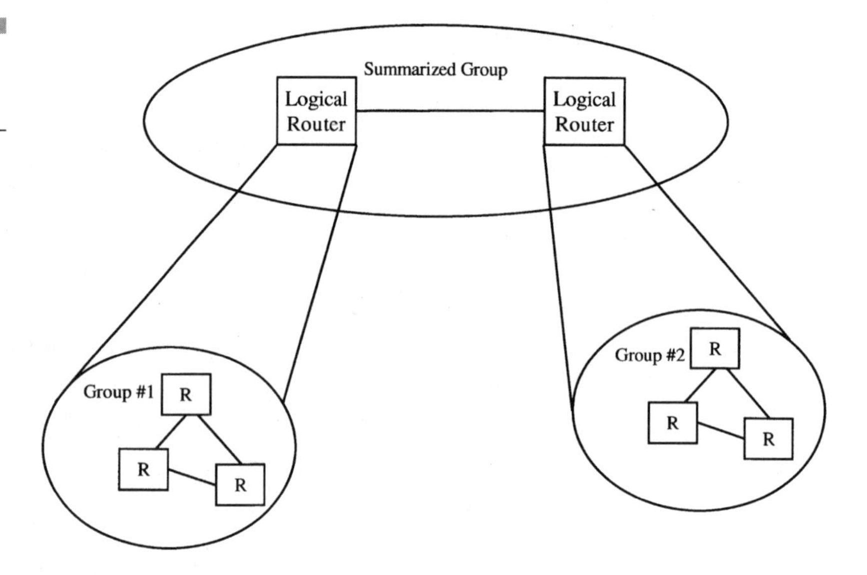

Figure 2-11
Link state routing hierarchy.

2.6.3 RIP

RIP is a common distance vector routing protocol that is implemented on just about any TCP/IP host or router. Its popularity can be traced back to the fact that it was distributed within some of the initial releases of UNIX in the mid-1980s. The primary functional attributes of RIP include the following:

- It has distance vector routing algorithm.
- It uses hop count as the metric.
- Routers broadcast the entire routing database every 30 s.
- The maximum network diameter of a network of routers supporting RIP is 15 hops.
- It does not support VLSMs.

RIP is quite simple to configure and runs on many small and medium-size corporate networks today and so is classified as an interior or intradomain routing protocol. RIPv2 was developed to address some of the shortcomings of RIPv1. RIPv2 is very similar to RIPv1 in operation and has been enhanced to support VLSM. This provides network managers who need more flexibility in managing the IPv4 address space with an alternative to OSPF for support of VLSM. RIPv1 is documented in RFC1058, and RIPv2 is documented in RFC1723.

2.6.4 OSPF

OSPF is the most well-known and deployed link state routing protocol today. OSPF is an interior/intradomain routing protocol and is supported on most if not all routers on the market. The primary functional attributes of OSPF are that they

- Include link state (Dijkstra) routing algorithm; also sometimes referred to as shortest-path first (SPF).
- Support multiple equal-cost paths to the same destination.
- Have VLSM support.
- Have two-level hierarchy.
- Generate link state advertisements only as a result of changes in network topology.
- Are extensible.

An example of an OSPF network consisting of several areas is shown in Fig. 2-12. An area in OSPF parlance is defined as a collection of routers and networks inside of an AS. An OSPF network must have an area 0 that is defined as the backbone area. If multiple areas are configured, all nonzero areas must attach to area 0 via an area border router (ABR). Routers inside an area exchange link state advertisements (LSAs) and build an identical map of the area called the link state data-

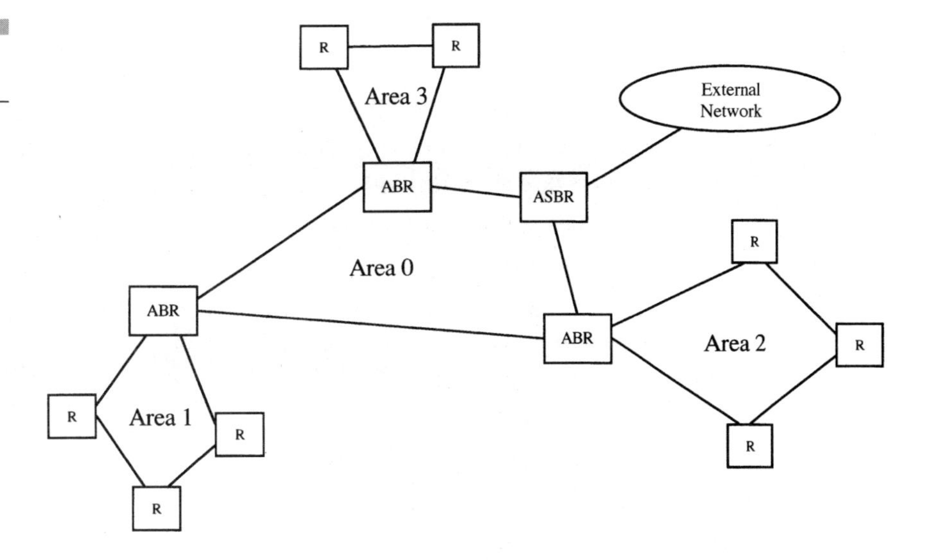

Figure 2-12
OSPF network.

base. Summarized information about specific topologies and networks is passed between areas through ABRs. Therefore routers maintain complete information about all networks and routers in their area and partial information about routers and networks outside of the area. To reach a network in another area, routers have enough information to direct the packets to the appropriate area border router.

OSPF is beginning to receive a lot of attention from network managers and developers for several reasons:

1. Larger networks consisting of many more routers are being built and deployed, and OSPF can scale much better than RIP and other distance vector routing protocols.
2. Additional functions and services are or will need to be deployed on these networks. As a link state routing protocol, OSPF possesses the ability to incrementally extend and enhance the function it provides by simply defining and adding new fields to carry new information in OSPF link state advertisements. Examples to date include multicast addresses and the opaque LSA, which is a placeholder to carry future advertisements.
3. The mystique and aura surrounding OSPF has begun to dissipate as more network engineers deploy and manage production networks running OSPF.

OSPF is documented in RFC2328.

2.6.5 BGP

BGP is classified as an exterior/interdomain routing protocol. Its primary objective is to communicate routing information between routers in different ASs. BGP is often referred to as a path vector routing protocol because BGP advertises reachability to a destination network by including a list of ASs that packets must pass through to reach the destination network. The path vector information is quite useful because loops can be avoided by simply looking at the AS numbers in the BGP routing update.

The primary functional attributes of BGP include

- ✓ Path vector routing protocol
- ✓ The support of policy-based routing by affecting which routes are selected and by controlling the distribution of routes to other BGP routers

- The use of TCP to reliably exchange routing information between BGP routers
- The support of CIDR aggregation and VLSM
- No restrictions on network topology

A network of several ASs running BGP is shown in Fig. 2-13. BGP routers in different ASs establish an external BGP (EBGP) relationship. BGP routers inside an AS establish an internal BGP (IBGP) relationship. To ensure that all BGP routers in the same AS maintain consistent routing information, each BGP router in the AS must establish an IBGP relationship with every other router BGP in the AS. However, to route packets inside of an AS, a normal intradomain routing protocol such as OSPF is used—not BGP.

At network startup, BGP neighbor routers (in our example let's use router 1 and router 2) open a TCP connection with each other, and the entire routing database is exchanged. After that, only changes in topology or policy are sent in BGP update messages. A BGP update message can either announce or withdraw reachability to a particular network. BGP update messages may also contain path attributes that are used by BGP routers to build and distribute routing tables based on specified policies.

The current version of BGP is version 4 (BGP4) and is documented in RFC1771.

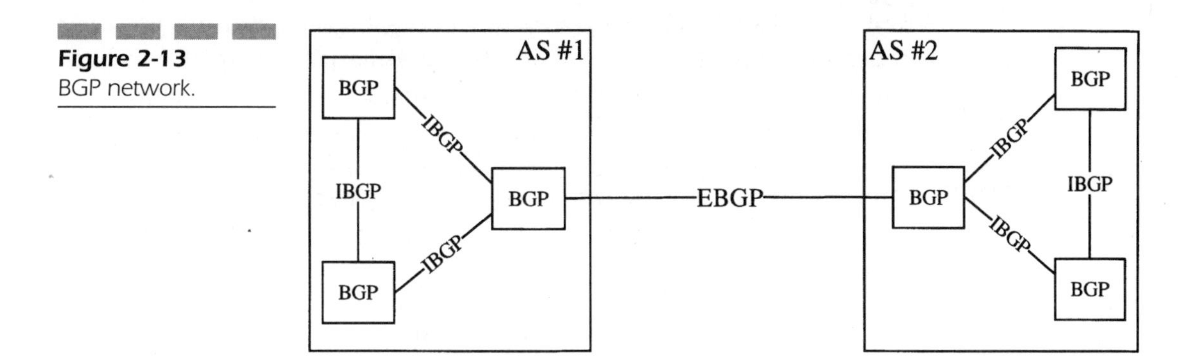

Figure 2-13
BGP network.

2.7 TCP and UDP

The TCP/IP protocol architecture supports two basic transport protocols: TCP and UDP. TCP stands for Transmission Control Protocol and supports a connection-oriented and reliable transport service between two TCP endpoints. UDP stands for the User Datagram Protocol and supports a connectionless and unreliable transport service between UDP endpoints.

Most traditional TCP/IP applications such as telnet and FTP as well as new ones such as HTTP use TCP as the transport service. The important functional attributes of TCP include

- *Connection orientation.* This means that TCP endpoints must establish a connection with each other before any data can flow. And only the TCP endpoints are concerned with connection state—the intermediate routers in the middle of the network are only concerned with forwarding IP packets.
- *Reliable data delivery.* TCP uses sequence numbers, explicit acknowledgments, and if necessary, retransmissions to ensure that the data sent from the source was delivered successfully to the destination.
- *Flow control.* This prevents a TCP sender from overrunning the capacity of the receiver, which is accomplished by means of a receiver window value that is sent from the receiver to the sender. This tells the sender how much data the receiver can handle. A TCP sender can only transmit at most a window's worth of data before waiting for an acknowledgment from the receiver to tell the sender to send more data.
- *Congestion control.* This prevents a TCP sender from overrunning the capacity of the links and routers in the network. After connection establishment, a TCP sender begins by sending a small amount of data into the network and waits for an acknowledgment from the receiver. With each acknowledgment, the TCP sender gradually increases the amount of data sent into the network until the TCP sender detects that a point of congestion is approaching. The TCP sender can detect when a congestion threshold has been reached by a number of factors including a timeout or the receipt of duplicate acknowledgments. The combination of flow and congestion control enables TCP hosts to quickly and fairly adapt their

sending rates to match the available capacity of the network and the receiver.

- *Support of point-to-point connections only.*

Work continues in the area of improving the performance and efficiency of TCP. RFC1323 defines a number of extensions that improve the performance and throughput of TCP over high-speed networks. RFC2018 defines a technique called TCP SACK (for selective acknowledgments) in which a window-size worth of data is divided into a number of blocks and acknowledged per block by the receiver. If some data are lost, only the missing (unacknowledged) blocks are retransmitted by the sender—not the entire window worth of data. Figure 2-14 illustrates the TCP connection setup, data exchange, and connection teardown between two hosts.

UDP supports applications that do not or cannot utilize the connection-oriented and/or reliability services of those provided by TCP. In some cases applications such as email or ones where only a few packets need to be sent need nothing more than UDP. In other cases applications may provide their own reliable delivery or flow control mechanisms. UDP has the following characteristics:

- Connectionless operation.
- Unreliability.

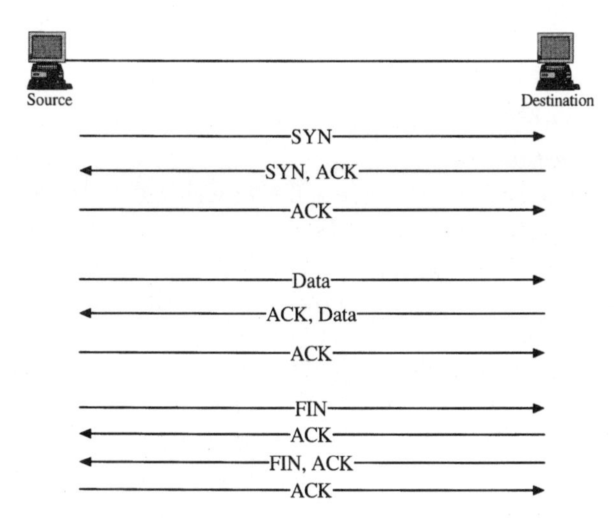

Figure 2-14
TCP connection.

- No flow or congestion control.
- Source and destination port numbers in the UDP header provide a simple multiplexing/demultiplexing service.
- Supports unicast and multicast transmissions.

MORE INFORMATION

Halabi, S., *Internet Routing Architectures,* Cisco Press, 1997.

http://www.ietf.org

Huitema, C., *Routing in the Internet,* Englewood Cliffs, N.J.: Prentice Hall, 1995.

Moy, J. T., *OSPF Anatomy of an Internet Routing Protocol,* Addison-Wesley, 1998.

Stevens, W. Richard, *TCP/IP Illustrated Vol. 1 The Protocols,* Addison-Wesley, 1994.

CHAPTER 3

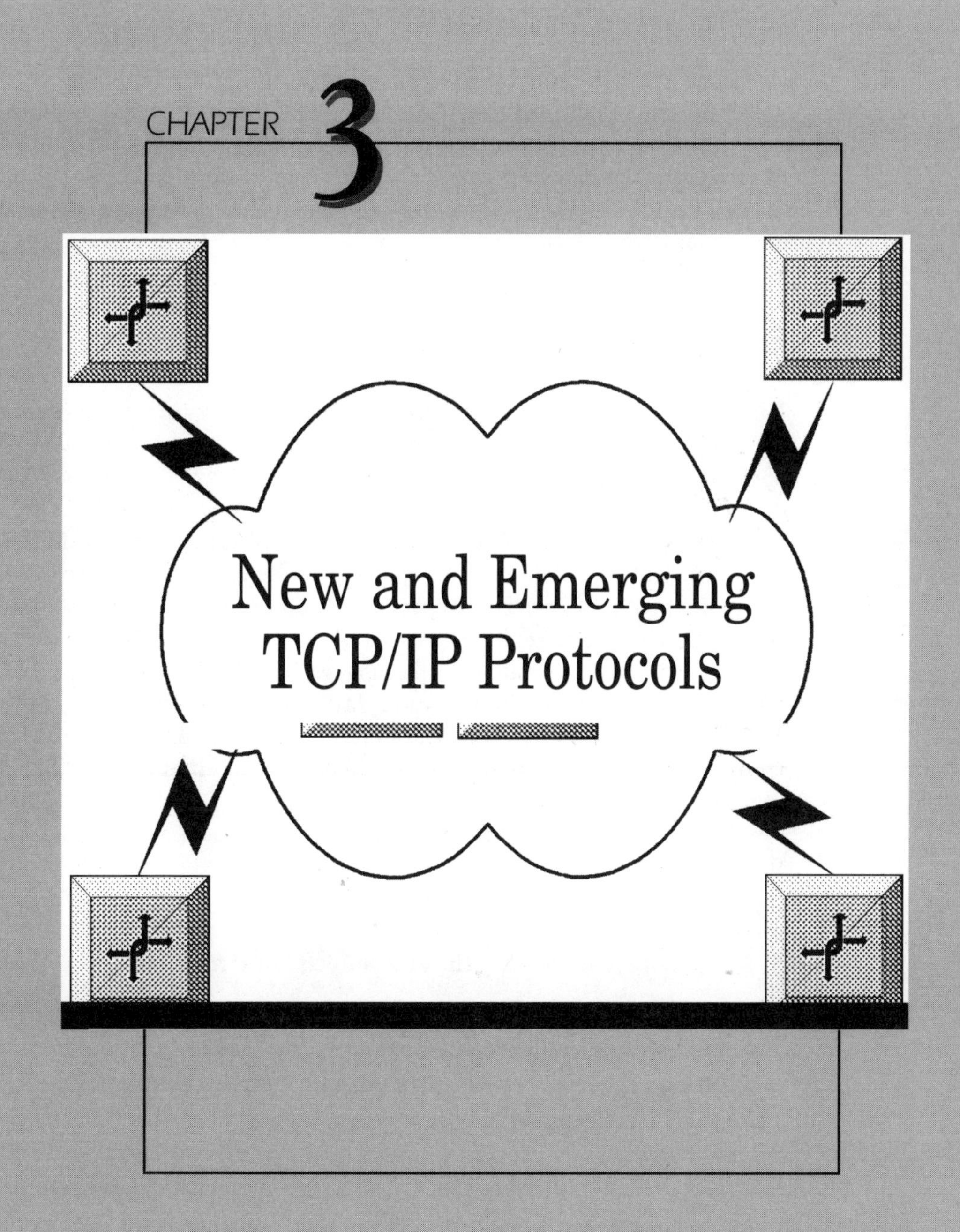

New and Emerging TCP/IP Protocols

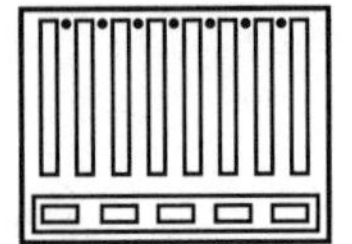

The previous chapter described the basic TCP/IP architecture and the functions and protocols that support IP routing. This chapter focuses on new and emerging protocols and functions that have been or will be added to the suite of TCP/IP protocols. The new protocols and functions discussed below add new services such as QoS, multicast, and security to the base internetworking functions of TCP/IP. In addition, a brief discussion of IPv6 functions and features is included.

3.1 IP Multicast

Up until now, all discussion has focused on the unicast aspects of TCP/IP. IP multicast is an additional component to the architecture that enables efficient group communications over IP-based networks. Applications requiring one-to-many, many-to-many, few-to-many, or few-to-few transmissions are ideal candidates for IP multicast. Examples of these applications include video conferencing, shared whiteboards, push technology (e.g., advertisements, info subscriptions, etc.), resource/server discovery, financial data distribution, and server replication.

The basic idea behind IP multicast is that multiple receivers would like to receive a copy of the same data sent by a source or group of sources. It certainly would be possible for a single sender or senders to transmit a separate copy of the data to each individual receiver but it is apparent that this is not very efficient for several reasons. The sender(s) must somehow learn and manage the individual host addresses of all of the receivers, and multiple copies of the same data would be flowing from the sender(s), consuming network resources in bandwidth and sender resources.

A more efficient model is the one defined by IP multicast:

- ✓ A group address is defined. This represents a session between one or more senders and one or more receivers.
- ✓ A receiver tells the router which multicast group it wishes to join (or leave) using the multicast address.
- ✓ A sender transmits a packet addressed to the multicast group address. It knows nothing about the receivers or where they are located and does not care. It only knows about the group address.
- ✓ Routers build a multicast delivery tree from the sender(s) that branches out to all networks that have at least one member of the group present. The routers forward the packets addressed to the

multicast group over the delivery tree until they reach the member networks.

The components of a network supporting IP multicast are shown in Fig. 3-1. The receivers tell the nearest router that they want to join a multicast group using a group membership protocol. A multicast routing protocol[1] operating on a network of routers constructs a multicast delivery tree that will deliver the packets of a multicast group to the group members. As far as the sender(s) are concerned, they simply need to transmit a single copy of the packet addressed to the appropriate group address, and the routers will handle the rest. This model can scale from a single receiver up to hundreds and thousands of receivers. If a receiver wishes to leave a multicast group at any time, it simply tells the nearest router on the network. The sender(s) is not involved and is not aware of group membership changes.

3.1.1 Class D Addressing

The Class D address space is defined for IP multicast group addresses. A multicast group address falls between the values 224.0.0.0 and 239.255.255.255. A portion of this address space is reserved for specific

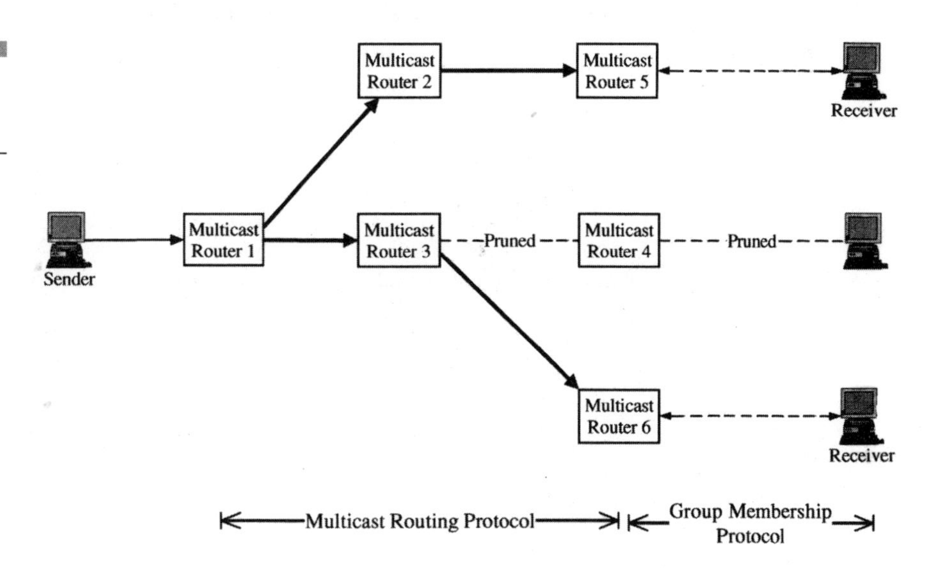

Figure 3-1
IP multicast components.

[1]Multicast and unicast routing protocols operate simultaneously in the same router.

group functions, well-known multicast applications, and administratively scoped multicast applications. The rest can be dynamically allocated when needed for multicast transmissions. The complete list of reserved multicast group addresses is available from the Internet Assigned Numbers Authority (IANA).[2]

IP multicast group addresses can be mapped to Institute of Electrical and Electronics Engineers (IEEE) 802 MAC multicast addresses. This is accomplished by taking the low-order 23 bits of the IP multicast group address and appending them to a special IANA designated prefix of 01-00-5E. For example, the multicast group address of 225.02.03.04 could be mapped to an Ethernet multicast address of 01-00-5E-02-03-04. Mapping IP multicast group addresses to IEEE 802 MAC-level multicast addresses enables multicast hosts to exploit the hardware multicast capability available in most network interface cards. The format of a class D address is shown in Fig. 3-2.

3.1.2 IGMP

The Internet Group Membership Protocol (IGMP) is the group membership protocol in the IP multicast model. IGMP is used by routers to solicit group membership from directly connected hosts and by hosts to inform a directly attached multicast-capable router that it would like to receive packets addressed to a particular multicast group address.

The basic IGMP message exchange is illustrated in Fig. 3-3. Routers first send out an IGMP Host Membership Query addressed to the "all-hosts group" address of 224.0.0.1. A host wishing to join a multicast group responds with an IGMP Host Membership Report addressed to the address of the multicast group. In the example shown, Host 1 responds to the query with a Host Membership Report specifying "Group 1." Host 2 does the same for "Group 2." The router only needs to receive one Host Membership Report per group. A message suppression

Figure 3-2 Class D multicast address.

	Leading 4 Bits	Remaining 28 Bits
Class D	1110	Multicast Group Address

[2] *http://www.isi.edu/iana/.*

Figure 3-3
IGMP flow.

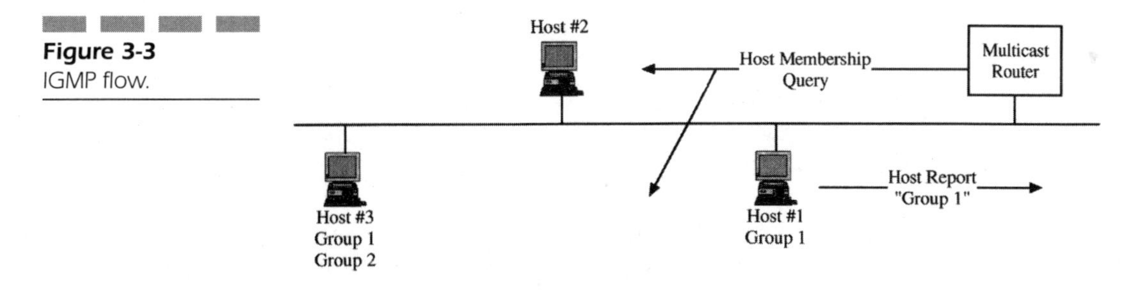

algorithm prevents other hosts on the same subnet from issuing host membership reports for a specific group if another host has already done so. This random delay interval prevents a flurry of reports for the same group flooding the LAN and the router interface with duplicate membership reports.

IGMP Version 1 is documented in RFC1112. RFC2236 defines IGMP V2, which introduces several new enhancements that improve the efficiency of IGMP. These include election of a multicast querier on a LAN (when two or more are present) and two new messages: Group-Specific Query and Leave Group. The two new messages will lower the "leave latency" by enabling a multicast querier to query for any hosts belonging to a specific multicast group, and hosts may leave specific groups immediately rather than waiting for a timeout period.

3.1.3 Multicast Routing

Two basic algorithms are used by multicast routing protocols to build delivery trees and forward multicast packets over these trees to group member networks. The first is called Reverse Path Forwarding (RPF) and the second is generally referred to as Core Based Trees (CBT).

The RPF algorithm is quite simple:

1. If a multicast packet arrives at a router interface that the router would use to send unicast packets back to the sender's network, forward the packet out all other interfaces.
2. Otherwise discard it.

This is illustrated in Fig. 3-4. Router A receives a multicast packet from source S.1 on interface w1. Router A looks at the source address in the packet (S.1) and then checks a routing table to see what interface it

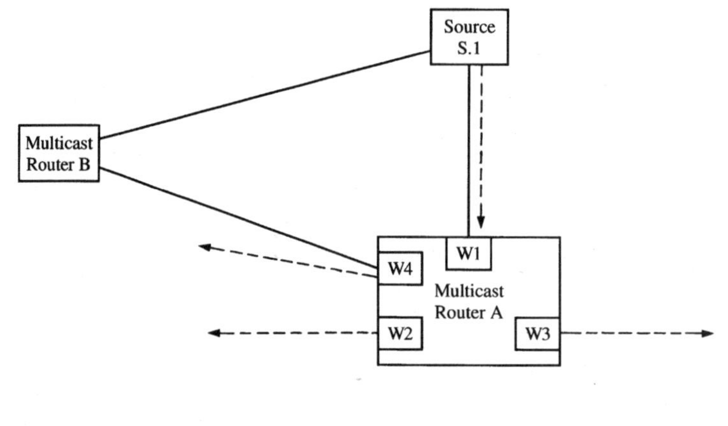

Figure 3-4 Reverse path forwarding.

would use to send unicast packets back to host S.1. Because it was received on the next-hop interface (w1) leading back to S.1, the packet is copied and sent out to all other interfaces. If the packet arrived on any other interface, router A discards it.

The end result of performing the RPF algorithm at each router is the establishment of a delivery tree or, to be more precise, a source-rooted shortest-path tree (SPT), that branches out to all routers and networks. However, the RPF algorithm alone can cause unnecessary packet duplication and the transmission of multicast packets on links with no active group members. The latter is because group membership as conveyed to a router by an interested host is not taken into account during the RPF process.

Several enhancements have been made to the original RPF algorithm to address some of these shortcomings. Packet duplication can occur on a link[3] connecting two routers that each have a separate path back to the sender (link connecting routers A and B in Fig. 3-4). This can be eliminated if each router can determine who its "on-tree"[4] upstream and downstream neighbors are. This determination can be made by several

[3]A link can also be a shared-media LAN, and the same mechanisms to prevent packet duplication are applied.

[4]On-tree means those routers that are on a delivery tree rooted at the source and branching out to all routers with active group members. The delivery tree is a set of routers and links that will forward multicast packets. Upstream is the direction toward the sender; downstream leads away from the sender.

techniques including examination of the routing tables, exchange of routing database messages, or perhaps sending or receiving an explicit message. The end result is that packets should not flow over links that are not part of the delivery tree.

Group membership information (or the lack of group membership information) should be used to truncate or prune router interfaces and links from the delivery tree if there are no group members present. Truncated reverse path broadcasting (TRPB) is another enhancement to RPF that removes router interfaces from the delivery tree if there are no active group members on the directly attached subnets. Reverse path multicasting (RPM) is a further optimization that enables a router to send a prune message to an upstream router if there are no group members present on directly attached interfaces or downstream from any of the router's interfaces. The SPT shown in Fig. 3-1 is rooted at the sender and branches out to all active group member networks. Observe that nonmember links and routers have been pruned as a result of RPM algorithms.

Multicast routing protocols that use RPF along with one or more of the aforementioned enhancements are generally classified as "broadcast and prune" protocols. This means that upon receipt of multicast data packets from the sender(s), routers will broadcast the packets out to all interfaces, resulting in a source-rooted shortest-path tree branching out to all routers. Once group membership is factored in, the delivery tree is dynamically pruned back to only those networks with group members. After some period of time the prune state times out and the broadcast process is repeated. The broadcast and prune approach to multicasting is simple to implement but can waste bandwidth and does not scale well because routers must maintain $O(S * G)$ information for each {source host, group address} pair.

CBT is the second algorithm used in some multicast routing protocols. CBT defines a core router that all downstream routers with group members send explicit join messages to. A single group-specific multicast delivery tree is created that is rooted at the core router and branches out to only those downstream routers with group members. The join and tree-building processes occur independently of the existence of any sender(s) or multicast data packets. The sender(s) for the multicast group forward all traffic to the core router, which places it on the group-specific delivery tree. This enables multiple senders of the same group to share a single delivery tree; therefore CBT multicast routing protocols are referred to as shared-tree protocols.

CBT scales much better than broadcast and prune techniques because the routers only need to concern themselves with maintaining a single core-rooted delivery tree for each group. This means they only need to maintain O(*, G), where G is the number of group addresses. Also, CBT does not periodically flood data into the network as do broadcast and prune protocols. The core router blocks multicast packets until it receives an explicit join message from the downstream router, and these join messages are only generated when a router receives an IGMP Host Membership report from a host.

3.1.4 Multicast Routing Protocols

Several multicast routing protocols have been defined and fall into one of two categories: dense or sparse mode. Dense-mode protocols work well in networks where there is a lot of bandwidth and where there is a reasonably high probability that group members are in close proximity to each other. Dense-mode protocols typically employ a broadcast and prune approach because it is simple and the network can absorb the overhead. A campus LAN and a corporate intranet are two examples where dense-mode protocols can work quite nicely. Sparse-mode protocols are designed for environments where bandwidth is not so plentiful and where group members are scattered across many networks and over many geographic regions. It makes sense here to employ the receiver-driven approach characterized by sending explicit joins to a single core router. An example of such a network is a very large corporate intranet or even the Internet.

The Distance Vector Multicast Routing Protocol (DVMRP) has been around for quite some time and is still in widespread use on the MBONE, which is the IP multicast backbone overlayed on the Internet. DVMRP uses RPM to forward packets and defines a capability to tunnel multicast packets through non-multicast-capable routers. MOSPF is a set of multicast extensions to OSPF and uses the OSPF topology database to compute source-rooted shortest path trees. PIM stands for Protocol Independent Multicast and comes in two flavors: dense mode (DM) and sparse mode (SM). PIM-DM is very similar to DVMRP without the dependencies on distance vector routing functions such as distance vector routing updates and the use of split horizon to signal the existence of an on-tree downstream router. PIM-SM can establish a shared-tree rooted at a rendezvous point (RP) router and dynamically switch over to a source-rooted shortest-path tree if better performance is required.

PIM and CBT are both distinguished by the fact that they are not dependent on any underlying unicast routing protocol functions. They can operate on the same router configured with RIP, OSPF, or any unicast routing protocol. DVMRP and MOSPF depend on certain unicast routing functions: DVMRP runs a RIP-like distance vector routing algorithm and MOSPF requires the presence of OSPF function.

Table 3-1 summarizes the functions and features of the various multicast routing protocols.

3.2 Integrated Services

The basic service model supported by an IP network consists of a single class of service: best effort. This means that irrespective of the type of traffic (e.g., voice, video, data) submitted to the network, once inside the network, information packets are serviced on a first-come, first-serve basis. This was and is quite satisfactory for data-only applications that have no specific delay or bandwidth constraints. However, the Internet and by default IP networks are quickly evolving out of necessity to support voice, video, and multimedia as well as traditional best-effort data applications.

TABLE 3-1 Multicast Routing Protocols

	Name	Dense/Sparse Mode	Algorithm	Environment	Documentation
DVMRP	Distance Vector Multicast Routing Protocol	Dense Mode	RPM	Campus, Intranet	RFC1075
MOSPF	Multicast Extensions to OSPF	Dense Mode	Link-State	Campus, Intranet	RFC1584
PIM-DM	PIM-Dense Mode	Dense Mode	RPM	Campus, Intranet	
PIM-SM	PIM-Sparse Mode	Sparse Mode	RPM/CBT	Intranet, Internet	RFC2117
CBTv2	Core-Based Trees, Version 2	Sparse Mode	CBT	Intranet, Internet	RFC2189

The IP Integrated Services Architecture (IntServ) is an enhancement to the existing IP network model that supports and provides both real-time and best-effort services. The motivations behind the development of this model are fairly straightforward:

- *Best-effort service is no longer good enough.* This is true from the perspective that different applications with different network requirements are being deployed, end users demand reliable and consistent performance, and network providers will not make any money offering just a single service.
- *Emerging multimedia application suite.* No longer confined to just supporting text-based applications, IP networks must be able to support the integration of different traffic types including video, voice, and data and to offer QoS guarantees.
- *Optimize network/resource utilization.* Rather than overprovisioning network capacity to support all real-time and best-effort traffic (possible in the LAN; not so in the WAN), it might be less expensive and more desirable to reserve only those resources needed to handle the real-time traffic. Best-effort traffic can flow over any remaining network capacity.
- *Offer and deliver "premium" services.* The IntServ model enables a network provider to offer and support special premium services, thus differentiating themselves from the competition.

The notion of supporting QoS over packet networks is not new, and a number of research papers, experiments, and prototypes have appeared over the years to support this capability. RFC1633 was the first formal document to come out of the IETF that described the requirements and basic components of the IntServ model. The IETF Integrated Services (IntServ) working group[5] was eventually formed with a charter of specifying "this enhanced service model and then to define and standardize certain interfaces and requirements necessary to implement the new service model." To date the IntServ Working Group has completed its work in the definition of the IntServ model, the basic components, and new traffic classes. Its efforts have also been coordinated with those of the separate RSVP Working Group.[6]

---

[5] *http://www.ietf.org/html.charters/IntServ-charter.html.*

[6] *http://www.ietf.org/html.charters/rsvp-charter.html.*

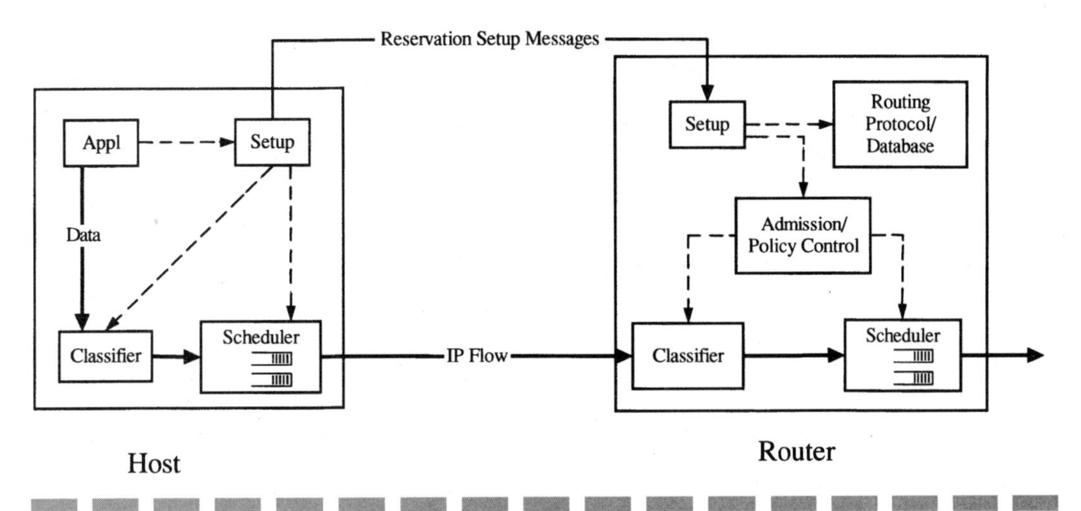

Figure 3-5 Integrated service model.

3.2.1 IntServ Model

The IntServ model is illustrated in Fig. 3-5 and consists of the following components:

- *Setup protocol.* A setup protocol enables hosts and routers to dynamically reserve resources in the network to handle the service requirements of a specific flow. RSVP is an example of a setup protocol. Other examples include ST2 + and Q.2931.
- *Flow specification.* Defines the QoS that will be provided for a specific flow. A flow defined in the context of the IntServ model is a series of packets flowing from a source to a destination with the same QoS. Conceptually and very loosely the flow specification could be viewed as a reserved set of network resources (e.g., minimum bandwidth) that the network must sustain to support the QoS of the flow. More specifically it includes objects that parameterize a policing function so that only traffic conforming to the service level defined in the flow specification will receive QoS. Nonconforming traffic will be treated as best effort.
- *Traffic control.* These are the components in the network devices (e.g., hosts, routers, switches) that control and manage the network resources needed to support the requested QoS.
  The traffic control components can be configured either by a sig-

naling control protocol such as RSVP or by manual configuration. The traffic control components consist of the following:

- *Admission control.* Determines whether the device (host, router, or switch) can support the requested QoS for a specific flow
- *Classifier.* Identifies and selects a service class based on the contents of certain fields in the packet header
- *Scheduler.* Provides the requested QoS service levels on the outbound link of the device

Observe from this model that the setup protocol is a control protocol only—it does not carry any data. The key components in the forwarding path of the IntServ model are the classifier and the scheduler.

3.2.2 IntServ Traffic Classes

The IntServ model currently supports three traffic classes: Guaranteed Service (GS), Controlled Load (CL), and, of course, best effort. GS[7] provides dedicated bandwidth, bounded delay, and no loss to queue overflows for conforming packets. GS was designed for applications with very stringent delay and bandwidth characteristics. Examples of applications that can benefit from GS include high-quality video conferencing, real-time financial transactions, and those transactions requiring a "circuit emulation" network service. GS is the only IP traffic class that supports a quantifiable QoS parameter (delay).

CL[8] service is more of a qualitative network service that approximates best-effort service over a lightly loaded network. CL makes no guarantees on bandwidth or delay, but unlike best-effort, CL will not noticeably degrade as the network load increases. CL was designed for applications that can tolerate some reasonable amount of delay or loss. Examples of such applications include low- to medium-quality audio/video multicasting or perhaps the provisioning of a tunnel across a private intranet that could carry delay-sensitive application traffic such as SNA or DEC Local Area Transport (LAT).

---

[7]RFC2212, "Specification of Guaranteed Quality of Service."

[8]RFC2211, "Specification of the Controlled-Load Network Element Service."

3.2.3 RSVP

RSVP[9] is the setup protocol component in the IntServ model. Essentially RSVP is a signaling protocol that enables IPv4 (and IPv6) applications to reserve resources for a stream of packets along a fixed path of network elements (hosts and routers). Although it is certainly possible for an application's QoS requirements to be satisfied by a traditional IP best-effort network with sufficient bandwidth and low delay or even by manually configuring the IntServ traffic control components along a path of routers, it is RSVP that gives an application the ability to dynamically request and receive QoS from the network and even to adapt QoS demands to changing network conditions.

Before describing the operation of RSVP, let's determine what RSVP is and is not. First, RSVP does not define or provide the QoS for a flow. The semantics of the QoS are defined in the form of "objects" by the IntServ architecture. An example of an IntServ object is the Flowspec (flow specification) that is used to parameterize the packet scheduler for a specific flow. The QoS for a flow is provided by the reservation state in the traffic control components, namely, the classifier (which identifies the packets of the flow) and the scheduler (which schedules the packets on the outbound link based on their QoS level). RSVP simply transports the IntServ objects opaquely and hands them off to the traffic control component at each host and/or router along the path. Second, RSVP is not a routing protocol. RSVP will work with any routing protocol such as RIP, OSPF, or BGP to route RSVP messages along a path from the source host to the destination host(s). Separating reservation setup from the routing process was a conscious decision by the designers of RSVP so as to provide maximum flexibility over any IP network by not linking its operation to that of a specific routing protocol. RSVP as part of the IntServ model is only concerned with the delivering QoS state to the host and routers along a path that is computed by normal routing protocols.

The operation of RSVP, like other (call or reservation) setup protocols is relatively straightforward. Resources are reserved for a particular call (flow) over a fixed path between a source host and destination host.[10]

---

[9]RFC2205, "Resource ReSerVation Protocol (RSVP)—Version 1 Functional Specification."

[10]The fixed path in this case is actually the routed path as computed by dynamic IP routing protocols that RSVP has reserved resources over. If dynamic IP routing computes a new path, RSVP must reserve resources along the new path.

But there are some distinct differences in the way RSVP operates versus other call/reservation setup protocols. These have mainly to do with the fact that it was designed to work over an IP connectionless network where network dynamics can suddenly alter the status quo as well as the fact that IP-based end users and applications are inherently heterogeneous. The unique capabilities of RSVP consist of the following:

- ✔ *RSVP is receiver-initiated.* The receiver of the flow, not the sender, is responsible for initiating and sending the reservation request into the network. In the case of a large multicast group, membership may be dynamic, and different receivers may have different network capabilities (some attached to 28.8-KB analog circuits, others attached to 100-MB Ethernets, etc.). Rather than burden the sender with all this information, it is the receiver that can best make a determination about its own QoS capabilities and follow up with the appropriate reservation request.
- ✔ *RSVP is soft state.* Reservation state along a fixed path will time out unless refreshed. Again, group membership in large multicast groups is dynamic, and one can avoid the overhead of explicit reservation tear-down procedures by simply timing out the reservation state. In addition, a topology change may result in a change in the best path leading to the destination.

The operation of RSVP in a multicast environment is illustrated in the conceptual diagram shown in Fig. 3-6 and consists of the following basic steps:

1. Sender begins transmitting packets to a group address.
2. Receivers join the group by sending IGMP host report messages to a directly attached router. With the assistance of a multicast

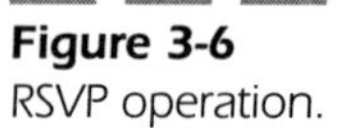

Figure 3-6
RSVP operation.

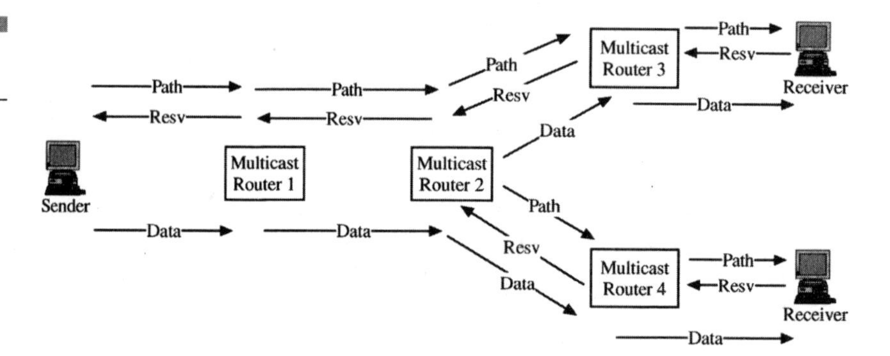

routing protocol, the routers create a delivery tree between the sender and the receivers.

3. An application on the sending host registers with and passes information through an Application Programming Interface (API) to the local RSVP entity. This information identifies the sending host, the destination (unicast or multicast address), and the characteristics of the flow that the sending application will generate.
4. The sending host generates an RSVP PATH message and sends it toward the receiving host(s). At each RSVP-capable router hop along the way, the PATH message stops, deposits path state in the routers (by leaving the IP address of the previous or upstream hop), and collects information about the QoS characteristics of the router (by updating an ADSPEC field). The PATH message marks the path between the sender and receiver(s) and collects information about the characteristics of path. Once the PATH messages reach their destinations, the receiver(s) have a pretty good idea about what they can request in terms of QoS based on the sender's transmission characteristics and the aggregate QoS capabilities of the route the PATH message traversed.
5. The application on the receiver host passes a reservations request over an API to the local RSVP entity. This in turn generates an RSVP RESV message that flows upstream toward the sender by following the reverse direction of the PATH messages. The RSVP RESV message contains a flow descriptor that is handed off to the traffic control component of each RSVP-capable router along the way. Admission and policy controls are invoked, and if the reservation request is accepted, the classifier and scheduler are updated. At branch points leading back to the sender (e.g., multicast router 2) RSVP may merge multiple downstream reservations into a single reservation, which reduces the amount of state that the router is required to maintain.
6. Once the RESV message reaches the sender, a fixed path with reserved resources in the direction of the sender to the receiver is established.
7. Because the RSVP state is soft, it will time out unless PATH and RESV refresh messages are periodically generated. Of course if a change in topology results in a new routed path between the sender and receiver, the whole PATH/RESV process must be repeated along the new routed path.

IP IntServ's and RSVP are described in RFCs 2205 through 2216. In summary, the important functional capabilities of RSVP are that

- There are support reservations for both unicast and multicast flows.
- Reservations are unidirectional only, which means they are only established in the direction of the sender to the receiver(s).
- Reservation requests are generated by the receiver.
- Reservations are soft state and will time out unless refreshed.
- It transports traffic and policy control information opaquely.

3.2.4 Other RSVP and Integrated Services Issues

The IntServ Model along with RSVP defines the machinery by which hosts and routers can support QoS. However, that alone may not be enough given the existing IP model to support true end-to-end QoS. In any typical network, hosts and routers are connected to each other over different data link technologies. Some such as ATM are capable of supporting QoS. With others such as shared Ethernet it may be very difficult to enforce a QoS for a specific flow. The IETF has formed a working group called the Integrated Services over Specific Link-Layers (ISSLL)[11] to tackle the issues of supporting IntServ and RSVP over different data link technologies such as ATM, 802 LANs, and low-bit-rate links.

Another issue involves the lack of QoS routing in IP routing protocols. The metrics used by these routing protocols consist of non-QoS variables such as the number of hops or the sum cost of the links to a destination network. Metrics such as available bandwidth capacity, delay, jitter, and other QoS attributes are simply not used. In the case of RSVP, the PATH is forwarded over a path computed by the routing protocol using non-QoS metrics. This could result in an attempt to establish a path over a suboptimal path, which in turn could lead to a reservation denial. Even if another path with sufficient QoS capacity is available, the routing protocols do not have any way of checking for or even making use of the alternate path. The problem definition and solutions for QoS routing are

---

[11]*http://www.ietf.org/html.charters/issll-charter.html.*

currently under examination by the Quality of Service Routing (QoSR)[12] Working Group of the IETF.

And finally, as noted in RFC2208, there are some concerns with the operation of RSVP regarding the following:

- *Scalability.* It may be difficult for current routers to support and maintain many small RSVP reservations or to classify flows over high-bandwidth connections. There is not yet a clear way to define reservations for an aggregate number of flows.
- *Security.* Protecting against a theft or hijacking of service is a big concern and should be addressed with an adequate, but as yet to be deployed, distributed key management system.
- *Policy.* Networks that require the delivery of service based on certain policies will need mature policy control mechanisms working together with RSVP entities.

3.3 Differentiated Services

Provisioning differentiated service levels over IP-based networks has been a very active area of research and standards development. Initially it was thought that ATM with its powerful QoS capabilities would solve the problem, but the ATM connection to every desktop and home has not (yet) materialized. Another very promising solution was the Integrated Services Model covered in the previous sections. But although theory, simulations, and some early testbed[13] results indicate that it can work, it is not likely that RSVP will emerge as the end-to-end QoS answer that everybody had hoped for. The reasons are quite simple: The overhead associated with maintaining per-flow state and signaling does not allow this model to scale. Therefore, there has to be something that falls between the best-effort FIFO service of the current Internet and the envisioned per-flow heavyweight mechanisms of RSVP and IntServ. The answer is called differentiated services (DiffServ). DiffServ uses packet marking and per-class queuing to support priority services over

[12] *http://www.ietf.org/html.charters/qosr-charter.html.*

[13] Baugher, M., and Jarrer, S., "Test Results of the Commercial Internet Multimedia Trials," *Computer Communications Review,* vol. 28, no. 1, Jan. 1998.

IP-based networks. DiffServ[14] is now a formal IETF working group and the group's work will result in a standard for differentiated services over IP networks.

3.3.1 DiffServ Model and Operation

The basic ideas behind DiffServ include the following:

- ✓ Define a small number of service classes or priorities. An example could be low, medium, and high priority. Another example could be best-effort and real-time. A service class may be associated with a traffic profile that consists of a minimum and maximum allowable bandwidth as well as a burst size or duration, or it simply could be a minimum amount of bandwith.
- ✓ Classify and mark individual packets at the edge of the network as belonging to one of the service classes. This may involve examining one or more fields in the IP and/or transport headers of each packet. One or more bits in the header of each packet may be marked to reflect the packet's service class or some other per-hop behavior such as drop preference.[15]
- ✓ The core network switches and routers service the packets based on the contents of the marked bits in the packet header.

This provides several significant advantages over the IntServ model. First, there is no requirement for per-flow signaling or state to be maintained in the network. Second, priority service can be applied to an aggregate number of individual flows that are mapped to a single service class. This enables network operators to easily provision and offer a small number of varying service levels to all potential end users and applications that may use the network. The aggregation of many individual flows to a few specific services allows this model to scale much more efficiently than do per-flow techniques. Another common name for this technique is COS.

Another important advantage of the DiffServ model is the fact that no change is required on the individual hosts or applications to receive a priority service. It is the job of the edge components to classify, mark,

---

[14] *http://www.ietf.org/html.charters/diffserv-charter.html.*

[15] Drop preference indicates packet's willingness to be discarded if congestion is encountered.

and possibly condition any packets entering the network. Similarly there are minor or even no changes needed to the core routing or switching elements. Most of those devices today support some sort of queue management. Adding a bit-level classifier to identify and place the packets in the appropriate service queue is a small upgrade when compared with an upgrade needed to support RSVP and IntServ traffic control. If the core network consists of frame relay or ATM, the edge components can simply place the packets on different per-priority VCs.

DiffServ provides flexibility and latitude in the kinds of services that network service providers may offer. The service bits in the packet header are only used to classify the packet. It is up to the service provider to define and provision what kind of service will be offered. Some service providers may offer bandwidth and delay guarantees for high-priority traffic; others may offer just minimum bandwidth. The only requirement is that the bits semantics are consistent so that multiple ISP networks interpret the priority bits in the same manner.

One other important advantage that is discussed in the next section is the notion of Virtual Private Networks (VPNs) and differentiated services. VPNs are formed by building secure tunnels through the Internet. Packets flowing over a secure tunnel may have all or portions of the transport header and payload encrypted. DiffServ can still provide priority service for these packets because it is only looking at bits in the cleartext packet header. This enables differentiated services to operate over VPNs.

DiffServ does have some issues that must be addressed. First, unless the network can be provisioned accordingly, the services offered by a DiffServ network may not approach the hard guarantees in bandwidth and delay that IntServ or ATM can deliver. It is really up to the individual networks to decide. Second, the edge components still require high-performance per-packet classifiers, which are not really any different from those found in the IntServ model. In addition, the issue of how classifier state will be managed on what could be a potentially large number of edge components needs to be addressed. Possibilities include manual configuration, Simple Network Management Protocol (SNMP), Lightweight Directory Access Protocol (LDAP), and RSVP. And finally it is important to be able price the service so that only a subset of the customers or subscribers will ask for it. If there is only a nominal differential in price between high-priority and best-effort services, presumably everyone would request high-priority services and that would defeat the purpose.

The DiffServ model consists of the following components:

- *DS byte.* The differentiated services byte is actually a revamped IPv4 TOS field and the newly defined IPv6 Traffic Class field. The bits in this field indicate the service the packet is expected to receive.
- *Edge components.* These devices are positioned at the ingress and egress of a DiffServ network. They are responsible for packet classification, marking, and conditioning (e.g., policing, shaping). An edge component can be a host, firewall, router, or some other edge device.
- *Interior components.* These devices can be core switches or routers that provide a per-hop service based on the contents of the DS byte. They should employ queue management techniques that control queue depths (e.g., Random Early Detection, RED) and schedule outbound transmission on a per-service basis (e.g., weighted fair queuing (WFQ), class-based queuing (CBQ), priority queuing, etc.)
- *Policy management.* Tools and people will be needed to administer the service policies and perhaps measure and enforce any service-level agreement made between the network and the end users. This is because someone will have to pay for receiving better service.

The specific elements contained within a DiffServ edge and core router are illustrated in Fig. 3-7.

The components and functions performed by the edge router consist of the following:

- *Multibyte classifier.* May look at one or more fields in the packet including the IP header, transport header, and even the payload
- *Policer.* May shape or discard packets so that they comply with traffic profile of the service to be delivered
- *Packet marker.* Marks one or more bits in the DS byte for those packets associated with a specific class or profile
- *Queue management/scheduler.* Manages queue lengths and schedules packets for transmission

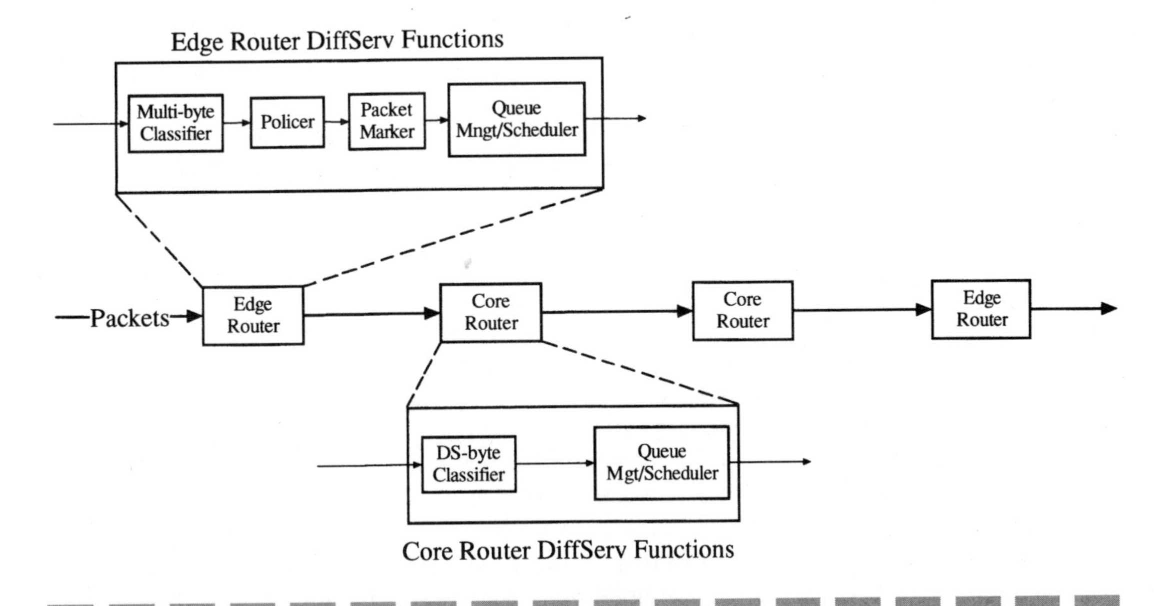

Figure 3-7 DiffServ edge and core router functions.

The functions of the core router are fewer and much simpler:

- *DS byte classifier.* Looks at the DS byte in the IP header
- *Queue management/scheduler.* Same as for the edge router

As illustrated in Fig. 3-7, DiffServ pushes per-packet classification and complexity to the edge of the network, leaving the high-performance switches and routers to do what they do best: forward packets at high speeds through the network.

3.3.2 DS-Byte Format

Although the IPv4 TOS byte was originally designed to establish transmission precedence and TOS route selection, it was never really deployed, or if it was, it was done in a proprietary manner between homogeneous devices. Given that the elementary function of DiffServ is to provide per-packet priority service, it was decided to resurrect the TOS byte and redefine it so as to meet the functional objectives of the DiffServ model.

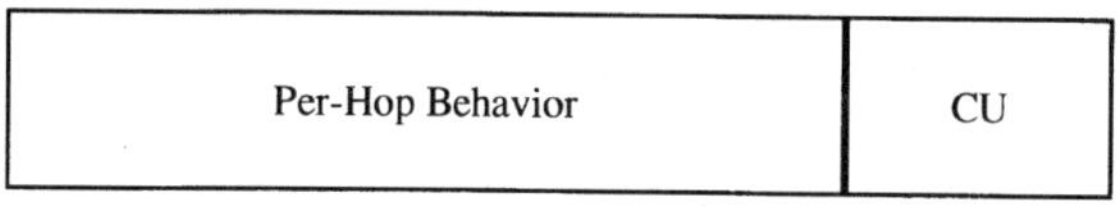

Per-Hop Behavior	CU

Figure 3-8
DS byte definition.

Several suggestions on how those bits should be allocated have emerged so far.[16] One preliminary suggestion that is a composite of earlier efforts is illustrated in Fig. 3-8. The first 6 bits define a per-hop behavior field that specifies the service or forwarding behavior that the packet should receive at a particular switch or router. The two CU bits are "currently undefined."

Work still remains in this area and it is likely that the DS byte definitions will change. Generally speaking, the final definition will include a per-hop behavior that specifies a few suggested services (e.g., default, high priority) but will leave enough room for implementation-specific treatment by network providers. There will also most likely be a drop preference bit that will be useful in controlling congestion through the use of active queue management techniques such as RED.[17] Other possibilities include an Explicit Congestion Notification (ECN) that would enable sources to adapt their transmission rates to available network capacity.

3.4 IP Security

Network security has always been a necessary and important element of data networks. In the "old days" it was the computer itself and the applications that were the resources. Protecting those resources was a relatively simple job. Confidential information was stored on big mainframe computers housed in monolithic glass houses. Password protection was used to authenticate and allow access to a closed set of end users who resided on a corporate network that was closed to the outside world. The Internet back then was a primitive and slow email and file transfer tool for various governmental and academic institutions.

The Internet and the computer revolution have changed all that. Now computers are fast, inexpensive, and mobile and typically store confidential information. While enabling ubiquitous communications on a

[16] *http://diffserv.lcs.mit.edu/.*

[17] *http://www-nrg.ee.lbl.gov/floyd/red.html.*

global scale, the Internet itself is about as unsecure as one can imagine. This means that end users transport data over and through network elements that are beyond their control or visibility. If the data are confidential in nature, they are exposed to possible theft, unauthorized inspection, deletion, or worse.

At the same time, private corporate networks must interact with and through the Internet. It is big business in the form of advertising, e-commerce, and overall global visibility. But doing so exposes corporate network resources to attacks from hackers, spoofers, pranksters, and other unscrupulous elements who may be doing it for laughs or may have some darker motives. Whatever the reason, network security is now more critical and necessary then ever before.

There are many ways of securing network traffic. One can erect a firewall at the edge of the private network so that unauthorized traffic is filtered out. Applications or transport protocols may use their own form of security. Another technique is to provide security at the IP layer or more specifically on a per-IP-packet basis. This makes sense for several reasons:

- ✔ The Internet and corporate intranets are IP-based. All traffic must travel through the IP layer. All traffic is carried in IP packets.
- ✔ It protects and isolates the upper-layer applications from security attacks.
- ✔ It compliments existing upper-layer security mechanisms.
- ✔ It is an enabler for the establishment of scalable and secure VPNs over the Internet

To address the need for security at the IP layer, the IETF formed the IP Security (IPsec) Working Group[18] Their efforts have resulted in a set of protocols, mechanisms, and services that provide network-layer security for IPv4 and IPv6. The services supported by the IPsec framework include the following:

- ✔ *Access control.*
- ✔ *Data origin authentication.* Validates the sender identity of each IP packet.

---

[18]*http://www.ietf.org/html.charters/ipsec-charter.html.*

- *Replay protection.* Prevents an attacker from stealing packets and playing them back at a later time
- *Data integrity.* Verifies that the IP packet was not tampered with while in transit
- *Data confidentiality (encryption).* Conceals some portion of the packet through encryption
- *Limited traffic flow management.* Hides the IP address of the original sender
- *Key management.* Manages the generation and distribution of encryption keys

The initial protocols defined by the IPsec framework are the Authentication Header (AH), Encapsulating Security Payload (ESP), and Key Management.

3.4.1 IPsec Security Associations

There are several pieces of information that must be made known to communicating parties using IPsec protocols. Each party must know the other's network address. And each party must agree on mechanisms such as cryptographic algorithms, shared keys, digital certificates, etc., to communicate in a secure manner across the network.

An IPsec Security Association (SA) is a unidirectional logical connection between two or more parties that defines a logical channel and the mutually agreed upon security mechanisms applied to data flowing over that channel. An IPsec SA provides a secure channel of communications for the exchange of data between IPsec parties.

IPsec SAs can be established between communicating parties in one of two modes: tunnel and transport. Tunnel-mode SAs establish secure channels between two intermediate devices (e.g., router, firewall). For example, a tunnel-mode SA could be established between the two routers. Tunnel-mode SAs encapsulate the original IP packet in a new IP header where the source and destination addresses are the tunnel endpoints. Security services are only applied to packets that are flowing between the two routers.

Transport mode SAs establish secure channels between two hosts, thus allowing security services to be applied on an end-to-end basis. It is also possible to nest a transport-mode SA inside of a tunnel-mode SA to enhance the overall security applied to packets flowing between different entities.

A specific SA is identified by the destination IP address, the security protocol (AH or ESP), and the Security Parameters Index (SPI) value. These values are carried in each IP packet and enable the receiving end system to select the SA that the packet will be processed under.

Examples of a tunnel- and transport-mode SA are illustrated in Fig. 3-9.

3.4.2 Authentication Header and Encapsulating Security Payload

The initial IPsec protocol suite consists of the AH and ESP protocols. AH supports data origin authentication and data integrity and has an option for replay protection. The integrity of the entire IP packet is verified except for the mutable fields in the packet header. Mutable fields such as the TTL and the header checksum may be altered at each hop. IPsec AH does not provide data confidentiality—the data payload still flows as cleartext. In addition, IPsec AH supports both tunnel and transport mode SAs.

IPsec ESP provides a higher degree of security that includes data origin authentication, data integrity, data confidentiality (encryption), and optional replay protection. ESP protects the entire payload by encrypting the data. If a tunnel-mode ESP SA has been established, the entire original IP packet including all header information will be encrypted, thus concealing the identity of the original data sender and destination.

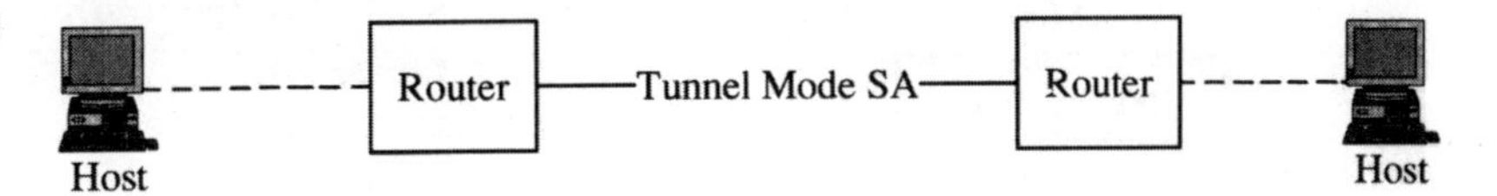

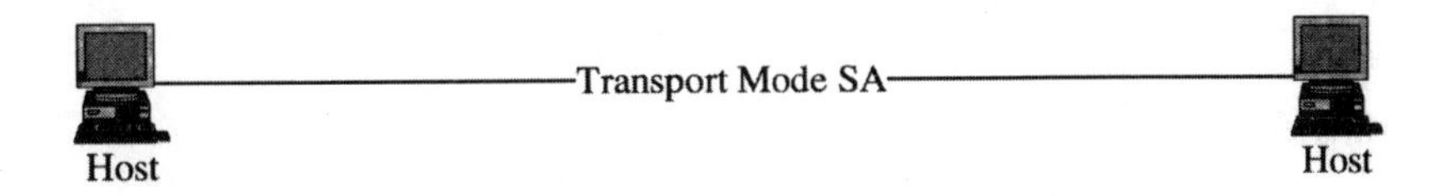

Figure 3-9 IPsec security associations.

The placement of the AH and ESP protocol headers within an IP packet for both tunnel and transport mode is shown in Fig. 3-10.

3.4.3 Key Management

The basic operation of cryptography requires that the communicating parties share a secret key. The secret key is used by the sender to encrypt the data and by the receiver to decrypt the data. If the secret key is lost or stolen, the data could become compromised and so a new key must be immediately provided. IPsec entities that wish to establish an ESP SA with each other must be provided with a shared secret key. The key can be manually configured, but as the number of SAs grows in the network, manual configuration imposes severe scaling limitations.

The IPsec working group is addressing the requirement for key management as defined in the Internet Security Architecture Key Management Protocol (ISAKMP). ISAKMP is not really a protocol but more of a framework that manages the process of securely establishing SAs in dynamic network environment. ISAKMP, or ISAKMP/Oakley as it is known, defines a framework for key generation, automatic key refresh, and key distribution. It is independent of any specific cryptographic algorithm and can operate to establish an SA over unsecured network links.

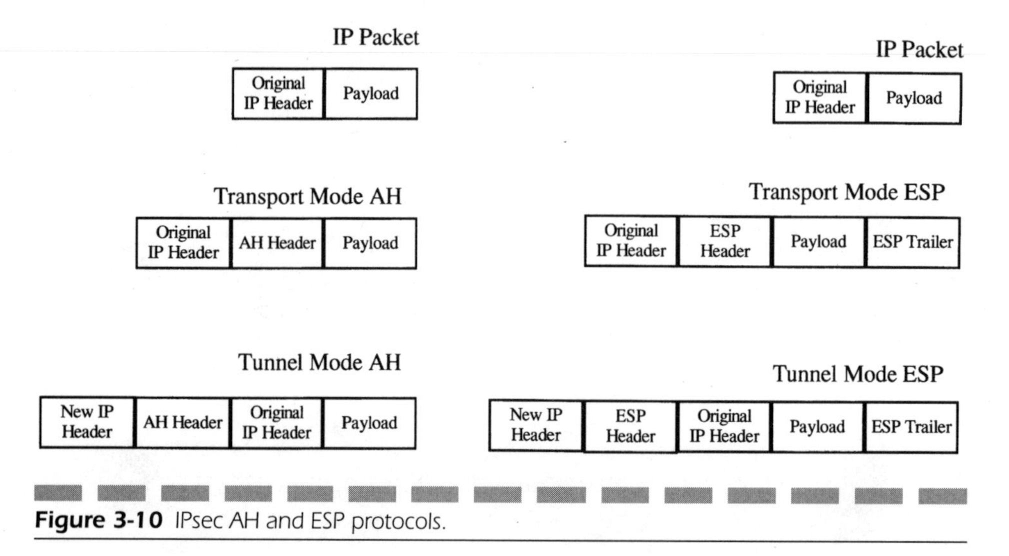

Figure 3-10 IPsec AH and ESP protocols.

3.5 RTP/RTCP

Streaming of audio and video flows over IP networks (e.g., RealAudio, InternetPhone, etc.) is now quite popular. The Real-Time Transport Protocol (RTP) is a new application-layer framing protocol (not really a transport protocol) that enables real-time applications to stream data over an IP network. Multimedia streams are first encoded (and compressed) and then encapsulated in RTP packets and transported via UDP/IP over unicast or multicast sessions. Timing, sequencing, and payload identification information is conveyed between RTP senders and receivers so that RTP receiver can begin "playing" the multimedia stream as soon as the first few frames arrive. RTP does not establish connections and does not guarantee that the data will arrive at the destination on time. RTP is a standard that will support all video and audio encodings (RFC1890) and will be incorporated into H.323 video-conferencing applications. The Real-Time Transport Control Protocol (RTCP) is a complimentary control protocol that is used to provide feedback and statistics to RTP session members.

3.6 IPv6

IPv6 is the next-generation version of IP, and among its many improvements is the address space that was increased fourfold to 128 bits. The increased IPv6 address space was implemented to address, what was in the early 1990s, the very real concern about the complete exhaustion of the IPv4 address space. Depending on who you talked to, it was projected that the entire 32-bit address space would be exhausted by the end of the decade if not sooner. That did not happen, but the IETF used the opportunity to engineer a new version of IP to address some of the real and perceived shortcomings of IPv4. A noninclusive sampling of some of those shortcomings and requirements follows:

- ✓ *Current IPv4 address space is too small.* Larger address space is needed.
- ✓ *Routing table size.* The number of networks in the Internet and the corresponding routing tables are growing quite large. A large address space allows more flexibility to create hierarchies and in

the spirit of CIDR aggregate many lower-level entries into a single higher-level entry.

- *Simplicity.* Many things about IPv4 are or seem complex. Fragmentation, options processing, and configuration are just a few examples.
- *Performance.* Large routing tables, fragmentation, and options processing all contribute to latency and delay in routers.
- *Ease of configuration.* IPv4 devices *must* be configured with a valid IP address or use Dynamic Host Control Protocol (DHCP) to retrieve one from a server.
- *Security.*
- *Extensibility.* This is the ability to seamlessly add new functions and features without requiring an overhaul of the existing protocol.
- *Coexistence with and a transition to IPv4.* It is safe to say that IPv4 devices and applications will be with us forever. Any new version of IP must work with and coexist with IPv4.

From these requirements and through the efforts of a good number of very smart and experienced network engineers came what is now called IPv6. IPv6 is the new version of IP and is not radically different from the overall operation of IPv4. Networks are still assigned a network address or prefix, IPv6 still forwards packets in a connectionless hop-by-hop manner, IPv6 routers still run routing protocols, a network device must be configured with a valid IPv6 address, and so on. Perhaps IPv6 can be best viewed as a simpler, more scalable, and more efficient version of IP.

The primary functions and features of IPv6 are

- *Expanded address space of 128 bits.*
- *Enhanced routing capabilities.* There are a number of little things here that should improve the performance and control of routing. Flexibility in creating address hierarchies is enabled with the larger address space. Source-routing is much more efficient using the routing extension header. The existence of a flow label field in the IPv6 header enables packets of a flow to be identified for special handling. The multicast address space is larger and includes a scope field to efficiently limit the range of a multicast transmission. Fragmentation is prohibited in routers, and the processing of options is much more straightforward and does not burden the

intermediate routers in many cases. Anycast addresses are defined in which a packet can be addressed to the "nearest" of one of several possible interfaces.

- *Simplified header format.* The IPv4 header contains 10 fields plus two addresses and possibly some options. The IPv6 header contains six fields and two addresses and possibly some extension headers.
- *Improved support for options.* Options, now called extension headers in IPv6, are encoded differently, which enables better performance and more flexibility to introduce future extensions.
- *Flow label.* A source can now label the packets of a particular flow that requires special service or handling by the intermediate routers. This will simplify the process of packet classification at each router.[19]
- *Transition mechanisms.* Defined within IPv6 are a number of mechanisms that will facilitate coexistence and interoperation with IPv4 hosts and routers.

The IPv6 header is 40 bytes long. Its format is shown in Fig. 3-11 and discussed below:

- *Version.* The current version is 6.
- *Traffic class.* This is DiffServ DS byte for IPv6 packets.
- *Flow label.* This is a nonzero value assigned by the source that identifies the packets of particular flow for special service or handling (e.g., QoS or nondefault special service). The flow label is 20 bits in length.
- *Payload length.* This is a 16-bit value that defines the length in octets of the payload plus any extension headers that follow the base IPv6 header.
- *Next header.* This defines the type of header immediately following, which could be an extension header, the transport layer header (e.g., TCP), or even the data payload. The next header is 1 octet in length.

[19]A common misperception is that IPv6 supports QoS. It only supports QoS to the degree that IPv4 can. IPv6 still requires IntServ or DiffServ mechanisms for service differentiation including QoS. The flow label and traffic class fields just offer a faster mechanism for identifying packets for special handling.

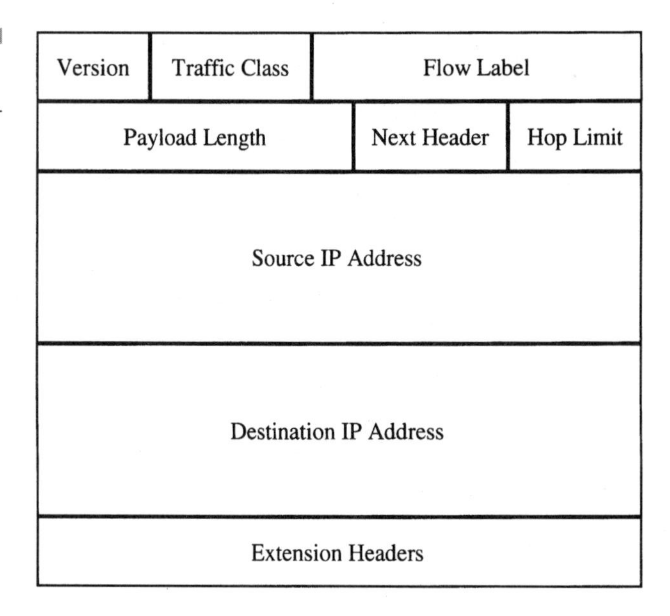

Figure 3-11
IPv6 header.

- *Hop limit.* This is an 8-bit value that offers the same function as the TTL field in IPv4.
- *Source address.* This is a 128-bit source IPv6 address.
- *Destination address.* This is a 128-bit destination IPv6 address.

After the 40-byte base IPv6 header and before the actual payload there may be one or more extension headers. Like IPv4 options, the extension header is a way to include information about any special processing that the packet may require. Most IPv6 extension headers are only processed by the destination and not by the intermediate routers, resulting in much better performance. The contents of the extension header are identified by the next header field that is present in both the base IPv6 header and in any of the extension headers. This concept is illustrated in Fig. 3-12, and the currently defined IPv6 extension headers are shown in Table 3-2.

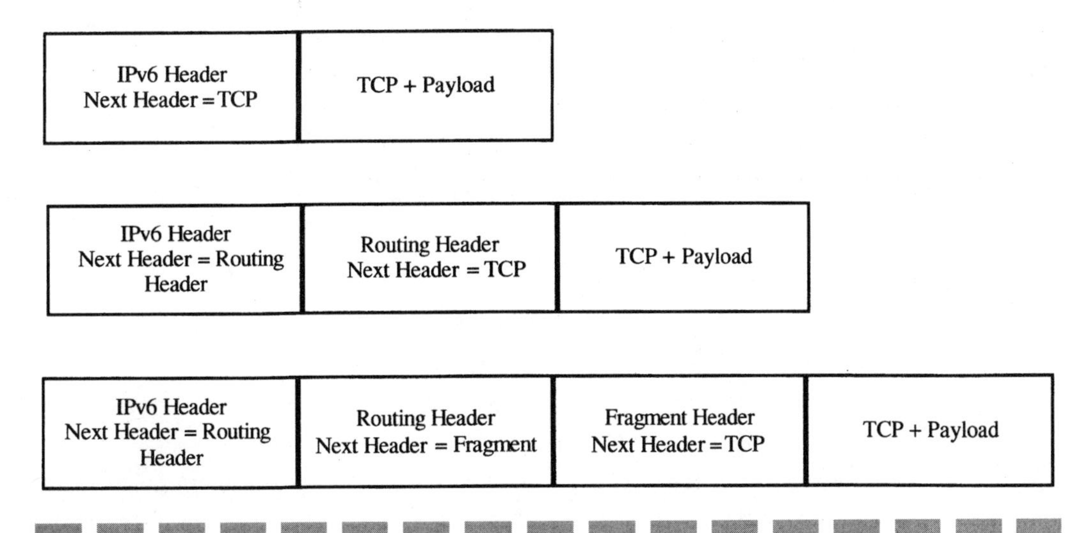

Figure 3-12 IPv6 extension headers.

TABLE 3-2

IPv6 Extension Headers

Extension Header	Description
Hop-by-Hop	Must be examined by every router along the path; examples are Jumbo Payload for IPv6 packets greater than 65,535 bytes in length and Router Alert, which instructs routers to more closely examine IPv6 payload (e.g., RSVP messages)
Routing	Defines source routes
Fragment	Is inserted by source host and permits sending of packets larger than the link MTU
Authentication	Ensures that the source address is authentic and that the packet has not been altered
Encapsulating Security Payload	Ensures confidentiality between sender and receiver
Destination	Processed only by destination host

MORE INFORMATION

Ferguson, P., and Huston, G., *Quality of Service,* Wiley Computer Publishing, 1998.

http://playground.sun.com/pub/ipng/html/ipng-main.html

http://www.ietf.org

http://www.ipmulticast.com

Huitema, C., *IPv6 The New Internet Protocol,* 2d ed., Englewood Cliffs, NJ: Prentice Hall, 1998.

Huitema, C., *Routing in the Internet,* Englewood Cliffs, NJ: Prentice Hall, 1995.

Keshav, S., *An Engineering Approach to Computer Networking,* Addison-Wesley, 1997.

Maufer, Thomas A., *Deploying IP Multicast in the Enterprise,* Englewood Cliffs, NJ: Prentice-Hall, 1998.

Stevens, W. Richard, *TCP/IP Illustrated Vol. 1 The Protocols,* Addison-Wesley, 1994.

Thomas, Stephen A., *IPng and the TCP/IP Protocols,* John Wiley & Sons, 1996.

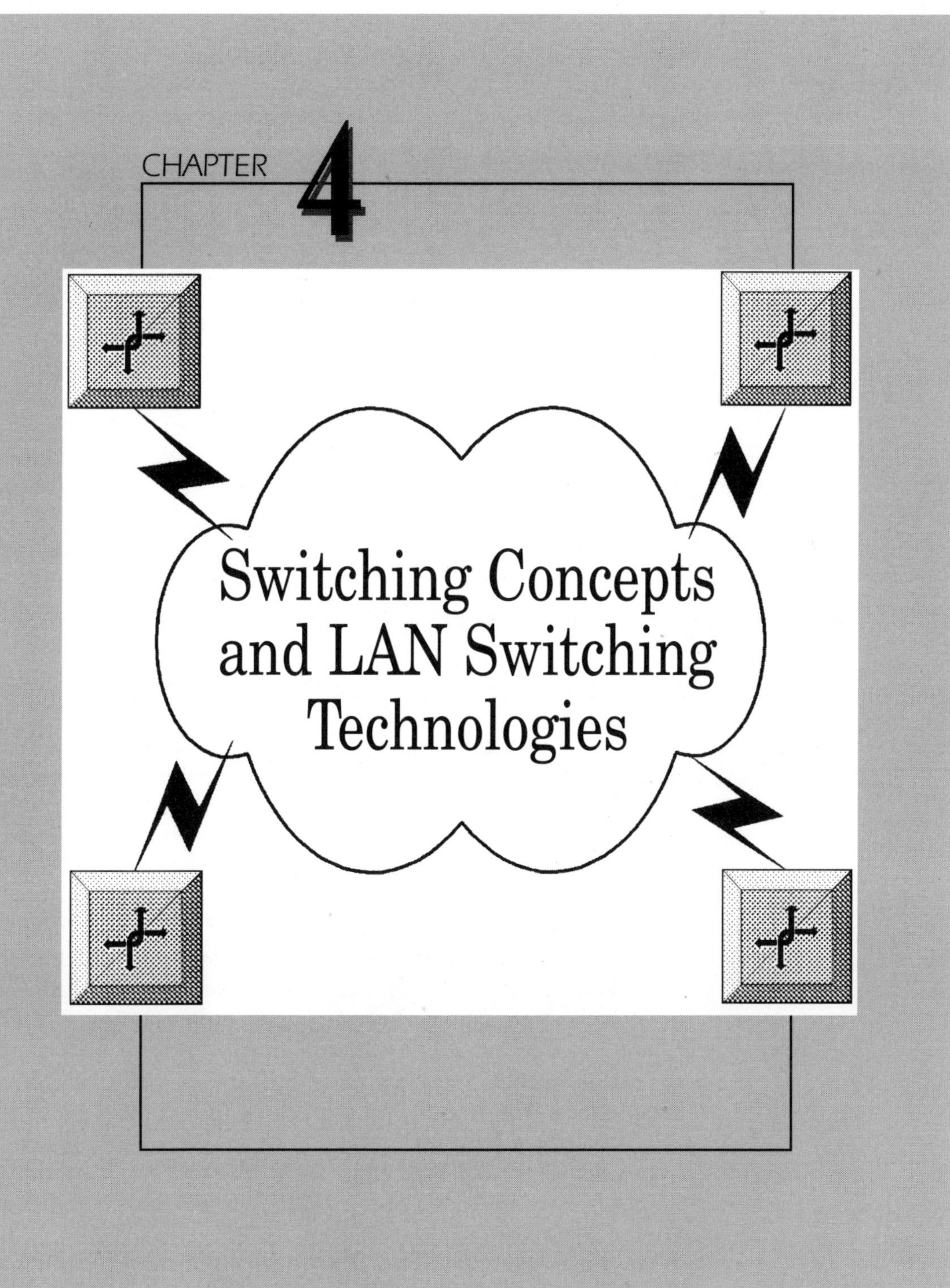
CHAPTER 4
Switching Concepts and LAN Switching Technologies

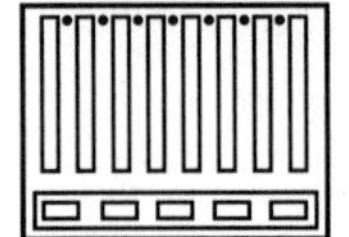

The previous chapters presented an overview of IP, or more specifically the architecture and protocols that compose TCP/IP. To further our discussion and contribute to the foundation behind IP switching, we will first discuss the topics of switch forwarding and switch path control. Switch forwarding techniques utilize information contained in the packet and maintained in the switch to forward the packet through the switch. Switch path control is used to create and maintain a switched path through a network of switches.

In data networking, switching can take on many different forms from defining a technique to rapidly move data packets between different LAN segments to being a full-blown end-to-end network architecture such as ATM. This chapter will review the basics of LAN switching. In the next chapter frame relay and ATM will be covered, and it is hoped that these two chapters will lay the groundwork for the IP switching solutions covered in the rest of the book.

4.1 Switching Concepts

Take the name of any network or application technology and append the word *switching* to it, and you will instantly invigorate it with an aura of high performance, low cost, and reduced complexity. Examples are Ethernet and Ethernet switching, Token Ring and Token Ring switching, and IP and IP switching. Network technologies that are already orders of magnitude faster than today's installed base sound better when invoked with the word *switching* (e.g., Gigabit Ethernet and Gigabit Ethernet switching). Heretofore distinct and seemingly unrelated computing elements were being combined with the word *switching* to form new and almost bizarre network technologies (e.g., application-layer switching).

What is switching and why is it so popular? The answer to the second question is simply price and performance. Switching technologies and in particular those associated with LANs offer much greater performance and capacity at much lower prices when compared to traditional shared LAN technologies. Now and for the foreseeable future, advances in silicon are placing more network processing on inexpensive chips, which drives the price down and boosts performance by orders of magnitude over older software-based processing. So the popularity of switching is rooted in the fact that it is fast and cheap.

So what exactly is switching? In the world of data networking, switching can take many different forms. LAN switching is used to move data packets between workstations on the same or different segments. WAN switching generally takes the form of a virtual connection that is provisioned between two endpoints such as a pair of routers. Whatever environment one may be operating in, all forms of switching share the following basic properties:

- *Operate at layer-2 and below of any protocol stack.* This means that LAN or WAN switching is transparent to IP or any other network-layer protocol or any of the applications running above it for that matter. Figure 4-1 illustrates two IP endpoints communicating over a layer-2 switched path.
- *Are performed in hardware.* Switches move data packets received on an input port to the output port for transmission without consulting a separate processor. Performing this function in hardware (or firmware) generally leads to better performance, lower latency, lower complexity, and lower costs.

The process of switching can be viewed from two different perspectives: local and end to end. From the local point of view, how does the individual switch decide, upon receiving a packet, which output port it will switch the data packet to? In other words, what information is available in the data packet and maintained in the switch that enables the switch to rapidly move data packets from an input port to an output port? This is called switch forwarding. The second perspective deals with the process by which a switched path between two endpoints is established and maintained through a network of switches. This is referred to as switch path control.

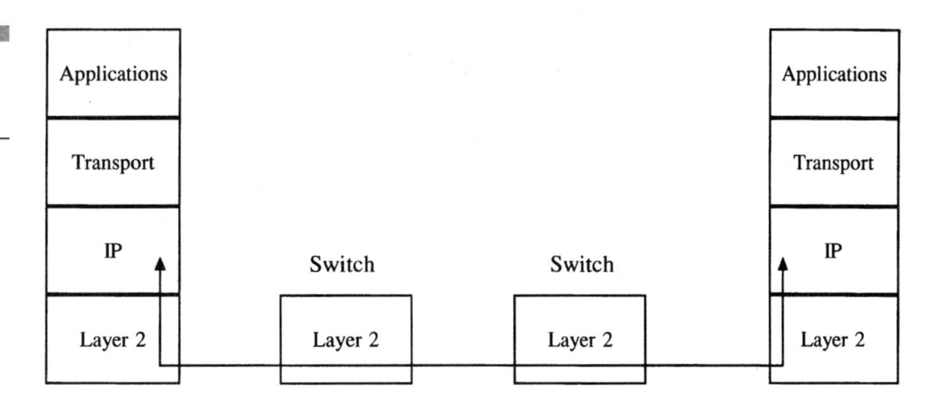

Figure 4-1 Layer-2 switching in the protocol stack.

4.1.1. Switch Forwarding Techniques

A switch is simply a box with some number of ports that different devices such as workstations, routers, and other switches attach to. When a packet arrives at an input port, the job of a switch is to either discard it or move it to the correct output port and send it on its way. The correct output port is determined by information contained in the packet and in some cases in the switch itself. The information contained in the packet consists of one of the following:

- ✓ Destination address
- ✓ Source-route vector
- ✓ Connection identifier

In the first case the packet contains a destination address field, which is simply the address of the endpoint (e.g., host) that the packet is destined for. Through one of several techniques discussed below, the switch has constructed an address/port table that consists of entries containing a destination address and the associated output port. When the packet arrives at the input port, the switch checks the destination address in the packet, matches it against an entry in the address/port table, and sends the packet on its way. Of course if there is no match, the switch may drop the packet or decide to flood it out *all* output ports except the one it arrived on in the hope it might reach its destination. If the destination address in the packet and the port that it came in on match the entry in the address/port table, the switch will drop the packet. This is based on the common sense assumption that a switch should not forward the packet out the interface it came in on.

Another observation is that the packet arrives at the switch in its native format, which implies that the sender can just transmit it, and the switches will handle the rest. This technique, common in Ethernet LAN switching, is illustrated in Fig. 4-2. The packet with a destination address of "aaa" arrives at the switch, the switch consults the address/port table, and forwards the packet to output port 4.

The next technique uses a source-route vector contained in the packet. A source-route vector is a sequence of topological elements in the network that the packet must "visit" on its way to the destination. The topological elements can range from the output port numbers of the switches along the path, the addresses of each switch along the path, or in the case of token-ring source routing, one or more ring-number-bridge number-ring number combinations. Again, through techniques dis-

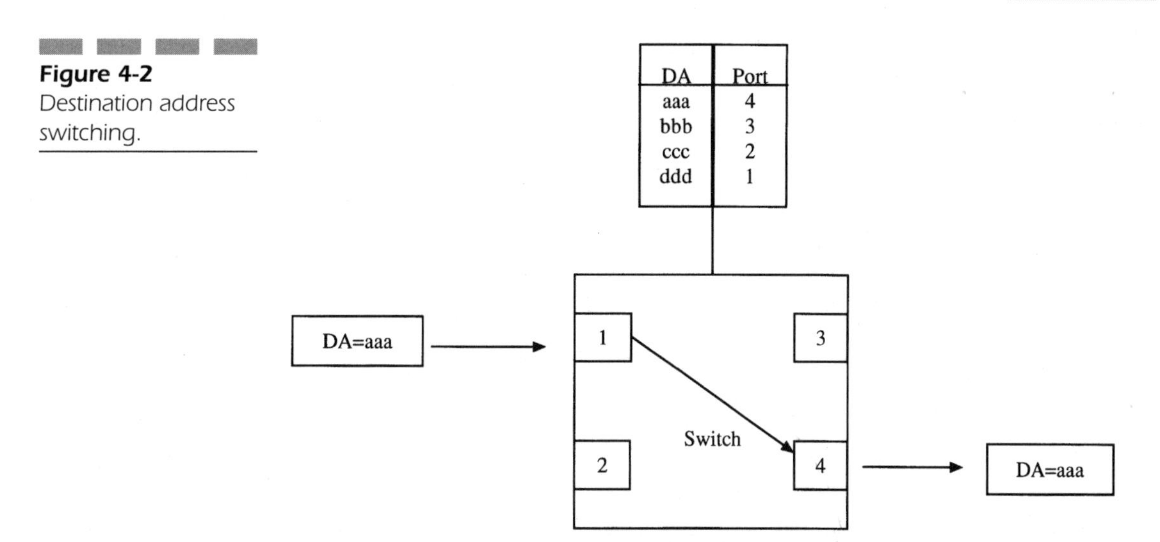

Figure 4-2 Destination address switching.

cussed below, a source-route vector is inserted into the packet, and the switch uses this information to direct the packet toward the appropriate output port. Note that the switch does not need to construct or maintain, per se, an additional switching table such as in the previous example. The source-route vector contains all the information about the path to the destination and the switch just needs to act upon it. Accepting the fact that it takes some work on the part of the network and sender to distribute the information necessary to assemble an accurate source-route vector, it is apparent that source-routing is a very efficient technique for switching packets through a network. Source-routing is used in token-ring and also by ATM PNNI Phase I when routing SVC setup requests. An example of source-routing using the output port as the vector element is illustrated in Fig. 4-3.

The third technique involves the use of a connection identifier contained in the packet to determine the output port. A connection identifier may also be referred to as a label. This is somewhat different from

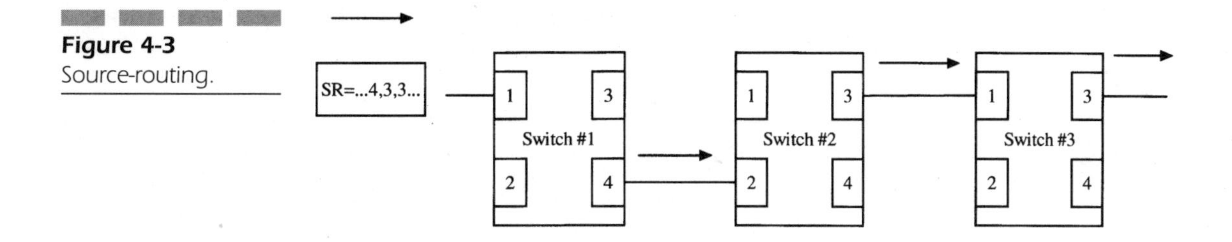

Figure 4-3 Source-routing.

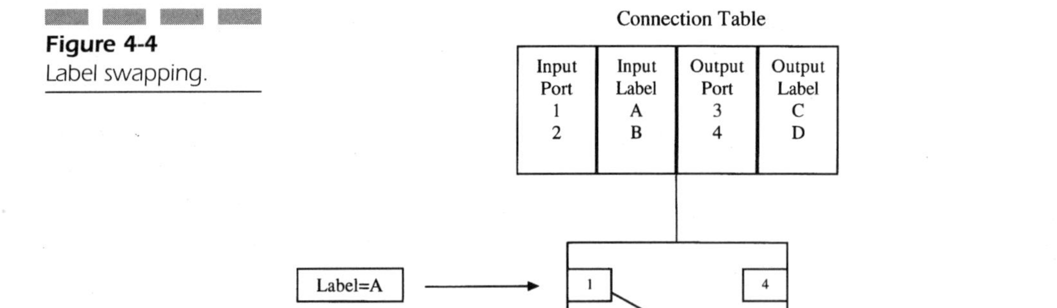

Figure 4-4
Label swapping.

the previous two techniques because in this case, the label in the packet header is swapped as it passes through the switch. Once again, through techniques discussed below, a label is appended to the packet, and the switch has constructed a connection table consisting of input port numbers, input labels, output port numbers, and output labels. When the packet arrives at the switch, the input port number and packet label are used to index an appropriate entry in the connection table. The label in the packet is promptly swapped, and the packet is directed out the appropriate output port. This can be done very rapidly, in hardware in most cases, because the switch merely has to extract a fixed-length label from the packet and compare it to entries in the connection table. Label swapping, as it is more generally known, is the forwarding paradigm used in ATM, frame relay, and several of the IP switching protocols that will be discussed later in the book.

It should also be noted that from a single switch perspective a particular label value is local to the switch and more specifically to the port (input or output) on the switch itself. This can be inferred from the fact that the label is swapped (rewritten) with a new value. However, to provide a complete explanation, it is also possible that the new label is the same as the old label. Even if that is the case, the labels still only have local significance.[1] An example of label swapping is shown in Fig. 4-4.

[1]*Local significance* may be a misleading term. In actuality, a pair of switches must agree on the use of a particular label value. In other words, the outbound port on the upstream switch and the inbound port on the downstream switch will use the same label value.

4.1.2 Switch Path Control

The switch forwarding techniques discussed in the previous section all depend on some form of switch path control to create and maintain information in the switches so that packets may be forwarded to the destination. Switch path control may involve an exchange of control messages between the switches in the network so that they can build a consistent set of switch forwarding tables. It could also involve the sending host generating a "path setup" message and sending that onto the network. The path setup message is used to discover a path through the network over which subsequent data packets can flow.

Switch path control is accomplished through one of the following techniques:

- ✔ Address learning
- ✔ Spanning tree
- ✔ Broadcast and discover
- ✔ Link state routing
- ✔ Explicit signaling

Address learning is probably the simplest technique that a switch can use to create and build a switch forwarding table. The switch promiscuously listens for *all* packets on a specific port and stores the source address of the packet along with the port number in an address/port cache. It also examines the destination address of the packet and attempts to match it against an entry in the address/port cache. If there is a match, the packet is switched to the appropriate output port. If there are no matches, the packet is flooded out all output ports except the one it came in on. And if the destination address is associated with the port that the packet came in on, the switch discards it. After some period of time, the switch may time out cache entries so that the information does not become stale.

Address learning is commonly used in Ethernet transparent bridges to build address/port caches. See Fig. 4-5. Observe that switch 1 has learned that stations aaa and bbb are located off of port 1 and that stations ccc and ddd are located off of port 3.

Address learning is quite easy to implement and requires no changes or processing on the part of the communicating hosts. However, it does require that the switches store a potentially large number of addresses in the cache. Address learning has one other major flaw: Loops can form

Figure 4-5
Address learning.

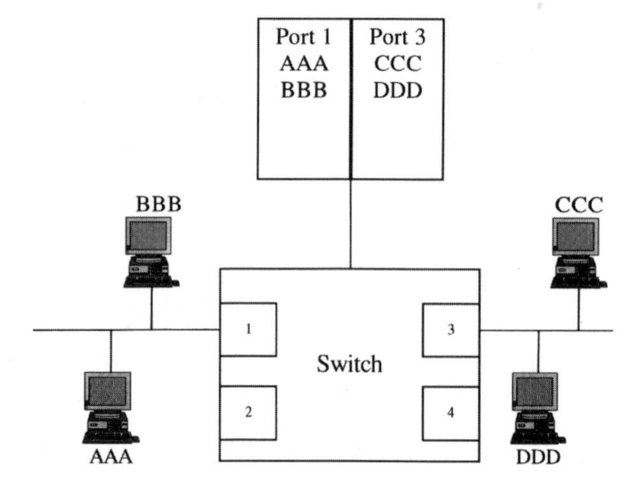

if there is more than one path between a source and destination host. A loop can form if a packet from a source LAN 1 crosses a bridge or bridges to a destination LAN 2 over a path 1 and returns over a separate path, path 2. The bridge(s) on path 1 believe that the source of the packet is on LAN 1. The bridge(s) over path 2 in the reverse direction believe that the source of the packet is on the destination LAN 2. With no way of resolving the apparent conflict, packets loop around the network indefinitely.

To get around the loop problem, the switches can establish a spanning tree. A spanning tree is a subset of links and switches that provide a loop-free path between any two stations or segments attached to the switched network. Switches build and maintain the spanning tree by exchanging control messages with each other that contain the identifier of a root switch and the distance from the root switch. Those switches and more specifically the ports on those switches with the least cost back to the root are placed on the tree. All other switches and ports are pruned, resulting in a set of switches and switch ports that can forward traffic in a loop-free manner. In other words some ports are in "forward" state while other switches and perhaps ports on the same switch are in "blocking" state. The combination of ports and links in the forward state comprise the spanning tree. A spanning tree for a collection of interconnected LANs is illustrated in Fig. 4-6.

The spanning tree algorithm is defined in the IEEE 802.1d LAN Bridging standard. It is used by Ethernet transparent bridges and Token-Ring source-routing bridges. Spanning trees are quite useful in extended LAN environments interconnected by bridges or switches. But

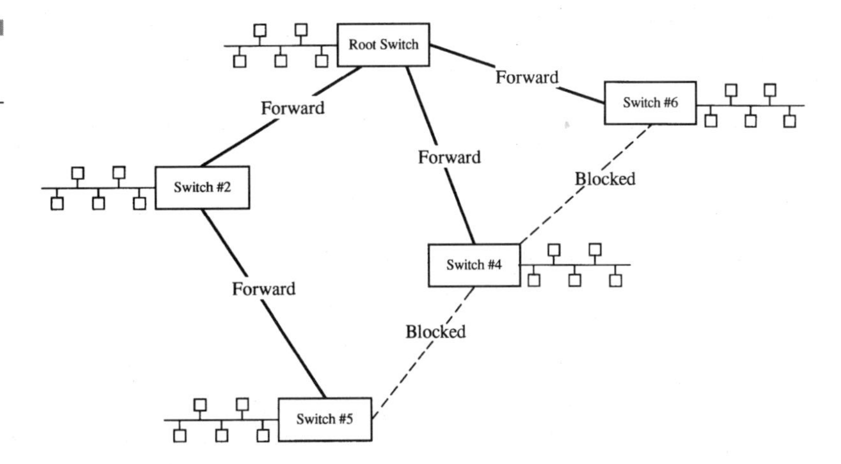

Figure 4-6
Spanning tree.

there are scalability and robustness limitations associated with spanning trees. First, depending on the topology, data traffic can concentrate on a small number of links and switches in the network. Even if alternate paths through active links and switches are available, the spanning tree will block traffic from flowing through them. This means that a spanning tree uses only a subset of the total available network topology (links and switches) to forward traffic. Second, there will be overhead associated with processing a large number of spanning tree control messages, particularly if the topology is large and less than stable.

Broadcast and discover is another technique that is commonly used in LAN switching (and bridging) to locate a switched path through the network. The idea here is to probe for one or more possible paths by sending a broadcast explorer packet to the destination, which will then return a packet over one or more possible paths. The returned packet contains a source-route vector the sender can use to forward all subsequent packets to the destination.

The concept of the broadcast and discover technique for establishing a switched path through a network is illustrated in Fig. 4-7. Observe that the explorer packet is flooded throughout the entire network. This is called an all-routes explorer (ARE). If a spanning tree is in place, the explorer packet may opt to follow the spanning tree path to the destination. This is called a spanning tree explorer (STE). The destination then returns a packet to the sender using possibly a broadcast packet, or it may decide to use a source route as shown in the figure. In this case the destination sends a packet back to the sender through switches 4, 3, 2, and 1, in that order. The sender will now use that path to forward pack-

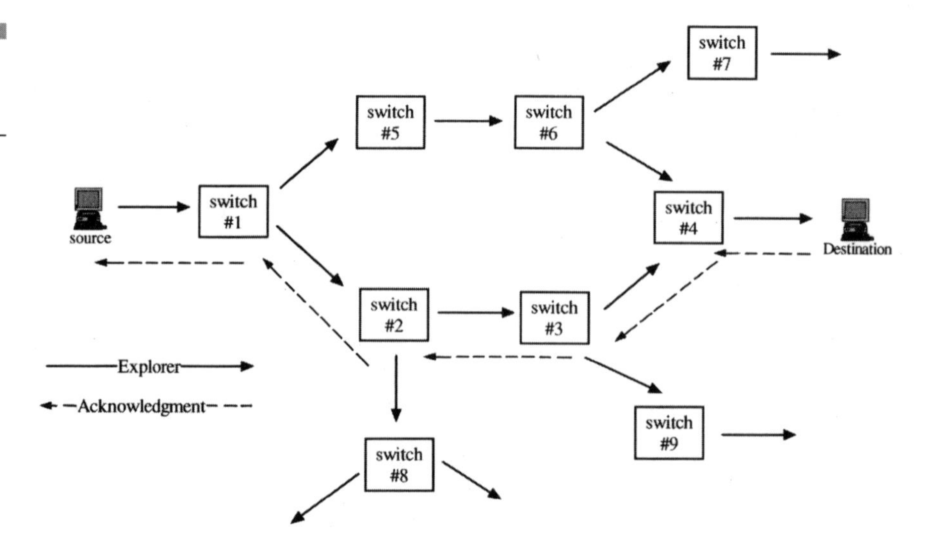

Figure 4-7 Broadcast and discover.

ets to the destination. If ARE frames are used, broadcast and discover techniques can provide some statistical load balancing over multiple paths between a source LAN and a destination LAN. Because source routing is used as the switch forwarding mechanism, there is very little overhead on the switch's part, but on the other hand, explorer packets do consume bandwidth and can lead to network congestion during network startup.

The fourth technique assumes that the switches are running a link state (or topology state) routing protocol. The switches exchange link state information with their neighboring switches and use this information to build an identical link state database of the network consisting of all active links and nodes. Because each switch knows what links and nodes are active in the network, a packet can be forwarded to the destination in a hop-by-hop manner (destination-based switching) or by using source-routing. In the later case, the ingress switch, upon receiving an indication that a source wishes to communicate with a destination, can consult the local link state database, assemble a source-route vector, append it the packet, and send it on its way. Note that the entire topology represented in the local link state database is examined each time a source-route vector is assembled, which is only during switch path setup.

Link state routing is used by IP routing protocols such as OSPF and Intermediate System to Intermediate System (IS-IS) to construct and maintain forwarding tables in routers. PNNI Phase I uses link state routing to forward ATM SVC setup requests through a network of ATM switches.

The last way to establish a switched path is with explicit signaling. Before any data packets can begin to flow over a dedicated switched path, the source must first send a connection setup request message through the network to the destination. This message ensures that there is an active path to the destination and possibly verifies that there is sufficient capacity over that path to handle the traffic. When the source receives an acknowledgment from the destination confirming that a path with sufficient capacity exists, data can begin to flow. The concept of explicit signaling is illustrated in Fig. 4-8.

The use of signaling to set up a switched path has its advantages and disadvantages. On the upside, signaling can confirm that there is, indeed, an available path to the destination before any data is sent. This obviates the need for a source to transmit successive data packets, wait for an acknowledgment, and then when none is returned, decide that the destination is unreachable. Another advantage to signaling is the fact that the route calculation is only performed at connection setup time and that subsequent data packets need only follow a dedicated switched path. No route calculation is needed once data starts to flow. Another way of looking at it is that the overhead of computing a correct path for a series of packets can be amortized over the life of the connection between the communicating parties.

A third reason why signaling might be attractive is that connection-specific parameters such as minimum bandwidth or maximum delay can be communicated into the network and processed by each switch along the path. The desired result is a switched path to the destination that meets the specific bandwidth and delay characteristics of the data flowing between the parties. And finally, explicit signaling could be useful as a method for accounting and for billing network usage on a per-connection basis.

The disadvantages of signaling consist primarily of the added delay and complexity that is introduced to the switching system. The additional delay is the connection setup latency or, put another way, the amount

Figure 4-8
Explicit signaling.

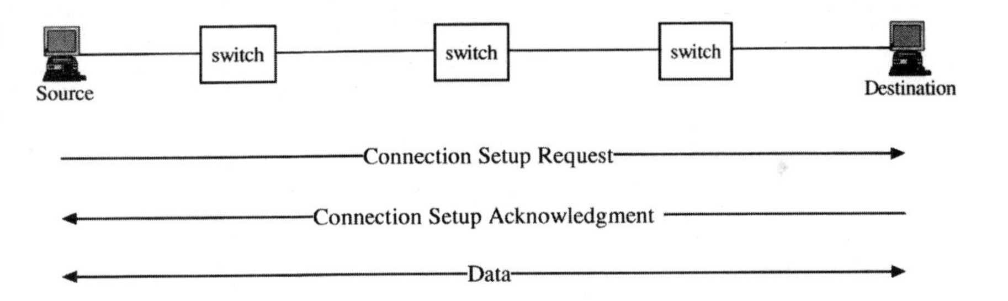

of time it takes to initiate and complete the connection setup process. More complexity is added to the network because of the fact that in addition to processing data, the network must now concern itself with processing signaling requests. And processing signaling requests breeds other control components such as admission control that all must participate in some manner during connection setup. And finally there is the question of scalability. There is a finite limit on the amount of processing that a system can reliably handle. In the case of a switched network that relies on the explicit signaling, there is a limit on the amount of the total switched paths supported and on the number of connection setups that can be concurrently handled.

Explicit signaling is used in the so-called hard-state connection-oriented switching solutions such as frame relay and ATM. It is also used in soft-state solutions such as RSVP and in the family of new IP switching protocols to be discussed later in this book. The difference between hard and soft state is that the former will not allow *any* data to flow until the connection setup process is completed, whereas soft state allows data to flow but over a separate path. Hard-state connections do not terminate unless requested to or by some unexpected network failure. Soft-state connections time out after some period of time unless refreshed by new connection setup messages and are considered more robust because data can continue to flow, even if the switched path has not been established.

4.2 LAN Switching

LAN switching is a natural evolution of traditional LAN technologies. Since the first colocated workstations, end users have found it convenient to attach to each other to share files and network resources such as printers. They connected to each other in the case of Ethernet by tapping into a common cable. This cable served as a common transport media to be shared by all attached workstations, but because it was shared, only one workstation at a time could transmit data. Protocols such as Carrier Sense Multiple Access with Collision Detection (CSMA/CD) and token passing were used to ensure that only a single workstation could be granted access to the shared media at any time. A shared media LAN is illustrated in Fig. 4-9. Three stations within the contention domain share a common transport media, but only one (station 1) can transmit data; the other two must wait.

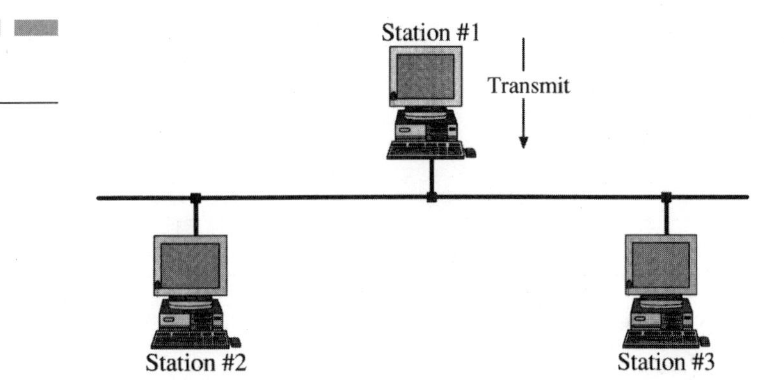

Figure 4-9
Shared LAN.

Although the protocols did not change, the physical LAN infrastructures consisting of workstations tapped into single-media, fixed-length cables gradually gave way to star-wired hubs. A shared LAN hub offered advantages in cost, media flexibility, and network management. Workstations simply plugged into a hub port and they were on the LAN. This simple deployment coupled with faster workstations, better network management tools, client/server application growth, and a move toward multiprotocol networking environments contributed to a tremendous growth in local area networking.

But at some point contention for shared resources by multiple workstations on the same LAN was going to become a problem. While one client is blasting away at a file server, using most of the available bandwidth, the other clients have to wait. The waiting contributed to lower network utilization and lower application throughput, and the problem was exacerbated with increased traffic volumes and more end users. There had to be a better way for LAN workgroups.

And that better way is to segment a LAN down to a few workstations or even a single one. The workstation therefore has access to the entire supported bandwidth (e.g., 100 MB with Fast Ethernet) and can transmit data without having to wait for any other clients to finish up. Extending the communications capabilities from half-duplex to full duplex enables a workstation to simultaneously send and receive data at the full bandwidth available. The notion of dedicated full-duplex communications is especially helpful to a server that must attempt to service multiple clients simultaneously. The concept of segmenting a LAN down to a single workstation is called microsegmentation and is illustrated in Fig. 4-10; it connects each workstation to a port on a LAN switch and gives each one full-duplex bandwidth.

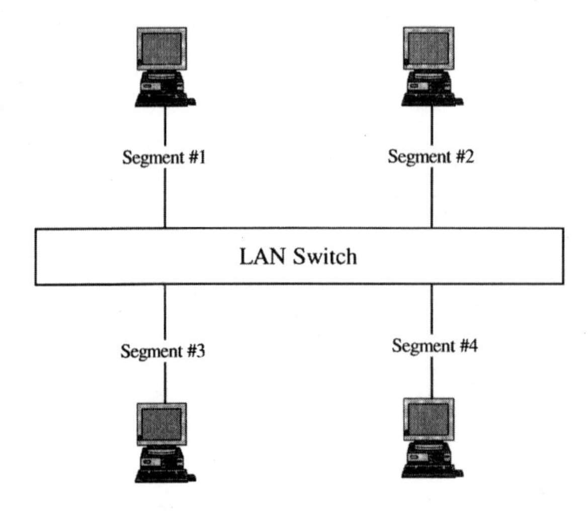

Figure 4-10
LAN switch.

Microsegmentation provides more bandwidth and better performance at the workgroup level. From a network perspective, LAN switching can offer similar gains in performance and capacity. First, LAN switching is a multiport extension to traditional two-port bridging. So a switch can be considered just a multiport bridge, but after factoring in more ports and a nonblocking architecture, one has a box that provides orders of magnitude more capacity and better performance. Second, high-speed trunks can interconnect switches to form a high-speed backbone for interworkgroup communications. And finally a network of switches can support one or more Virtual LANs (VLAN). A VLAN is a broadcast domain whose members communicate with each other as if they shared the same segment even though they may be many switches and links away. VLANs are useful for administrative, security, and broadcast control reasons and are discussed in more detail in Sec. 4.2.3.

4.2.1 Cut-Through Forwarding

The process of switching a packet from an input port to an output port is based on some piece of information contained in the packet itself. As discussed in Sec. 4.1.1, it could be a destination address, a label, or a source-route vector. LAN switches typically switch packets based on the destination address. Furthermore, a LAN switch can process a packet using two possible internal forwarding techniques: cut through and store and forward.

In cut-through mode, a switch begins forwarding the packet as soon as the destination address is examined and verified. The forwarding of the first part of the packet can begin even as the remainder, or "tail," of the packet is being read into the input port switch buffers. The advantage to this mode of forwarding is lower latency, which can be useful when supporting delay-sensitive traffic on the LAN. The disadvantage of cut-through switching is that all packets, valid or invalid, are forwarded to the output port.

The concept of cut-through switching is illustrated in Fig. 4-11.

4.2.2 Store and Forward

Store and forward is the other per-packet forwarding mechanism used by conventional LAN switches. In this mode of operation, the LAN switch reads the entire packet into a buffer before deciding where to send it. This enables the switch to validate the length of the frame and run a cyclic redundancy check (CRC). If the frame is too big, too small, or fails the CRC, the switch can discard it without forwarding it to the next station or segment. However, this error checking performed by the switch on the entire packet can increase latency. The amount of time spent in the switch is directly proportional to the length of the frame.

The store-and-forward technique is used by LAN switches that move packets from one media type to another. For example an Ethernet LAN switch with an ATM uplink receives the entire packet before commencing the segmentation process, chopping it up into 53-byte cells, and switching them over the uplink.

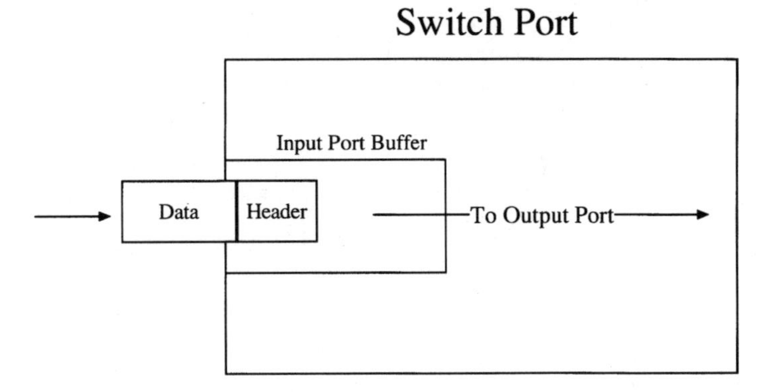

Figure 4-11 Cut-through forwarding.

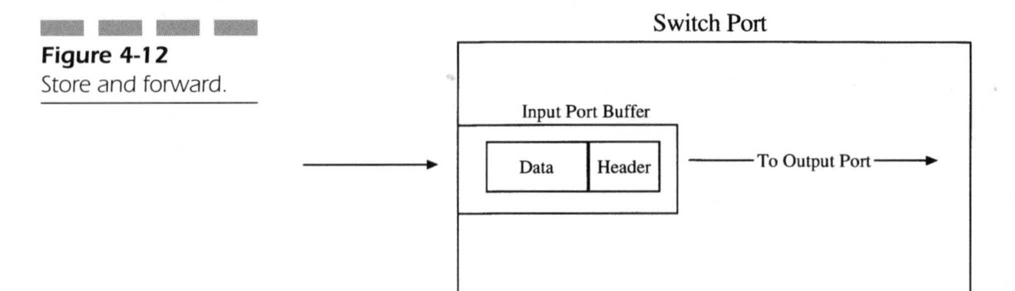

Figure 4-12
Store and forward.

The concept of store and forward is illustrated in Fig. 4-12.

4.2.3 Virtual LANs and IEEE 802.1Q

A VLAN is an important and useful application associated with LAN switching. By definition, a VLAN is a broadcast domain whose members use LAN switching to communicate as if they shared the same physical segment. The members of the VLAN grouped into the broadcast domain are under no physical or geographical restrictions—they may be physically attached to separate switches in a network of switches. Broadcast, unknown, and data packets exchanged between members are confined to that VLAN. Another definition of VLAN is that it enables a single switched infrastructure to be divided up into a smaller set of broadcast domains.

The criteria for joining a VLAN can be based on any number of different factors. Common examples are

- Physical port
- MAC address
- Layer-3 unicast address
- Multicast address
- Some number of different policies such as time of day, application, MAC address, etc.

The concept of a VLAN is illustrated in Fig. 4-13. Observe that the members of a VLAN may be physically attached to the same or different switches. Using LAN switching as the transport vehicle enables the members of the same VLAN to communicate at near or actual wire-speeds with very low latency. Membership control and topological inde-

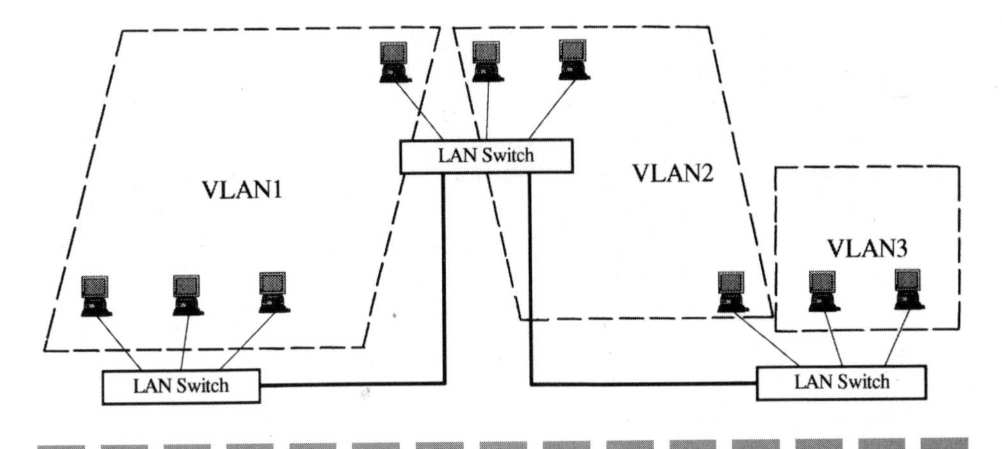

Figure 4-13 Virtual LANs.

pendence coupled with high-speed intra-LAN connectivity provide a number of advantages:

- *Configuration.* Adds, moves, and changes can be an operational and administrative burden for network managers. A dynamically defined VLAN in which some client parameter (e.g., address) determines membership simplifies this task. A client can simply move from one switch port to another and remain on the same VLAN.
- *Security.* VLANs can protect critical resources by requiring some form of authentication before joining.
- *Network efficiencies.* Traffic flows on different VLANs are essentially firewalled from each other. This prevents traffic from one VLAN from arriving on another one and possibly contributing to unnecessary workstation and segment utilization. Inter-VLAN traffic must flow through a bridge or a router. It is also a useful technique when grouping a high-performance server with a number of affined clients.
- *Broadcast containment.* A properly configured and operational VLAN should prevent or minimize broadcast leakage from one VLAN to another.

Unless the members of a VLAN are physically separated (e.g., partitioning the ports of a switch), switches must have some way of identifying and then forwarding packets to the correct VLAN. This is accomplished by the use of a tag that is appended to each packet and is used to determine which VLAN the packet belongs to.[2] VLAN tags enable switches to multiplex traffic from different VLANs over the same physical switched network. After a network of VLAN switches is configured, the first packet sent by a workstation arrives at an ingress switch port. Unless the membership criteria is based on a physical port, some portion of the packet's contents will be examined and compared with a VLAN configuration database. A tag or VLAN identifier is then appended to the packet. The VLAN identifier and destination address are used by the VLAN switch to forward the packet to the appropriate same-VLAN destination. At the same time, this information is used by a VLAN switch to filter out nonmember traffic.

An example illustrating the operation of VLAN switching is shown in Fig. 4-14. Switches 1 and 2 have been configured to build VLAN A with stations aaa and ccc as members and VLAN B with stations kkk and jjj as members. Observe that the configuration information must be distributed to every switch that might have a possible member of either of

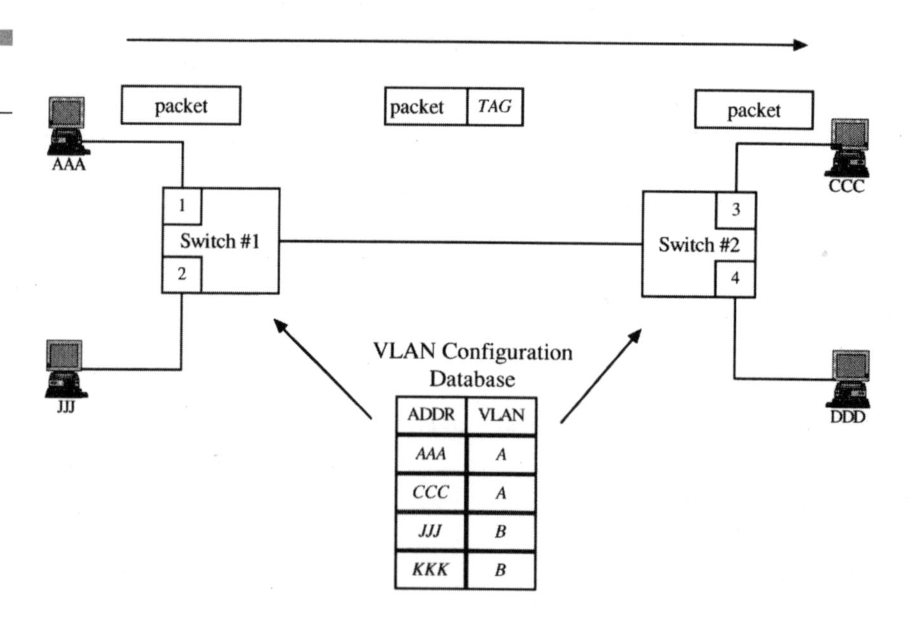

Figure 4-14 VLAN switching.

[2]The use of a tag in the VLAN definition should not be confused with the tags used with Cisco Tag Switching. Tags in the VLAN context identify packets belonging to a VLAN. Cisco Tag Switching uses tags to speed up the IP forwarding process.

these VLANs. Now assume that the switches have learned of members associated with VLANs on their respective ports through either a static or dynamic learning process. When station aaa forwards a packet to station ccc, it arrives at port 1 of switch 1, which consults the configuration database, notes that station aaa belongs to VLAN A, tags the packet with a VLAN identifier, and then forwards the packet to station ccc. The tagged packet crosses the link connecting switches 1 and 2, arrives at switch 2, which notes that the packet belongs to VLAN A . Switch 2 then removes the VLAN tag and forwards the packet out port 3 to station ccc.

Until recently, switch vendors used their own proprietary methods to tag packets. This introduced interoperability problems because switches from different vendors might use different methods to tag packets. One technique employed by several vendors was the use of the IEEE 802.10 Secure Data Exchange as a VLAN identifier. IEEE 802.10 was developed as a way to provide authentication and encryption services for MAC-layer devices. Although it served the purpose, the use of a developed standard for purposes other than what it was designed for was frowned upon.

The IEEE thus embarked on defining a formal VLAN standard, and its efforts have resulted in the IEEE 802.1Q specification for Virtual Bridged LANs. The 802.1Q specification sets out to accomplish the following:

- ✔ Define an architecture for the provisioning of VLAN services over existing IEEE 802 Bridged LANs.
- ✔ Define the frame format for carrying VLAN tags in Ethernet/IEEE 802.3 and IEEE 802.5 Token-Ring frames.
- ✔ Define the protocols and mechanisms by which configuration and membership information can be communicated to VLAN-aware devices.[3]
- ✔ Define the criteria and procedures for forwarding frames through a network of IEEE 802.1Q VLAN-aware devices.
- ✔ Ensure full interoperability and coexistence with non-VLAN-aware devices. A non-VLAN-aware device is one (e.g., workstation or bridge) that does not transmit or receive tagged packetxs or those

---

[3]VLAN-aware devices can be workstations, bridges, switches, or a router interface attached to a VLAN network. VLAN-aware devices can transmit, receive, and interpret VLAN-tagged packets.

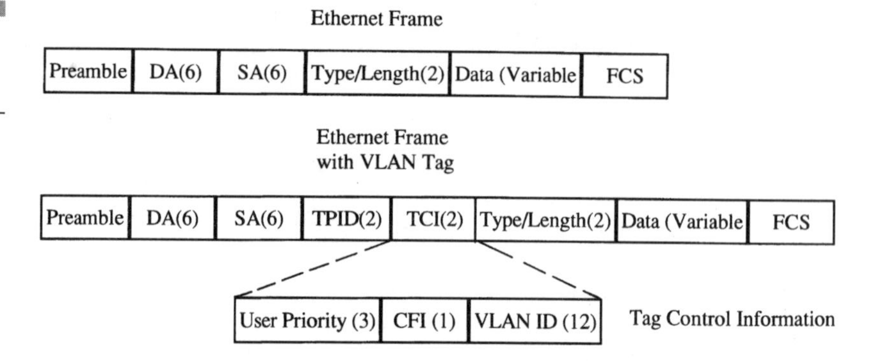

Figure 4-15 Ethernet VLAN frame format.

containing a VLAN Identifier (VID) and does not understand VLAN membership information.

The format of the tag header differs depending on whether it is carried in an Ethernet or a Token-Ring frame. The placement and format of the tag header inside an Ethernet frame is shown in Fig. 4-15. The tag header adds 4 bytes to the original Ethernet frame, thus extending the maximum Ethernet frame size to 1518 bytes. This is greater than the 1514-byte value specified in the IEEE 802.3 standard but is expected to be amended so that Ethernet frames with a VLAN tag and length of 1518 bytes will be supported.

The 4-byte tag header consists of

- *Tag protocol ID (TPID).* The 2-byte TPID consists of a hexadecimal value of 81-00 and must be different from any value used in the Ethernet Type field.
- *Tag control information (TCI).* The TCI consists of a 3-bit User Priority field that is used to indicate the frame's priority as it is forwarded through switches supporting the IEEE 802.1p specification, explained below. The 1-bit Canonical Format Indicator (CFI) indicates whether the MAC address information is in canonical format, and the 12-bit VID defines the VLAN that the frame belongs to.

The placement and format of the tag header inside a Token-Ring frame is shown in Fig. 4-16. This is different from the format of the Ethernet frame in that an 8-byte Sub-Network Access Protocol- (SNAP) encapsulated TPID is used rather than the 2-byte value for Ethernet. Together with a 2-byte TCI value, the VLAN tag header adds 10 bytes to the length of a token-ring frame.

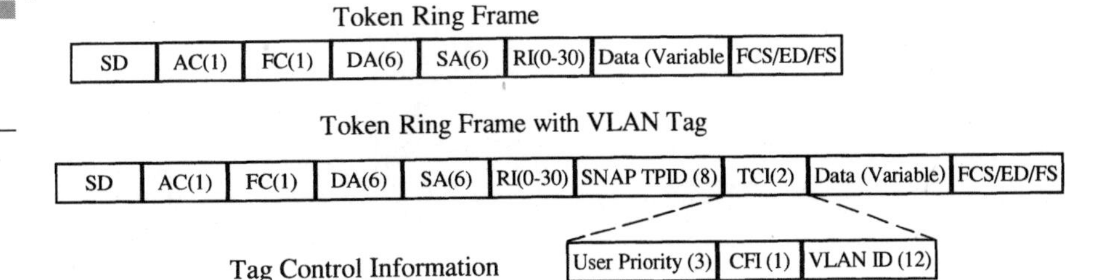

Figure 4-16 Token-Ring VLAN frame format.

The values contained in the 8-byte SNAP-encapsulated TPID are as follows:

- Bytes 1 through 3 carry the basic SNAP header, which consists of a hexadecimal AA-AA-00.
- Bytes 4 through 6 carry a SNAP protocol id, which consists of a hexadecimal 00-00-00.
- Bytes 7 and 8 carry the aforementioned TPID and have a value of hexadecimal 81-00.

The operation of the IEEE 802.1Q specification is best illustrated by examining the network in Fig. 4-17. The workstations connected to the Ethernet segments attached to VLAN switches 1, 2 and 4 are on access links. An access link is a LAN segment that is used to attach non-VLAN-aware workstations to a port on a VLAN switch. Access links can be a single segment or may consist of a number of segments and/or workstations connected by non-VLAN-aware bridges and switches. An access link carries no tagged packets.

A trunk link is one that carries tagged packets only, or in other words, those containing VIDs. A trunk link can only support devices that understand VLAN frame formats and membership. The most common implementation of a trunk link is one that connects two VLAN switches together. But a trunk link can be also be a shared LAN segment with many VLAN switches and/or VLAN-aware workstations attached to it.

The third type of link defined in the IEEE 802.1Q specification is a hybrid link. A hybrid link can transport both tagged and untagged packets and supports both VLAN-aware and non-VLAN-aware devices. In Fig. 4-17 the hybrid link is attached to VLAN switches 3 and 4, non-VLAN-aware workstations that comprise members of VLAN C, and a VLAN-aware workstation that is part of VLAN A. Observe that a hybrid link can act as both an access link for some VLANs and a trunk for others. It should be noted that for members of the same VLAN attached

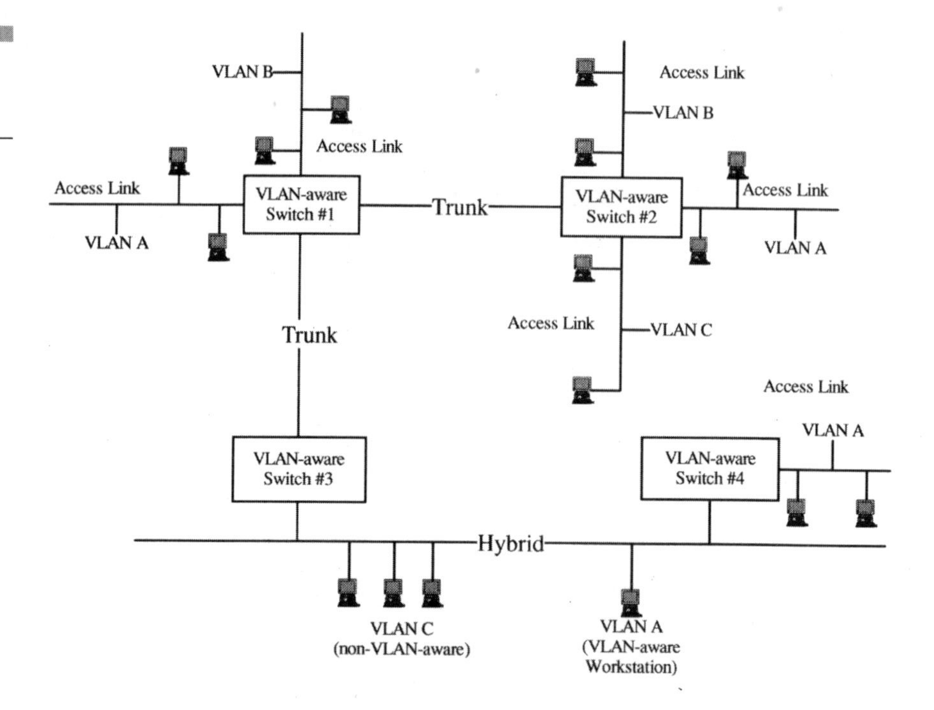

Figure 4-17 IEEE 802.1Q VLAN network.

to the same hybrid link, all packets must either be tagged or marked with the same VID. For example, if non-VLAN-aware members of VLAN B were attached to the hybrid link shown in Fig. 4-17, they would not be able to communicate with the VLAN-aware workstation that is also a member of VLAN B.

Reliable and scalable operation of a VLAN requires that consistent membership be maintained across all VLAN-aware devices in the network. Conceivably, in small environments it is possible to manage this information on a per-switch basis. However, in networks that have a large number of VLAN-aware devices and a large number of individual VLANs and where VLAN membership is dynamic (frequent joins and leaves), some form of dynamic VLAN membership mechanism is required. IEEE 802.1Q defines the Generic Attribute Registration Protocol (GARP) VLAN Registration Protocol (GVRP). GVRP is used by VLAN-aware devices (e.g., workstations and switches) to dynamically register and distribute VLAN membership information across a network of VLAN-aware devices. This enables a collection of VLAN switches to establish and update which VLANs are active and through which switch ports they can be reached.

GVRP is an application that uses GARP, which is a receiver-driven protocol for registering attributes across a network of GARP-capable devices. The attributes can consist of multicast addresses or, in the case of VLANs, membership information from VLAN. The concept of GARP (and by extension GVRP) is illustrated in Fig. 4-18. In this example VLAN switch 4 has detected the presence of members of VLAN A attached to one of its ports. VLAN switch 4 will then generate a GVRP message and pass it to its neighboring VLAN switch. Upon receipt of the GVRP message, each switch will update its VLAN forwarding tables to associate the outbound switch port the message came in on and the VID of VLAN A.

4.2.4 IEEE 802.1p

Closely aligned with the development of the IEEE 802.1Q specification is the ongoing work to define a specification for the expedition of real-time traffic and dynamic multicast filtering over IEEE 802 LANs. This is defined in the IEEE 802.1p specification that is incorporated into a revision of the IEEE 802.1D specification for MAC-level bridging.

IEEE 802.1p extends the capabilities of basic LAN bridges by adding the following:

- Mechanisms for identifying and prioritizing data frames as a means of expediting delay-sensitive traffic flows in a bridged LAN environment
- Provisioning of dynamic filters to support group multicast communications over a bridge LAN environment

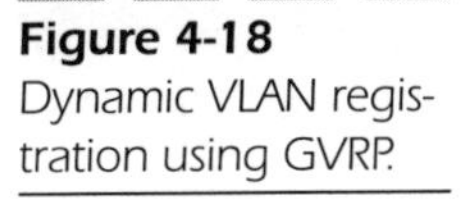

Figure 4-18 Dynamic VLAN registration using GVRP.

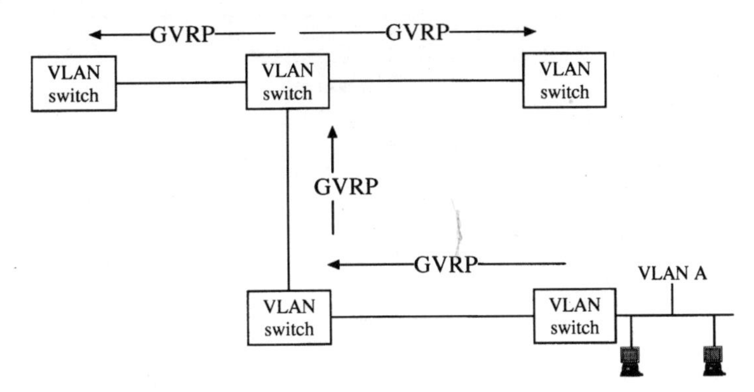

The ability to support different traffic classes and thus provide different service levels over 802 LANs is accomplished by marking "priority" bits in the frame itself and then having the switch place the frame in one of several queues for transmission based on the contents of the frame's priority bits. IEEE 802.1p switches make use of the 3-bit user priority field that is located in the TCI field of a VLAN tag header. The 3 bits contained in the user-priority field mean that up to a maximum of eight different traffic classes or levels of priority can be defined.

Having a user-priority field in each data frame controls and possibly minimizes the overall transmission delay as the frame passes through the network. The total delay is the sum of the queuing and the access delays. The queuing delay is the amount of time that the frame sits in a queue until it is selected for transmission on the outbound link. The access delay is the amount of time that the frame must wait until it has access to the LAN and thus can be transmitted. Some LAN technologies, like Ethernet, have no control over this process. User priority is a technique that can be used to provide access priority to those LAN technologies that do not support this capability.

The user priority can be set by either the data traffic source or by an 802.1p switch. The switch may do so based on some policy or configuration information. For example, all data frames coming in on, say, port A will be marked with a high priority. Figure 4-19 illustrates the components and operation of an 802.1p switch.

When the frame arrives at a switch port, various criteria related to the MAC addresses, filters, and correct LAN operation are applied to the frame. If the frame fails any of these checks or cannot pass through a filter, it is discarded; otherwise it is placed in one of several priority

Figure 4-19
IEEE 802.1p switch.

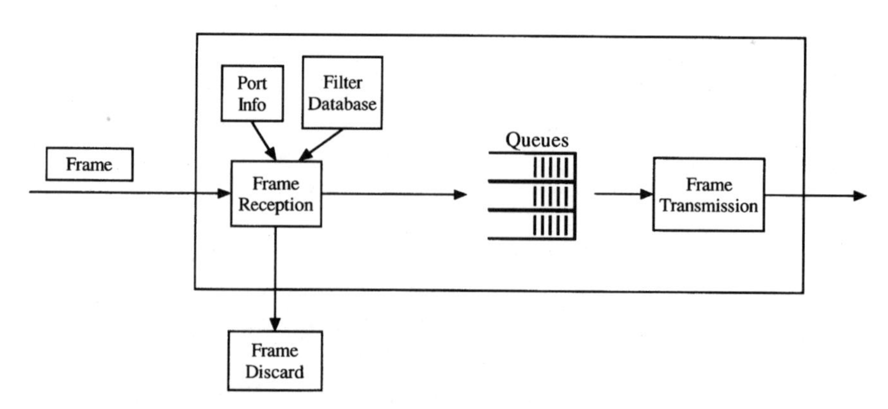

queues based on the contents of the user-priority bits. If a user-priority field is not present in the frame (the frame came in from a non-802.1p-capable workstation), the switch may select a default value and encode it in the frame.

Historically, LANs have treated multicast data as broadcast packets that were flooded to every workstation and out every bridge or switch port. This is not an efficient use of network resources because only a subset of the LAN population may wish to receive the multicast data frames. The second goal of the IEEE 802.1p specification, therefore, is to define techniques that efficiently support multicasting in a LAN environment.

This goal is accomplished by creating a bridge multicast filtering database in each switch and then updating that database using a dynamic group membership protocol. The switch then is able to keep track of which multicast groups are active and which ports have members. For those ports with no members, the switch filters out multicast traffic.

The dynamic group membership protocol specified in IEEE 802.1p is the GARP Multicast Registration Protocol (GMRP). GMRP is an application that enables end systems and switches to register and deregister multicast group membership information and allows that information to be disseminated throughout the network. GMRP uses GARP to accomplish this, as shown in Fig. 4-18. Potential group members register group addresses with the port of an adjacent switch. The group address is propagated throughout the network of switches, resulting in a loop-free spanning tree. The network is then able to accommodate dynamic multicast groups while minimizing the impact on nongroup resources.

4.3 LAN Switching Network Examples

LAN switching comes in several flavors, and its deployment depends on the needs of the end users and applications as well as the actual capacity and performance gain that one might expect. The two cases that will be illustrated are workgroup and backbone switching.

Workgroup switching is a simple and cost-effective technique to increase the bandwidth and performance of end users who work frequently with nearby or local computing resources. The idea behind

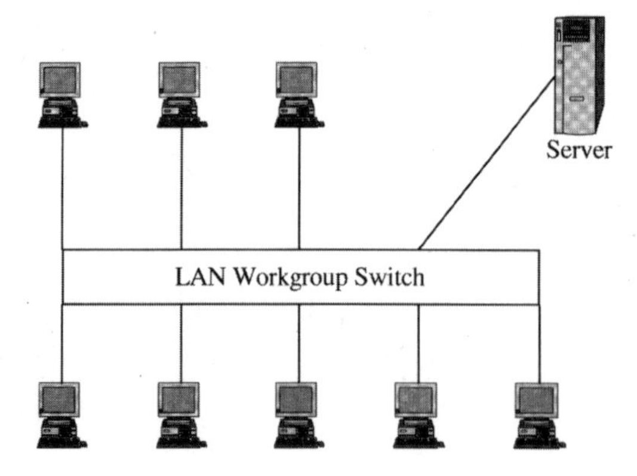

Figure 4-20
LAN workgroup switching.

workgroup switching is shown in Fig. 4-20. Clients who do most of their work with a colocated server may have once shared a LAN segment (Ethernet or token-ring). Replacing the shared media technology with a LAN workgroup switch removes contention for access to the LAN—each client now enjoys dedicated access to the network with no change required to the workstations, LAN adapters, servers, or applications. In addition, a server can be outfitted with a high-speed full-duplex LAN adapter (e.g., Fast Ethernet), which will further improve service to the clients. Given the low costs of workgroup switching, it almost does not make sense to do anything else.

Workgroups constitute a small portion of the extended LAN environment. Fully extending the performance and capacity gains of LAN switching to the entire LAN population requires interworkgroup traffic to be switched. This is especially true when server resources are not colocated with the clients. In fact this is probably the most common case. Therefore deploying a switched backbone makes the most sense because it provides connectivity and a high-speed transport between workgroups.

There are many different possibilities and scenarios for building a LAN backbone. One option might be to collapse all workgroup switches and servers into a single backbone switch(es) in sort of a hub and spoke topology where the backbone switches are the hub. From the perspective of the workgroup clients, this provides a direct high-speed link to the servers attached to the backbone but does not offer any path redundancy in case the workgroup-to-backbone link fails.

Another option to consider is to form a network of backbone switches and attach the workgroups into the backbone. The backbone provides

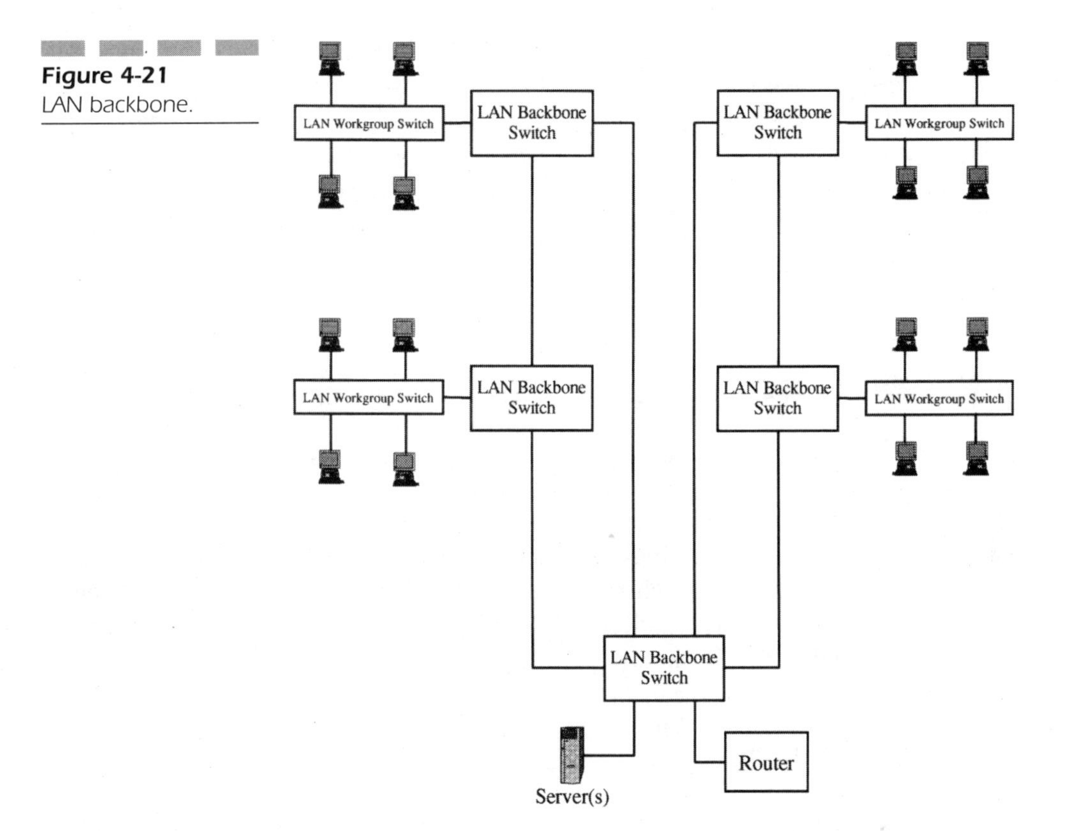

Figure 4-21
LAN backbone.

multiple switched paths between any two points in the network. This provides an added degree of redundancy and reliability in case of a link or switch failure. Figure 4-21 illustrates a partial mesh switched backbone. For illustrative purposes the colocated workgroup and backbone switches are shown as two separate boxes, but it is expected that real deployments may use a single box to perform both functions. Observe that there are multiple switched paths for traffic flowing from any workgroup down to the servers. If a single workgroup/backbone switch fails, only that workgroup will be affected. However, just because there are multiple possible paths does not mean that traffic can make use of them. Recall from Sec. 4.1.2 that a spanning tree is used to form a loop-free forwarding path for data traffic through a network of bridges or switches. The spanning tree in this environment would limit the forwarding links and switches to a subset of the network topology. This can be a problem if the spanning tree forwarding path is congested—there is no way to make use of other available links and switches. To get around this limitation, ATM is used as a backbone to interconnect LAN-based workgroups. ATM uses connection-oriented mechanisms to first locate any available path that can accommodate the bandwidth and other requirements of the traffic flow. So although it is certainly conceded that LAN switching is the best choice for workgroups, ATM serving as an interworkgroup backbone can be quite useful.

MORE INFORMATION

Draft Standard P802.1Q/D8, IEEE Standards for Local and Metropolitan Area Networks: Virtual Bridged Local Area Networks, December 1997.

Information technology—Telecommunications and information exchange between systems—Local and metropolitan area networks—Common specifications—Part 3: Media Access Control (MAC) Bridges: Revision (Incorporating IEEE P802.1p: Traffic Class Expediting and Dynamic Multicast Filtering).

Sackett, G., *IBM's Token-Ring Networking Handbook,* New York: McGraw-Hill, 1993.

Saunders, S., *The McGraw-Hill High-Speed LAN Handbook,* New York: McGraw-Hill, 1996.

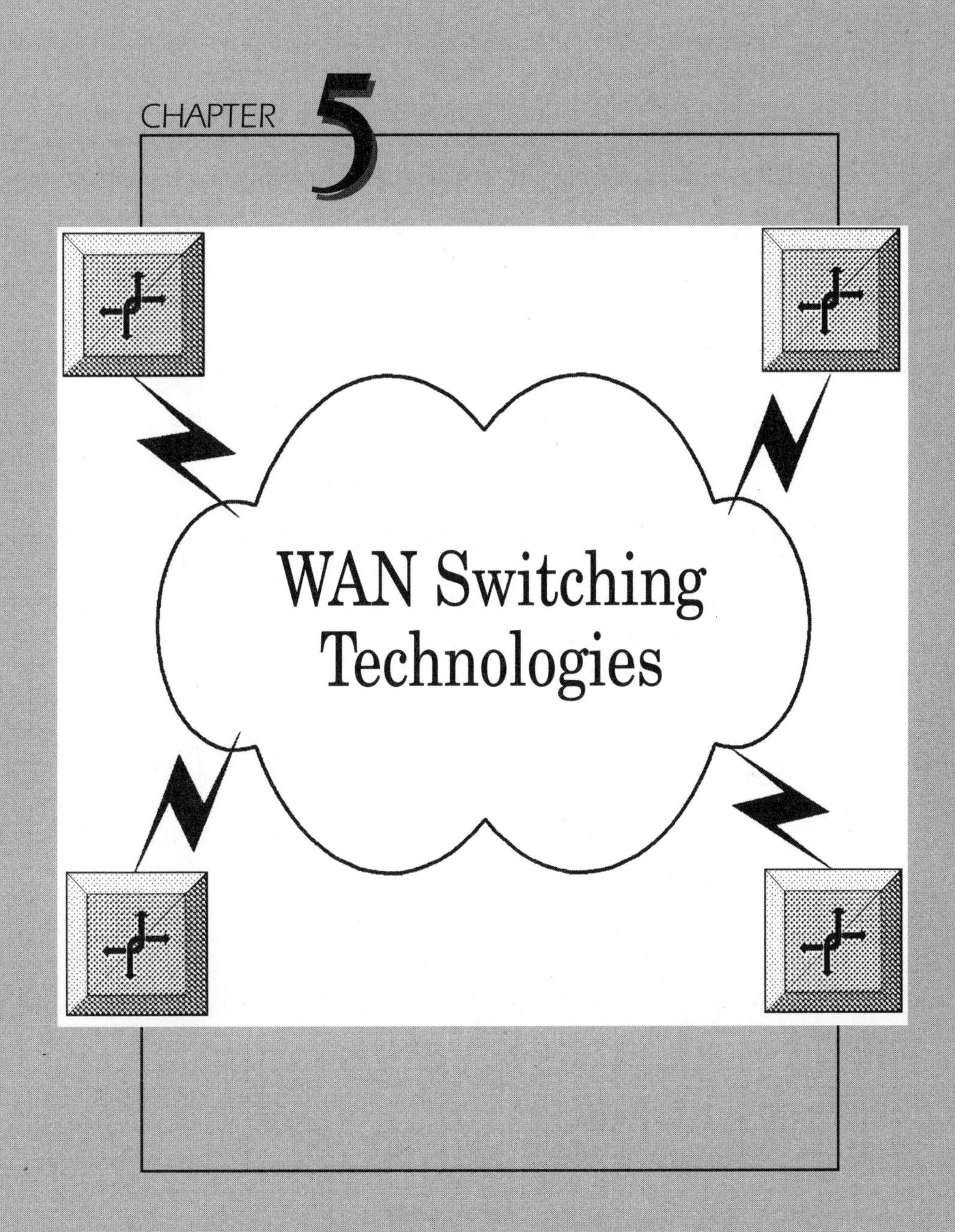
CHAPTER 5
WAN Switching
Technologies

Now that TCP/IP and the concepts of switching and LAN switching technologies have been covered, it is time to focus in on the technology of switching in the WAN. Frame relay and ATM are the two primary technologies utilized by carriers, ISPs, and corporate intranets to switch data and some video and voice over the WAN.[1]

5.1 Frame Relay

Frame relay is a connection-oriented fast-packet switching technology that is prevalent in many public and private WANs today, including the Internet. Devices connected via frame relay establish one or more VCs with each other and communicate through a network of frame relay switches. Frame relay is quite popular as a WAN networking service because many VCs can be multiplexed over the same physical network, and VCs can be provisioned to meet the bandwidth and protocol requirements of the individual subscribers.

5.1.1 Architecture

Frame relay technology evolved from experiences gained from previous technologies as well as improvements and advances in current technologies. From its predecessor, X.25, frame relay picked up on the notion of a VC and a per-VC connection identifier. But X.25 was developed and deployed at a time (1970s-1980s) when line speeds were not fast (compared to today) and the line quality was sometimes questionable. Therefore it became necessary to engineer error checking and reliability mechanisms on a per-hop basis. These techniques eventually delivered the data safely to the destination but at a significant cost in latency and network delay.

With the advent of the LAN and the requirement to interconnect LANs over WANs, new techniques that could handle greater traffic volumes at variable rates were needed. Conceivably, X.25 bandwidth and performance could have been cranked up to meet this demand but that really became unnecessary because digital communications (e.g., fiber) offered much greater capacity through much cleaner pipes. It was no

[1]ATM spans both the local and wide area network. This chapter will introduce the basics of ATM that can be applied to both environments. The specific use of ATM in the LAN environment will be covered in later chapters.

longer necessary to have the reliability and error-checking mechanisms located at each hop in the network. In addition it was also deemed important to allocate bandwidth on demand so that the network could support the increasingly variable and random nature of data traffic. And finally, like X.25 but unlike SNA and IP, a new technology had to be transparent to network-layer protocols.

Frame relay addresses these requirements and has proven to be a versatile and popular technology for supporting data communications over WANs. It takes advantage of the improved quality and higher speeds afforded by better communication link technologies such as fiber. Error checking is still done on a per-hop basis, but error recovery has been pushed to the edge of the network. Therefore much better performing and lower-cost packet-switching machinery can be placed in the middle of the network. Any intelligent functions such as flow control or error recovery are performed by the end systems attached to the frame relay network.

The primary functional attributes of frame relay consist of the following:

- ✔ Variable frame size.
- ✔ Variable bandwidth (DS0 up to OC3+).
- ✔ Supports PVCs and SVCs.
- ✔ Port sharing—multiple VCs can originate/terminate from the same physical device port.
- ✔ Bandwidth Management. A frame relay VC can be defined with traffic parameters that govern the amount of data that can flow through the network under normal and heavy conditions. A committed information rate (CIR) defines the amount of traffic in bits per second that can normally flow through the network in an interval of time. The committed burst size (Bc) defines the maximum amount of data in bits per second in an interval of time that the network can support and the excess burst size (Be) defines the maximum amount of data that the network can attempt to deliver.
- ✔ Supports multiple network layer protocols (e.g., IP, IPX, SNA, etc.) as well as bridged LAN traffic.
- ✔ Label-swap forwarding based on the Data Link Control Identification (DLCI) connection identifier.
- ✔ Congestion control. Frame relay supports explicit congestion notification. In addition if the information transfer rate exceeds a

defined threshold, some packets may be marked as eligible to be discarded or will be discarded all together.

- Interworks with ATM.
- Supports data and can accommodate voice.
- Is supported on routers, switches, frame relay access devices (FRADs), and other WAN networking devices. Frame relay is not really a host-based technology although its use on a host is not precluded.

The basic architecture of a frame relay network is illustrated in Fig. 5-1. Frame relay switches form the backbone of the network and are connected to each other by frame relay network-to-network interfaces (NNIs). Devices performing a frame relay terminating equipment (FRTE) function are attached to ports on the frame relay switches over a Frame Relay User-to-Network Interface (UNI) connection. Examples of devices that attach to frame relay switches include routers, FRADs, and possibly hosts. FRTEs are responsible for flow control, error recovery, and reliability. Frame relay switches are responsible for multiplexing, frame delimiting, CRC checking, and congestion notification. Communication through a network of frame relay switches is accomplished by establishing a permanent or switched virtual connections between FRTE endpoints. In Fig. 5-1 a PVC has been established between FRTE 1 and FRTE 2.

Specifications related to frame relay functions have been developed over the years by various standards bodies including the American

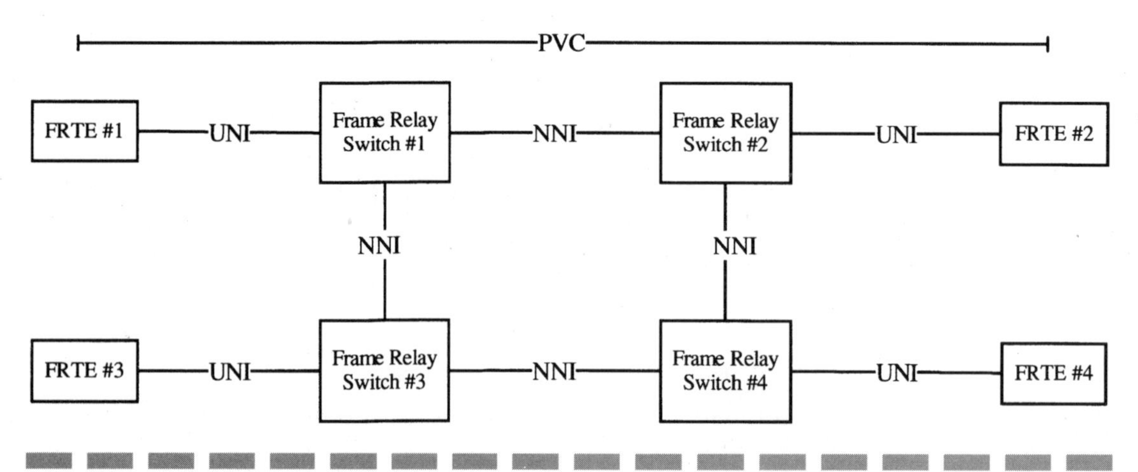

Figure 5-1 Frame relay architecture.

National Standards Institute (ANSI) and the International Telecommunications Union (ITU). The best source for an understanding of basic frame relay functions and specifications is provided by the Frame Relay Forum[2] in the form of Implementation Agreements (IAs). The Frame Relay Forum is a consortium of vendors, developers, carriers, and end users that are committed to the promoting and deploying frame relay technologies. Table 5-1 describes the current frame relay IAs.

5.1.2 DLCI

Frame relay VCs are identified by DLCIs. A DLCI is a multibit field contained within a frame relay protocol data unit (PDU) that identifies the VC that the packet belongs to. More generally, a DLCI is a label affixed to the packet that corresponds to a VC that traverses the switch. The VC will have end-to-end significance (between a pair of switches), but the DLCI will only have local significance, because it is swapped as it passes through each switch in the network. Therefore one can say that frame relay switching employs a label-swapping forwarding mechanism.

TABLE 5-1

Frame Relay Forum Implementation Agreements

Standard	Description
FRF.1	Frame Relay UNI
FRF.2	Frame Relay NNI
FRF.3	Multiprotocol Encapsulation
FRF.4	Frame Relay SVC
FRF.5	Frame Relay/ATM Network Interworking
FRF.6	Frame Relay Customer Network Management
FRF.7	Frame Relay PVC Multicast
FRF.8	Frame Relay/ATM Service Interworking
FRF.9	Frame Relay Data Compression
FRF.10	Frame Relay Network-to-Network SVC
FRF.11	Voice over Frame Relay
FRF.12	Frame Relay Fragmentation

[2] *http://www.frforum.com.*

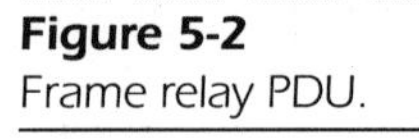

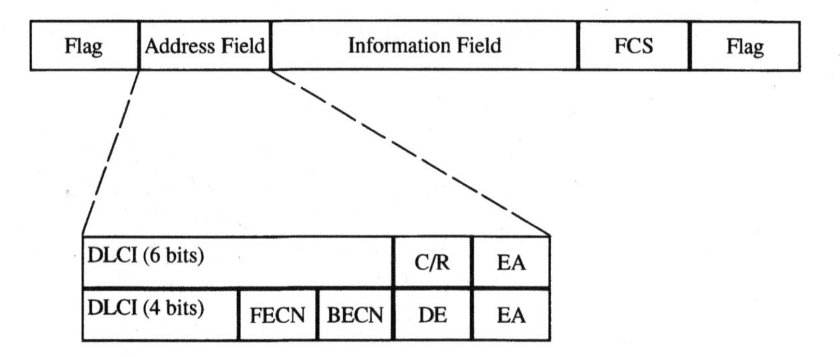

Figure 5-2
Frame relay PDU.

A DLCI can be at least 10 bits in length and is contained in a 2-byte address field. Larger DLCI fields contained in larger address fields have been defined. Figure 5-2 shows a basic 2-byte address field containing a 10-bit DLCI field. The fields in the frame relay PDU consist of the following:

- *Flag.* Value of a hexadecimal 7E delimits frame relay PDUs.
- *Address field.* Field of 2, 3, or 4 bytes that contains the DLCI. Most implementations use the basic 2-byte format. This is also referred to in some documentation as the Q.922 address.
- *DLCI.* Unique 10-bit value that identifies the frame relay VC the packet belongs to.
- *Command/response (C/R).* Bit indicating whether the frame is a command or response.
- *Extended address bit (EA).*
- *Forward explicit congestion notification (FECN).* A 1-bit field used to notify the end user that congestion has been experienced in the direction the frame was sent.
- *Backward explicit congestion notification (BECN).* A 1-bit field used to notify the end user that congestion has been experienced in the reverse direction from which the frame was sent.
- *Discard eligibility (DE).* A 1-bit field that indicates a discard preference when congestion is encountered.
- *Frame check sequence (FCS).* A 2-byte field used to perform a CRC check on the data, excluding the flags and FCS value itself.

5.1.3 Congestion Control

Congestion control is generally defined as the mechanisms that enable network components and end systems to gracefully adapt to variations in available network capacity. Congestion occurs when the amount of traffic injected into the network exceeds the capacity of the links and switches. The objective of congestion control therefore is to have the traffic source(s) reduce the amount of traffic that is being sent into the network so that it will not exceed the available network capacity. The traffic source(s) can detect the existence of congestion through implicit or explicit means. An example of implicit congestion notification might be packet loss or delay—the traffic source could infer from this that congestion is present in the network and that it should reduce its sending rate until the congestion clears up, which is how TCP works. An example of explicit congestion notification is an explicit signal received by the traffic source from the network indicating the presence of network congestion.

Frame relay uses explicit notification to control congestion. The explicit congestion notification signals are carried in the FECN and or BECN bits of the address field on a frame relay PDU. The FECN bit is turned on by a frame relay switch when congestion is encountered by the packet as it travels from the sender to the receiver. The BECN is turned on by a frame relay switch when congestion is encountered in the reverse direction. Receipt of a FECN or BECN indication may cause the traffic source to reduce its CIR or perhaps mark additional packets as eligible for discard.

5.1.4 RFC 1490

The versatility of frame relay is demonstrated by the fact that it can transport both routed (layer-3) and bridged (layer-2) data traffic. RFC1490 defines extensions to the information field that support the encapsulation of routed and bridged packets. This makes it ideal to carry LAN interconnect traffic over a WAN. The reason that RFC1490 is important is that it enables multiple protocols (e.g., IP, IPX) to be carried over the same PVC rather than requiring a separate PVC for each protocol. This obviously has the benefit of requiring fewer PVCs to be provisioned, thus reducing costs and resource consumption both in the network and at the frame relay device such as a router.

There are several different formats defined in RFC1490 that are used to indicate the presence of either routed or bridged traffic and what protocol is being carried within the packet. Figure 5-3 illustrates an example of the routed and bridge format. In the routed format example, the RFC1490 header follows the Q.922 address field. The RFC1490 contains a control field, a packet assembler/disassember (PAD), and a Network Layer Protocol Identifier (NLPID), which identifies the protocol that is contained in the information field. The NLPID value for IP is x'CC'. Many protocols including bridged data do not have an assigned NLPID value; in that case it is necessary to define additional fields. In the bridged format example, a 5-byte SNAP encapsulation header is used. It consists of a 3-byte OUI field and a 2-byte protocol identifier (PID) field. The combination of the OUI and PID fields and their respective values indicates that the information field may be carrying, for example, a bridged IEEE 802.3 Ethernet frame. If so, following the SNAP header, a MAC address may also be included.

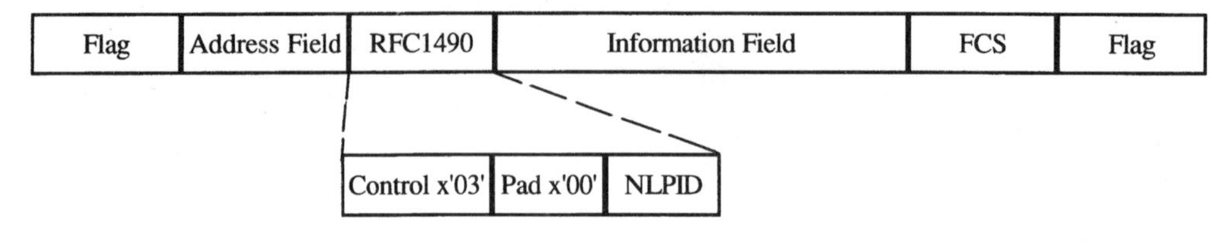

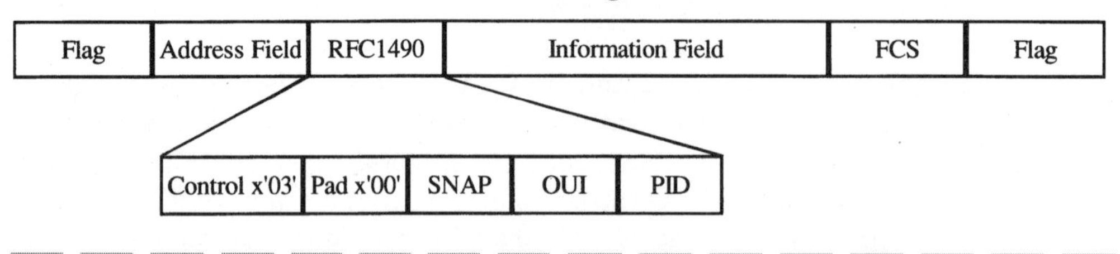

Figure 5-3 RFC1490 (*a*) routed and (*b*) bridged formats.

5.1.5 Frame Relay Network Examples

Frame relay is a powerful and versatile technology for interconnecting data devices over a WAN in a cost-effective manner. A few examples will help illustrate its usefulness in achieving this and some of the concepts that have been discussed.

Figure 5-4 shows three routers interconnected in a full mesh over a frame relay network. Observe the following:

- A single frame relay interface at a source can support multiple PVCs going to different destinations. This port-sharing capability reduces ports costs in both the routers and switches in the network.
- The DLCI configured on each router has local significance only. In other words, the DLCI is not the same at each end of the PVC.
- Each PVC can be configured with different traffic parameters (e.g., CIR), thus tailoring the service that is specific for the traffic flowing between any two routers.

Another illustration of frame relay's versatility is its ability to interwork with ATM. Figure 5-5 illustrates two examples of where frame relay and ATM can work together. In the case of Frame Relay/ATM Network Interworking, the ATM network is serving as the backbone

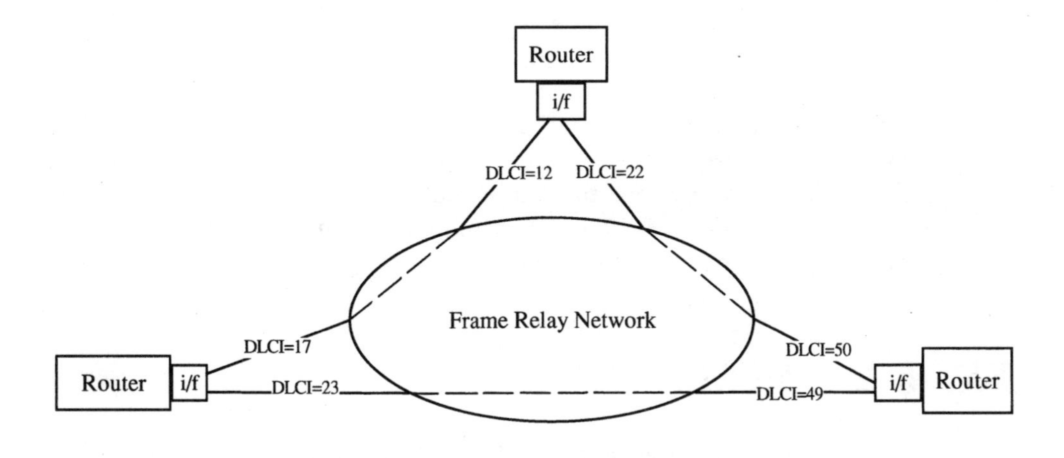

Figure 5-4 Routers connected to a frame relay network.

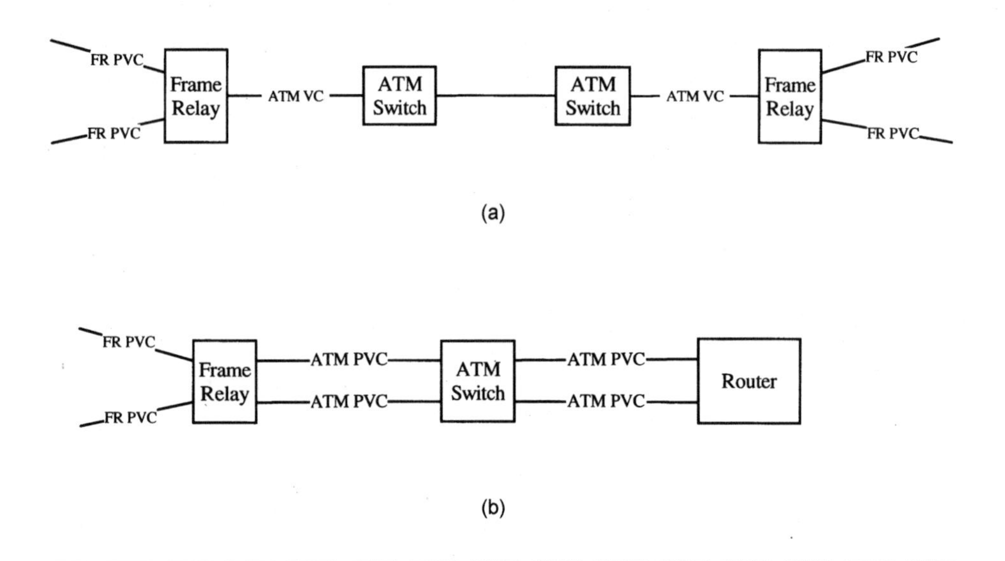

Figure 5-5 Frame relay/ATM (*a*) interworking and (*b*) service interworking.

between two frame relay switches. The frame relay PVCs come into an ingress frame relay switch that is outfitted with an ATM uplink. The frame relay traffic then rides the ATM network to an egress frame relay switch where the traffic is then placed back on frame relay PVCs for continuation onto its destination. The ability to aggregate and transport frame relay PVC traffic onto an ATM backbone is a very common use of ATM technology by ISPs. The second example, Frame Relay/ATM Service Interworking, is a more specialized use of ATM. In this case, frame relay PVCs are mapped to individual ATM PVCs that then terminate in a router or possibly a host. A possible application of this configuration might be the case where a number of remote sites interconnected by a frame relay network need to communicate with resources on a high-performance ATM network. The job of mapping frame relay PVCs to ATM PVCs is performed by an interworking function located in a switch.

5.2 ATM

ATM is a connection-oriented cell-switching technology that supports data, voice, and video over the same physical infrastructure. Devices and applications use ATM to establish virtual connections with each other through a network of ATM switches. Unlike frame relay or LAN switching, ATM offers a number of services, including QoS, that make it ideal for supporting real-time applications as well as normal data. As will be described later on, ATM is also a very powerful layer-2 switching technology that IP and other network-layer protocols can take advantage of.

5.2.1 Architecture

The architecture of ATM is rooted in the work performed in the late 1980s to broaden and enhance the capabilities of ISDN. Broadband ISDN was developed with the goal of enabling high-speed, digital, connection-oriented services that could support a number of different applications including data, real-time transmissions, interactive video, voice, and so on. ATM was selected as the transfer mode for Broadband ISDN for several reasons including the fact that it is transport independent, bandwidth can be allocated on demand, it is simple to provision semi-permanent connections, and the hardware and associated performance was envisioned to scale. Given these characteristics and the fact one can statistically multiplex a large number of data, voice, and video connections through a common infrastructure, it is no wonder that ATM was viewed just a short few years ago as *the* next-generation network. It probably won't turn out quite that way, but ATM can and will play an important role in the networks of the future.

Without going in to a whole lot of unnecessary detail, the architecture of an ATM system is depicted in Fig. 5-6. Although all networking archi-

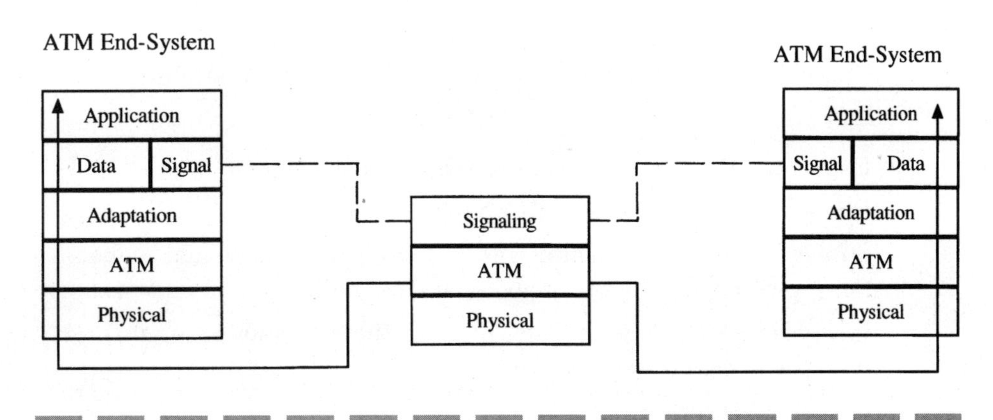

Figure 5-6 ATM architecture.

tectures possess some control plane functions, those in ATM stand out because the ATM SVC, a commonly used function, requires explicit signaling for connection setup and teardown. Therefore ATM end systems—hosts, routers, edge devices, or whatever—have both a signaling plane and a control plane. Signaling is used to build and tear down connections in an ATM network. Once the connection is established, data passes down from the application to the adaptation layer. The adaption layer adapts the higher-layer application data to the underlying ATM services. Several different types of adaptation layers have been defined and their use depends on the ATM service offered. Another function performed at the adaptation layer is segmentation and reassembly. This involves segmenting the payload into 48-byte payload segments at the source and reassembling it into a 48-byte variable-length payload at the destination.

After the data in the form of 48-byte payload segments are passed down to the ATM layer, 5-byte cell headers are appended to each 48-byte payload, forming a cell, and then each cell is passed down to the physical interface. The physical interface consists of any number of different link-layer technologies including Synchronous Optical Network/Synchronous Digital Hierarchy (SONET/SDH), unshielded twisted pair (UTP), T3, and T1.

At each intermediate switch along the path the individual cells are switched based on the contents of the cell header, using label swapping as the forwarding mechanism. The cells are not reassembled into packets until they reach their destination.

The interface between an end system and an adjacent switch is the UNI and the interface between adjacent ATM switches is the NNI. That distinction is important because there are different protocols, messages, and cell formats that flow over the UNI than flow over an NNI.

The primary functional attributes of ATM can be summarized as follows:

- ✔ Uses connection-oriented cell-switching technology.
- ✔ Has variable bandwidth (DS1 up to OC48+).
- ✔ Supports permanent, semipermanent, and dynamic virtual connections and permanent and switched virtual paths.
- ✔ Includes port sharing; multiple VCs originate at and terminate from the same physical device port.
- ✔ Uses fixed-length 53-byte packets called cells.

- Supports per-connection QoS for different traffic classes (real-time, variable rate, best-effort).
- Supports both point-to-point and point-to-multipoint connections.
- Label-swaps based on VPI/VCI field in ATM cell header.
- Is supported on hosts, routers, switches, edge devices, etc.
- Interworks with frame relay and LANs.
- Supports LAN Emulation, IP over ATM, and native ATM applications.
- Supports data, voice, and video.

The ATM Forum,[3] a consortium of vendors, developers, carriers, and customers that was formed in 1991 has for the most part driven the ATM standardization process. Table 5-2 lists some of the implementation standards based on many of the original ITU Broadband ISDN specifications that came out of the ATM Forum.

TABLE 5-2

ATM Forum Standards

ATM Forum Implementation Agreement	Description
UNI 3.0/3.1/4.0	Different releases of the ATM User-to-Network Interface signaling specification
Traffic Management 4.0	ATM Traffic Management Mechanisms including ABR Rate-Based Flow Control
PNNI Phase 1	Dynamic Interswitch Signaling and Routing Protocol for SVC setup requests
IISP	Interim Interswitch Signaling Protocol; Static Routing for SVC setup requests
LAN Emulation (LANE) V1.0	Services and protocols for emulating Ethernet and Token Ring on top of an ATM network
Multiprotocol over ATM (MPOA) V1.0	Services and protocols for operation of a distributed virtual router over an ATM network

[3] *http://www.atmforum.com.*

5.2.2 Virtual Paths and Virtual Connections

ATM supports two types of connections: VCs and virtual paths (VPs). A VC is defined as a dedicated logical channel between two or more end systems that transports cells. A VC originates and terminates within ATM end systems and can be configured with specific traffic and QoS characteristics such as bandwidth and delay. PVCs are statically configured and initialized when the network is brought up. Soft PVCs are semipermanent PVCs that can be configured and established on demand between multiple end systems while the network is active. And finally VCs that are dynamically established via signaling are called SVCs. Cells belonging to a specific VC are identified by a VC identifier (VCI) in the header of the cell.

A VP is a group of VCs, and each VC is associated with a single VP. Cells belonging to a specific VP are identified by a virtual path identifier (VPI) contained in the header of the cell. VPs can be permanent (PVP) or switched (SVP). VPs are an effective technique for switching large numbers of VCs around a network because the switching (label-swapping) is based on the VPI in the cell—the VCI values are ignored. PVPs are also useful in supporting SVC-based communications over WAN ATM networks that do not support SVCs. This is accomplished by tunneling the SVC setup requests through a PVP. Multiple VPs and VCs inside those VPs can flow over a single physical link.

A physical link, VP, and, VC are illustrated in Fig. 5-7. It should be noted that a VCI is only unique inside of a VP. Therefore, it is possible for two different VCs inside different VPs to share the same VCI value.

5.2.3 Cell Format

ATM cells are 53 bytes in length and consist of a 48-byte payload and a 5-byte header. The formats of the UNI and NNI cells are shown in Fig. 5-8

Figure 5-7 ATM virtual paths and virtual connections.

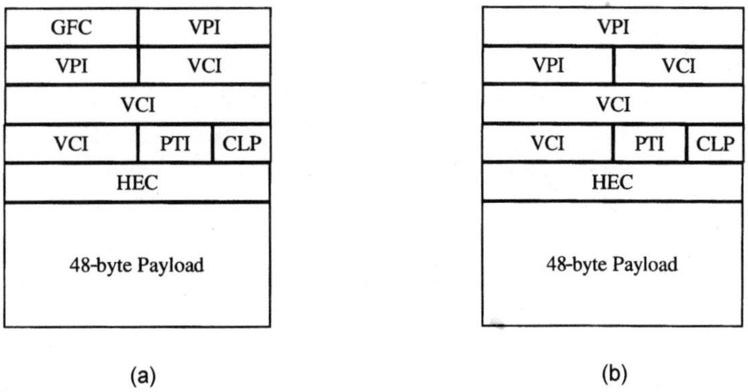

GFC	VPI	
VPI	VCI	
VCI		
VCI	PTI	CLP
HEC		
48-byte Payload		

(a)

VPI		
VPI	VCI	
VCI		
VCI	PTI	CLP
HEC		
48-byte Payload		

(b)

Figure 5-8
ATM cell format. *(a)* UNI. *(b)* NNT.

and the fields are described below. Again, UNI-formatted cells flow between an end system and a switch, and NNI-formatted cells flow between two switches.

- *Generic flow control (GFC).* A 4-bit value originally intended as a flow control indicator between the end system and an ATM switch but not implemented. Only exists in UNI format
- *VPI.* This field is 8 bits on the UNI side and 12 bits on the NNI side. The additional bits for the VPI on the NNI side are taken from the GFC field, which does not exist on the NNI side.
- *VCI.* This field contains 16 bits.
- *Payload type indicator (PTI).* A 3-bit field that indicates whether the payload contains user data, network management data, or traffic management information. One bit in the PTI field, the AAL-indicate bit, is set to zero for all cells in a packet except for the last one and thus is used by the receiving end system to identify the end of one packet and the beginning of a new one.
- *Cell loss priority (CLP).* A 1-bit value that discriminates between cells that are compliant with the traffic contract and those that are not. Cells that are not compliant are eligible to be discarded if congestion occurs.
- *Header error check (HEC).* This is used by ATM physical layer to detect and correct errors in the cell header.

5.2.4 ATM Addressing

ATM end systems and switches use a 20-byte addressing structure that is loosely modeled on the OSI Network Service Access Point (NSAP) structures. There are several formats to use depending on whether the ATM end systems are in a public or private network. Regardless of the format, the ATM addresses consist of an address prefix, which can be up to 13 bytes in length, a 6-byte end system identifier (ESI), and a 1-byte selector field, which only has local interface significance. The format of a 20-byte ATM address is shown in Fig. 5-9.

5.2.5 AAL5 Format

ATM supports different adaptation layers that provide specific services depending on the application. AAL1 has mechanisms to convey clocking between a source and destination and is used to support real-time services such as circuit emulation and voice. AAL3/4 supports the ability to multiplex and interleave cells from different sources over the same link and is used to support connectionless networking technologies such as Switched Multimegabit Data Service (SMDS). Both AAL1 and AAL3/4 extract some additional overhead from the 48-byte payload. Because there is already the so-called cell tax of roughly 9.4 percent (5/53) to deal with, use of payload bytes for anything but payload reduces the efficiency of ATM even further.

Originally termed the Simple and Efficient Adaptation Layer (SEAL), AAL5 has emerged as the most commonly used AAL in just about all ATM implementations. AAL5 requires no additional overhead from the 48 byte payload and is quite simple to implement. This is because there are no additional headers or trailers to process other than a 32-bit CRC value that is used to provide error protection at the frame level. In addition, a large percentage of all ATM implementations carry data (IP specifically) that are connectionless and best effort in nature. So as far as IP over ATM is concerned, there are really no extra requirements made for AAL functionality other than minimal processing, minimal overhead, and low cost to implement. AAL5 fits the bill on all of these.

Figure 5-9 ATM address.

ATM Address Prefix (13)	ESI (6)	Sel (1)

Figure 5-10
AAL5 format.

User Information (1-65,535) bytes	Pad (0-47) Bytes	CPCS UU 1 byte	CPI 1 byte	Length 2 bytes	CRC 4 bytes

An AAL5 frame is illustrated in Fig. 5-10. The user information can range from 1 byte all the way up to 65,535 bytes. A PAD field that follows is used to pad the user information out so that it is a multiple of 48. Following the PAD field is the 8-byte AAL trailer, which contains primarily a length field and the CRC check.

As will be shown throughout the rest of the book, every internetworking over ATM solution (except LANE 1.0) uses AAL5 as the adaptation layer.

5.2.6 UNI Signaling

The ability to establish a virtual connection on demand between end systems is one of the most attractive features of ATM. Network resources in the form of bandwidth, buffers, labels, and so on are allocated on demand, used for the duration of the connection between the communicating ATM end systems, and then released when they are no longer needed so that others may use them. This is all performed in a dynamic fashion, so there is no requirement to perform any preconfiguration such as is the case with PVCs. Another attractive but subtle feature of SVCs is that the resources needed to support the connection are not taken away from any existing connections. In other words, if a request for an SVC is sent into the network and there is insufficient capacity to support the connection, the SVC request is denied. This is analogous to a telephone busy signal. And finally each individual SVC request can contain specific traffic and QoS parameters that are specific to the data flow between ATM end systems.

UNI signaling is the means by which an ATM end system conveys an SVC setup request between one or more ATM end systems. More specifically, UNI signaling, consists of the protocols and messages that are exchanged over the UNI interface between an ATM end system and an ATM switch. UNI signaling messages do not carry any data but instead carry information elements containing connection-specific information such as the ATM address of the destination, traffic descriptors consisting of average and peak bandwidth, QoS parameters such as maximum allowable jitter, and so on. The information elements for the various UNI specifications are described in the ATM Forum UNI specifications.

UNI signaling currently supports the establishment of point-to-point and point-to-multipoint SVCs. Point-to-point SVCs can be bidirectional and point-to-multipoint SVCs are unidirectional from the sender out to all receivers. An example of the UNI message exchanges needed to build a point-to-point SVC is illustrated in Fig. 5-11.

UNI 3.0 and 3.1 specify that the SVC requestor must know a priori the addresses and the destinations for both point-to-point and point-to-multipoint SVCs. UNI 4.0 added Leaf-Initiated-Join (LIJ), which is the ability of a leaf node[4] to dynamically join an existing point-to-multipoint SVC. Further enhancements to UNI signaling might take the form of a new lightweight signaling protocol and the ability to establish multipoint-to-point SVCs.

5.2.7 ATM Service Categories and Traffic Management

ATM supports many different application types that place different demands on the network to function in an optimal manner. For example, real-time applications require minimal latency, low jitter, bounded delay, and minimum bandwidth. On the other hand best-effort applications use any and all bandwidth that is available but are capable of adapting

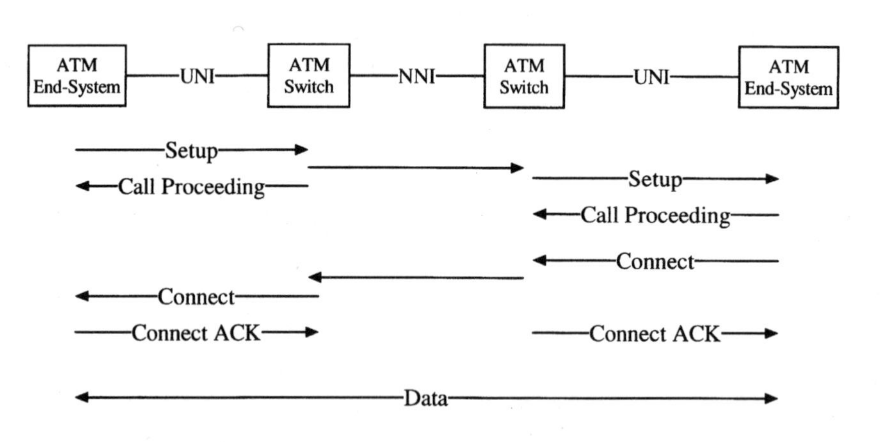

Figure 5-11 *ATM UNI signaling.*

[4]A leaf node is an ATM end system connected to the receiving end of a point-to-multipoint virtual connection.

their information transfer rate to variations in available network capacity with no disruption to the application. In between are applications whose traffic demands vary over different time scales. These applications may or may not need specific guarantees from the network pertaining to delay, jitter, bandwidth, and so on.

To support the wide spectrum of possible application behaviors, the ATM architecture has defined a set of service classes or categories. ATM service categories define and support within an ATM network a small group of services which any application can fall into. Defining a small group of services that covers all possible applications makes the specification easy to follow and build on and makes it simple for ATM end systems to understand what they can and cannot ask for from the network.

The ATM service categories shown in Table 5-3 are defined in more detail in the ATM Forum TM 4.0 specification.[5]

Depending upon the services requested by an end system from the network, ATM can employ a number of traffic management techniques. Traffic management optimizes network resources and avoids or limits network congestion. Traffic management also helps enforce and deliver

Table 5-3

ATM Service Categories

ATM Service Category	Description
CBR	Constant Bit Rate: for applications such as Circuit Emulation or real-time video that require a fixed amount of bandwidth and some clock synchronization between end points.
Rt-VBR	Real-time Variable Bit Rate: for applications that transfer real-time data at variable bit rates such as video.
Nrt-VBR	Non-real-time Variable Bit Rate: same service as rt-VBR, except delay is not bounded.
ABR	Available Bit Rate: provides best-effort service for applications that can adjust their transfer rate based on network feedback.
UBR	Unspecified Bit Rate: purely a best-effort service with no flow control.
GFR	Guaranteed Frame Rate: similar to UBR with a minimum cell rate.

[5]"ATM Forum Traffic Management Specification V4.0," ATM Forum, af-tm-0056.000, April 1996.

the QoS on the connection that the end system and network contracted for.

The TM 4.0 specification defines the ATM traffic management functions in more detail. Some of the more well-known traffic management mechanisms are

- ✓ *Connection Admission Control (CAC).* CAC checks to see if resources are available for a connection request and either permits the connection setup process to continue or denies the request.
- ✓ *Usage Parameter Control (UPC).* Otherwise known as traffic policing, UPC polices the traffic at network ingress for conformance with the traffic contract agreed upon by the connection source and the network. UPC may tag or discard nonconforming cells.
- ✓ *Selective cell discard.* Cells with CLP = 1 may be discarded by a congested component (e.g., switch).
- ✓ *Traffic shaping.* May alter the characteristics of cell traffic so that it conforms more closely to the traffic contract.
- ✓ *Frame discard.* When congestion occurs in a switch, cells are dropped. If one cell is dropped, frame discard drops all the remaining cells in the frame. Variations on this function appear in the literature under the Early Packet Discard (EPD) family of congestion control techniques. Frame discard is useful as a means of limiting unspecified bit rate (UBR) traffic congestion.
- ✓ *ABR rate-based flow control.* ABR sources can dynamically adjust their transfer rate based on network feedback. ABR sources periodically generate special Resource Management (RM) cells that traverse the network toward the destination. Intermediate switches may mark congestion indicators in the cells or include an explicit rate (ER) value. When the RM cell is turned around by the destination and returned to the source, the source may adjust its transfer rate based on the congestion indicators or change it to the value of the ER field contained in the RM cell.

5.2.8 PNNI

Many believe that because ATM is a connection-oriented switching technology, routing is not required. This is true for cells flowing over an established PVC or SVC. The VPI/VCI, or more generally the label contained in the cell header, is not an entity that can be routed or addressed. It simply identifies the connection that the cell belongs to, has local significance only, and most likely will be swapped as the cell is forwarded through a switch. However, before an SVC connection is established, the SVC request must travel from the source to the destination. Contained within that SVC request is a field that can be routed and addressed, the ATM address of the destination. In addition the SVC request may contain traffic descriptors such as peak and sustained cell rates (PCR, SCR) and QoS parameters such as Cell Loss Ratio (CLR), maximum Cell Transfer Delay (CTD), and so on. Therefore the ATM routing system must identify and select a path to the destination that is able to support and sustain the bandwidth and QoS requirements specified in the SVC request.

This is not as simple a task as it might seem. Consider that the path calculations performed by your favorite IP routing protocol only take into account a single dimension of metrics (e.g., fewest number of hops, low cost) when computing a path to the destination. Routing of an ATM SVC setup request must factor in reachability to the destination, available capacity along the path, the offered traffic load to be placed on the path, and whether the path can meet the QoS requirements of the connection. And given that there might be a large number of SVC setup requests to manage over a large ATM infrastructure, any ATM routing protocol must be scalable and self-configuring and must place as little load on the network as possible. At the same time it must accurately track and disseminate the status, quantity, and quality of network resources to all switches in the ATM network.

PNNI Phase I is the signaling and routing protocol developed by the ATM Forum to dynamically route SVC setup requests over a network of private ATM switches.[6] PNNI Phase I contains a routing component that distributes topology state information to all ATM switches in the network and a signaling component that makes use of the topology state information to compute a route and forward the SVC setup request through the network from the source to the destination.

PNNI Phase I borrows some of the mechanics from previous routing protocol efforts to achieve its goals of scalability and efficiency. Specifically PNNI makes use of the following:

---

[6]"ATM Private Network-Network Interface Specification V1.0," ATM Forum, af-pnni-0055.000, March 1996.

- *Link state routing.* ATM switches advertise the topology state information to other switches and use this information to build a topology state database. Topology state routing means that the switch will advertise information on the status of the links and the switch itself. The contents of the topology state database are then used compute a path for the SVC setup request to follow.
- *Source-routing.* PNNI limits resource consumption during SVC setup by having the source or ingress ATM switch compute the entire path through the network based on the topology state database. The source ATM switch then appends a source-route vector called a Designated Transit List (DTL) to the SVC setup request message and sends it on its way. Each intermediate switch just follows the DTL and need not concern itself with path calculations. Observe also that during SVC setup, the entire topology database at the ingress switch is consulted to determine an optimal path through the network for that specific connection. Unlike spanning tree or IP routing protocols, PNNI can build connections that utilize the entire network topology of links and switches.
- *Hierarchical routing.* PNNI must scale, and to do so one must introduce hierarchy. A collection of ATM switches that share a common ATM address prefix and topology state database is called a peer group. The contents of the topology state database is then summarized by a single peer group leader (PGL) and advertised in aggregated form to other PGLs in other peer groups. The PGLs of individual peers groups can actually form their own higher-level peer group and as members of this higher-level peer group they are called logical group nodes (LGNs). The LGNs are not actually physical switches themselves but rather serve as abstract representations of the topology of the lower-level peer groups. The concept of the PNNI hierarchy is illustrated in Fig. 5-12.

The hierarchy limits the amount of topology state information that is advertised around the network by presenting some of it in summarized or aggregated form. In the example in Fig. 5-12, the switches in peer group 1 share a single topology database that reflects the actual topology of that peer group but may only possess summarized knowledge of switches in external peer groups such as peer group 2.

PNNI Phase I is the standard ATM routing protocol used in both private and some public ATM networks. PNNI extensions

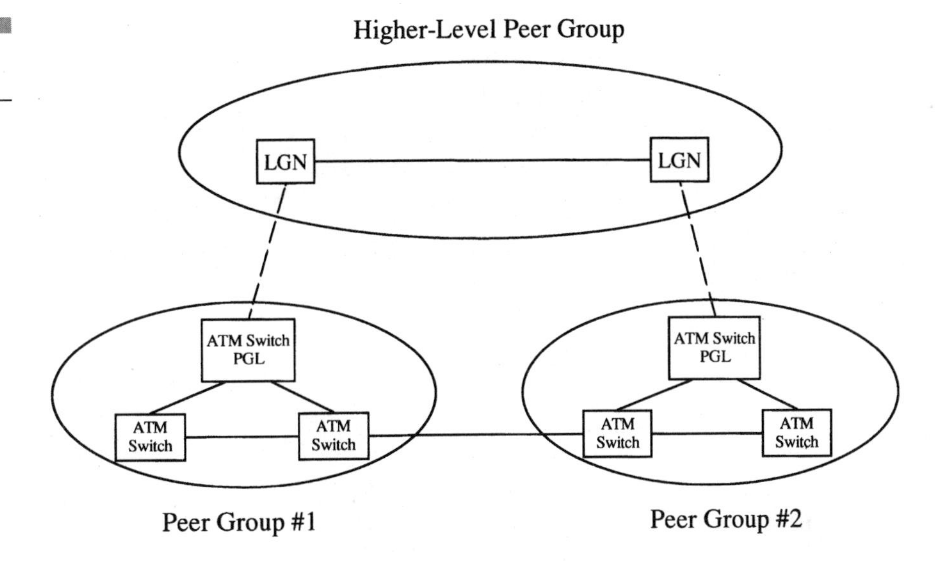

Figure 5-12
PNNI hierarchy.

(I-PNNI,PNNI-augmented routing [PAR], proxy-PAR) to support the integration and cooperation with IP routing protocols will be discussed in Chap. 14.

5.2.9 ATM Network Examples

ATM can support a diverse set of applications and topologies and in fact can take on several identities. It can operate as a complete end-to-end network architecture with its own set of applications, addressing, routing, and data transfer. That is not very common today, although some niche applications such as high-quality MPEG2 video operate ideally in this mode. ATM can also operate as a high-speed, high-performance layer-2 infrastructure for data networking applications either on an end-to-end basis or serving as a backbone in a campus environment. A combination of the two is illustrated in Fig. 5-13.

Observe that some workstations along with a couple of servers are directly connected to ATM switches. Utilizing existing applications and networking protocols such as IP or IPX, the workstations are able to use the underlying ATM network as a high-speed transport for client/server application flows. At the same time, the same ATM network could serve

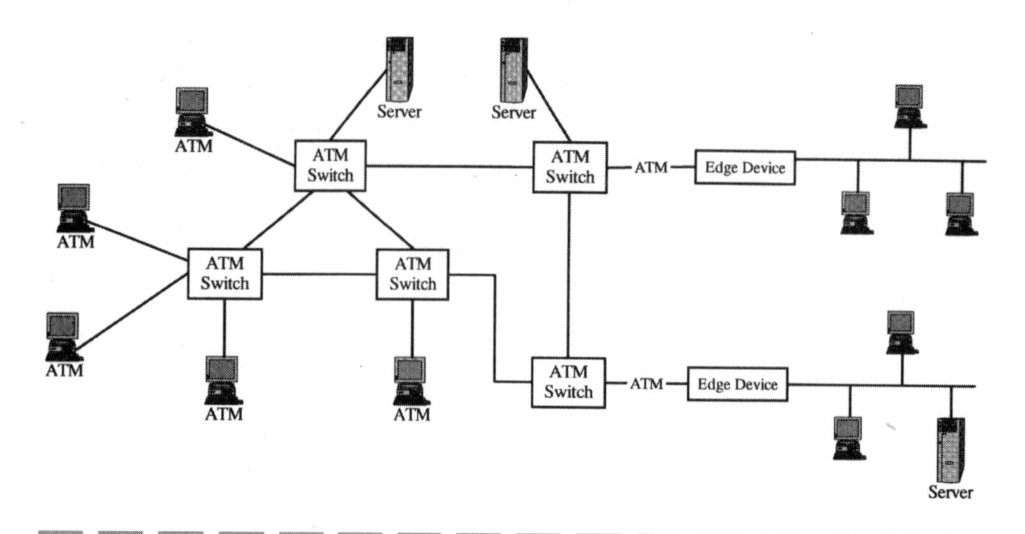

Figure 5-13 ATM campus network.

as a backbone between the two Ethernets shown on the right side of the diagram. Presumably the Ethernet-attached workstations and servers could interoperate with the ATM-attached devices in a transparent manner. All of this is quite possible and actually quite common. The underlying IP over ATM protocols such as LAN Emulation and Classical IP that facilitate this scenario will be discussed in later chapters.

ATM is also used as a backbone technology for ISPs. Perhaps the best illustration of this concept is shown in Fig. 5-4 in which one could substitute the frame relay network for an ATM network, or using the Frame Relay/ATM Network Interworking scenario shown in Fig. 5-5, ATM could serve as the core of the network backbone. ATM is quite useful for ISPs and carriers because of its bandwidth management functions as well as the ability to provision virtual connections over explicit paths and thus achieve some level of load balancing.

MORE INFORMATION

Black, U., *ATM: Foundation for Broadband Networks,* Englewood Cliffs, NJ: Prentice Hall, 1995.

Dutton, H., and Lenhard, P., *Asynchronous Transfer Mode (ATM) Technical Overview,* Englewood Cliffs, NJ: Prentice Hall, 1995.

Dutton, H., and Lenhard, P., *High-Speed Networking Technology: An Introductory Survey,* Englewood Cliffs, NJ: Prentice Hall, 1995.

Miller, M., *Analyzing Broadband Networks,* M&T Books, 1994.

Spohn, D., and McDysan, D., *ATM Theory and Application,* New York: McGraw-Hill, 1995.

CHAPTER 6

The Concept of IP Switching

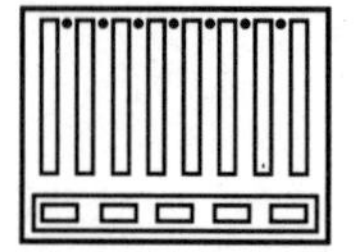

In the previous chapters we reviewed in some detail the suite of TCP/IP protocols and some of the important and popular underlying switching mechanisms and protocols that are prevalent in both local and wide area networks today. We now bring together these two topics and discuss the concept of IP switching. IP switching can be defined as the protocols and mechanisms that utilize layer-2 switching to accelerate the forwarding of IP packets through a network. The most popular switching mechanism for IP switching is ATM, although other layer-2 switching (e.g., frame relay) technologies could work. This chapter introduces some of the important concepts and definitions behind IP switching and discusses some of the motivations behind its development.

6.1 Definitions and Terminology

IP switching is a class of protocols and mechanisms that utilize layer-2 switching as the primary forwarding mechanism for transporting IP packets through a network.[1] IP switching takes advantage of the high bandwidth and low latency of switching to transport a packet as quickly as possible through the network. Even though all IP switching solutions, and there are several, do support IP routing and forwarding as we know it, they all attempt to redirect all or some number of packets through a switching component, thus obviating the need to perform any layer-3 processing.

To best understand the functions and operation of IP switching, let's first define several important concepts.

6.1.1 IP Switch

Generally speaking, an IP switch is a device or system that can forward IP packets at layer 3 and possesses a switching component that enables packets to be switched at layer 2 as well. An IP switch possesses mechanisms to classify which packets will be forwarded at layer 3 and which will be switched at layer 2 and then to redirect some or all packets over

[1]Although we will discuss IP switching techniques that bypass intermediate routing hops, it should be noted that this definition is inclusive of other forms of ATM switching that facilitate IP connectivity and performance. Classical IP over ATM and LAN Emulation are two such examples that will be discussed in later chapters.

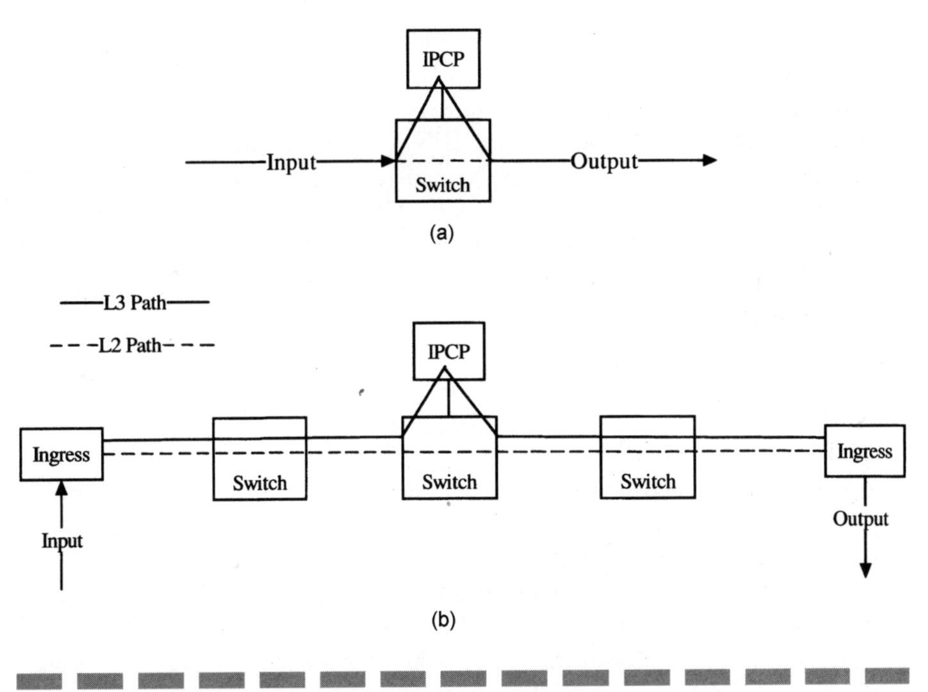

Figure 6-1 IP switching systems. (*a*) IP switch device. (*b*) Virtual IP switch.

a layer-2 switched path. Most IP switches utilize an ATM switching fabric, but other layer-2 switching technologies can be used as well.[2]

Two examples of IP switching systems are shown in Fig. 6-1. The IP control point (IPCP) in both examples runs a typical routing protocol (e.g., RIP, OSPF, BGP), provides a default layer-3 hop-by-hop routed path, and may directly or indirectly communicate with the switch component(s) to redirect IP packets through the switching component(s). Like normal ATM switches, the switch components maintain a connection table of input ports, input labels (VPI/VCI), output ports, and output labels. The top diagram shows an IP switch and the bottom shows a virtual IP switch.

There are several differences between the two models. One lies in the scope of the layer-2 switched path. With the IP switch, the switched path

[2]Both general and specific discussions of IP switching solutions assume the use of ATM as the switching component because that is what most of the solutions use. Other layer-2 switching technologies are used as well. Specific cases (e.g., tag switching, MPLS) where ATM is not necessarily required will be pointed out.

is contained within a single device and switch component that is under the direct control of a single IPCP. Multiple adjacent IP switches would have to cooperate in some manner to establish an end-to-end switched path that traversed more than one IP switch. With the virtual IP switch, an end-to-end switched path may span multiple switches that are all under the indirect control of a single IPCP. Another difference is the location of the input and output "ports." In the IP switch device configuration, the input and output ports of the IP switching system are located on the same box. With the virtual IP switch, the input and output components can be located on the same box or on physically separate ones.

Another difference between the two models is the use of, and coexistence with, ATM Forum UNI/PNNI signaling and routing protocols. A single IP switch depends on the IP network topology and routing protocols to select a forwarding path through the network and then uses a specialized control protocol exchange between itself and adjacent IP switches to map the hop-by-hop forwarding path to a layer-2 switched path. A virtual IP switch also uses special control protocols to initiate the process but then relies on the ATM network topology, signaling, and routing protocols to select and build a layer-2 switched path. In this case standard ATM Forum UNI/PNNI signaling and routing protocols are used to accomplish this task.

It should be noted that an IP switch is capable of running ATM Forum protocols from an end system as well as from a switch perspective. IP switch UNI signaling and VC management would be needed if two IP switches wanted to communicate through a network of intermediate ATM switches. Another scenario might be the case where the switch component of an IP switch is running ATM Forum protocols (e.g., PNNI) in a "ships in the night" configuration. This enables a single device to provide IP routing and forwarding, IP switching, and ATM switching services all in one box. One can assume that with practical deployments of IP switching services that the ATM switch component will indeed support ATM Forum protocols to give the network provider several different service options to offer.

The term *IP switch* was first popularized by Ipsilon when it burst onto the scene in 1996. The company's device consisted of an IP switch controller (e.g., router) attached to a separate ATM switching fabric running the IFMP and GSMP protocols.[3] The Toshiba CSR and the IBM

---

[3]IFMP is documented in RFC1953. GSMP is documented in RFC1987/2297.

Integrated Switch Router (ISR) have similar configurations and use special control protocols, but they also include specific support for and coexistence with ATM Forum protocols. MPOA is an example of a virtual IP switch and in fact uses the term *virtual router* to describe the function that it is providing.

6.1.2. IP Switch Ingress and Egress

An IP switching system provides default layer-3 IP routing and forwarding and accelerated layer-2 IP switching services. The beneficiaries of these services can be individual applications or collections of end users grouped by LAN segment, network, or a shared destination. Whatever the criteria is for determining who or what can receive IP switching services, one must have a way of entering and leaving the IP switching system. Therefore to round out the necessary components of an IP switching system, one needs an ingress and egress component.

IP switching ingress and egress components are positioned at the edge of an IP switching system. An ingress and egress component can consist of a piece of code running in a workstation, added function in an existing edge device or router, or a specialized black box. The important functions of an IP switching ingress and egress consist of the following:

- ✔ Provide normal default IP forwarding for traffic entering and exiting the network.
- ✔ Provide media translation (e.g., Ethernet to ATM) services for packets entering and exiting the IP switching system.
- ✔ Participate in the control procedures for establishing, maintaining, and removing a layer-2 switched path between the appropriate ingress and egress.
- ✔ At the ingress, classify eligible packets and then forward them onto the layer-2 switched path. This generally involves inspection of some number of fields in the packet header to determine if the packet should be placed on a switched path and if so, which one, a standard routing table lookup, insertion of a label into the packet, or, in the case of an ATM switching system, segmentation and transmission over a virtual connection.
- ✔ At the egress, receive packets over the layer-2 switched path and perform standard IP forwarding procedures as the packet exits the IP switching system.

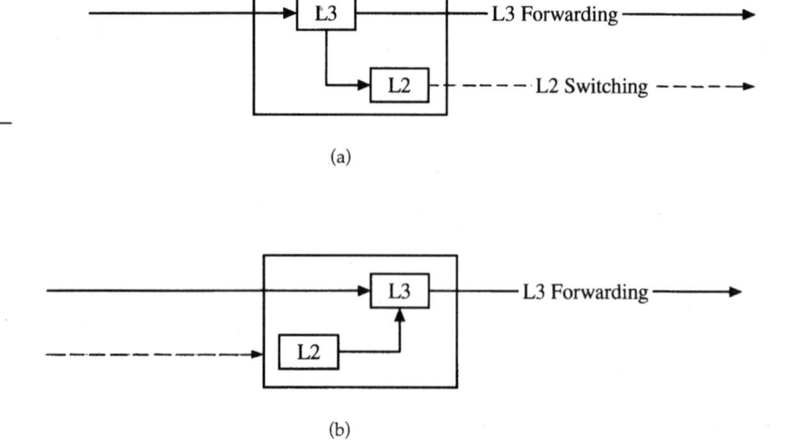

Figure 6-2 IP switching (*a*) ingress and (*b*) egress functions.

The concept of an IP switching ingress and egress is illustrated in Fig. 6-2. Assume up front that the IP switching system has determined that some packets can be forwarded over a layer-2 switched path and that such a path has been established between the network ingress and egress components. Packets arriving at the network ingress are classified for either default layer-3 IP forwarding or layer-2 switching based on some IP-layer criteria such as source/destination IP addresses or destination network prefix. Packets classified to be layer-2 switched are forwarded over the switched path. All others are forwarded at layer 3 using normal IP forwarding procedures. When the packets arrive at the network egress, either at layer 3 or layer 2, they will be passed to layer 3 for normal IP processing and forwarded to the destination.

An IP switch does not operate in a vacuum or in standalone mode. To be useful it must have a complimentary function at the network ingress and egress that enables packets to enter the IP switching system, transit over a layer-3 routed or layer-2 switched path, and then exit.

6.1.3 Shortcut Path

In a normal IP routed environment, packets from a source network are forwarded hop by hop through a series of routers to a destination network. Associated with each router hop are necessary functions such as a routing table lookup, header checksum, TTL decrement, media translation, and so on. This introduces some amount of aggregate delay and latency for packets traversing the routed path. One way to reduce delay and latency is to bypass intermediate routing hops where possible.

The layer-2 switched path alluded to in the previous sections is called a shortcut path. A shortcut path, by definition, is a source-to-destination virtual connection that bypasses intermediate layer-3 routing hops.[4] A shortcut path can be established between two hosts, two edge devices (e.g., routers), or a combination of the two. It all depends on the placement of the aforementioned ingress and egress functions, granularity of the traffic to be placed on the shortcut path, and the IP switching protocols used to build it.

The basic properties of a shortcut path are

- It bypasses intermediate layer-3 routing functions.
- The trigger to build a shortcut path can be based on data traffic or control traffic. Data-driven shortcut paths can result in some data traffic being routed at layer-3 prior to the IP switching system, establishing a shortcut path. Control-driven shortcut paths are established based on the appearance of control traffic (e.g., routing table update) so that all subsequent data traffic would flow over the shortcut path.
- If the shortcut path between an ingress and egress does not exist or suddenly goes away, traffic can still reach the destination by traveling over the routed path.
- It may follow the same physical path (pass through the same nodes and links) as the routed path (IP switch), or it could pass through a separate layer-2 switch topology (virtual IP switch). In the case of the virtual IP switch, the shortcut path could be topologically independent of the IP routed path (see Fig. 6-1 for an example).
- An ingress-to-egress shortcut path can be built in one "end-to-end" pass, or it can be built by concatenating multiple smaller shortcuts.
- It may originate from a network ingress or egress or may "chaotically" fall into place based on concurrent local shortcut setup actions performed in each IP switch.
- It can be point to point, point to multipoint, or multipoint to point.

[4]A shortcut path can be distinguished from a normal path by the fact that it is bypassing or shortcutting router hops. A normal path is layer-2 path (e.g., ATM SVC) established between two devices on the same subnet.

- It can provide some enhanced QoS or COS beyond what would be available over the routed path.

It is quite clear that there are many different criteria and techniques for establishing shortcut paths through a network. Later sections discuss how and when shortcut paths are built and managed.

6.2 Motivations

The development of IP switching technologies was motivated by a number of factors, some business-related and others technical. Technically speaking, there was a big motivation to develop faster and less expensive IP forwarding devices because the Internet and probably many intranets exist in a state of constant or near congestion. This should not be surprising because most IP networks are actually engineered for congestion. It would be cost prohibitive to provision a network to handle 100 percent of the traffic at burst-rate speeds, so they are built with considerably less capacity with the hope that most end users can get most of their traffic through the network most of the time. If not, TCP hosts adapt the sending rate to whatever network capacity is available, congestion clears up, and people can continue with their business. Of course if the hosts are not using TCP (streaming audio/video comes to mind) and continue to inject packets into an already clogged network while ignoring or not understanding implicit or explicit congestion indicators, the problem may not disappear. In fact it could worsen.

According to standard network theory, performance bottlenecks can be reduced or eliminated by speeding things up. Increasing the forwarding rate of data packets can have a sort of positive domino effect on network performance in general. Faster transmission rates mean that packets will spend less time waiting in queues. The queues will empty out faster, leaving more available queue capacity, for transient bursts. Instead of discarding packets due to insufficient queue capacity, resulting in retransmissions and worse, the traffic bursts can be handled. Overall network performance will improve and so on.

Now things are not really so simple. There will always be points of congestion in any network, and it has been suggested that network switching or routing nodes take a more active role in queue scheduling and management. "Congestion-ignorant" hosts will have to somehow be policed and so on. But one will definitely need more bandwidth and faster forwarding devices to handle the growth in traffic and to minimize congestion.

Using ATM technology was and is viewed by many as the ideal solution for bypassing the performance bottlenecks associated with current router (circa up to 1997) technology. A standard ATM switch offers multigigabit forwarding rates, which are orders of magnitude faster then existing routers. ATM also offers excellent price performance and port speeds up to OC12 and higher and was a primary focus of over a decade of industry research and standards development.

Somehow using ATM technology in the IP routed Internet environment was at the forefront for those that began investigating IP switching solutions. The technical reasons why IP switching came about include some of the following:

- ✔ Routers are bottlenecks and cannot support traffic at OC3 + forwarding rates.
- ✔ Routing tables are too big and lookups take too long. Improvements can be made if many routes are mapped to a smaller group of labels (e.g., VPI/VCI) and a lookup consists of indexing VPI/VCI tables in hardware.
- ✔ Next-generation IP networks and applications need the bandwidth management, performance, and QoS that ATM can offer.
- ✔ A simple way of supporting connectionless IP traffic over a connection-oriented network was needed.
- ✔ A simple way of constructing shortcut paths for applications that need it was needed.
- ✔ ATM Forum UNI/PNNI signaling and routing was considered too "heavyweight." Something simpler that would be palatable to the "IP-centrists" was needed.
- ✔ ISP core networks already had or were planning to deploy ATM. A simple and scalable way of supporting large amounts of IP traffic over the ATM VC-oriented network was needed.
- ✔ Developed platforms and protocols needed to support emerging and advanced IP functions such as IP Integrated Services (RSVP) and traffic engineering (explicit routing).

There are other reasons without doubt. But the key technical motivator behind the development of IP switching was a desire to significantly boost performance by utilizing switching and specifically ATM switching as the forwarding mechanism for some or all of the traffic.

The business reasons for doing anything new in the networking industry these days are obscured and hidden behind the puffed-up, grandiose, and exaggerated claims generated by the overpaid marketing and consultant-public relations arms of the organization or venture capitalist firm. With some hindsight though it is not too difficult to recognize a few reasons why some established players and startups invested so much in IP switching technology and continue to do so:

- ✓ Justify investment in ATM
- ✓ Reposition ATM as an "IP-friendly" technology
- ✓ Exploit the real and perceived performance problems of the Internet and intranets with integrated platforms that provide IP routing and high-capacity ATM switching
- ✓ Offer new technology to compete against an installed base of routers
- ✓ Offer new technology to confuse the market and protect the installed base of routers by obscuring their deficiencies

Who knows for sure. But the idea in this business is to make money, and some did not and others still might. It is likely that IP switching in various forms will be a very useful technique that both public and private network operators will deploy for added capacity and special services.

6.3 IP Switching Addressing Models

IP and ATM addressing were discussed in previous sections. A network supporting IP switching services with ATM technology must support at minimum the IP address space. The IP address may be bound to an ATM address for IP packets with a specific source or destination, or a combination of IP addresses may be mapped to a connection (by virtue of labeling the traffic with a VPI/VCI). Therefore there are two basic addressing models for IP switching: separated and IP-VC.

6.3.1 Separated Addressing

The separated addressing model specifies that both an IP and ATM address space should be maintained within the network. Thus a router or host that is attached to the ATM network must be configured with both an IP and an ATM address. This is not different from any LAN-attached IP host that is configured with both an IP and a MAC address.

Unless a PVC has been previously established, a source IP host must know the ATM address of the destination IP host that it wishes to communicate with. This IP-to-ATM address resolution is required so that the source IP host may initiate an SVC connection request to the destination IP host. A source IP host may know this information a priori through configuration or some other means, or it may have to query an address resolution server that keeps a table of IP and ATM address mappings. In practice, separate routing protocols are needed to advertise both IP and ATM destination prefixes, although it is possible that a single routing protocol could support both (e.g., I-PNNI [Integrated PNNI]).

The separated addressing model has the following characteristics:

- ✔ Each IP-addressable ATM-attached device must be configured with an IP and ATM address
- ✔ Separate routing protocols for IP (e.g., OSPF) and ATM (e.g., PNNI) are typically supported in this environment.
- ✔ In practice dynamic resolution of IP-to-ATM addresses requires an additional query-to-server mechanism, which can add delay to the time it takes when establishing either a normal path (e.g., between two hosts on the same subnet) or a shortcut path between a source and destination.

Separated addressing is used in the IP switching overlay model that is described below. MPOA is an example of an IP switching system that uses separated addressing.

6.3.2 IP-to-VC

The term *IP-to-VC* is an attempt to describe an addressing model where just the IP address is maintained, and packets are placed on a separate shortcut path, a VC, based on the contents of IP-layer information in the packet header. In this model, no ATM addresses are used to facilitate IP switching communications. Separate control protocols described in later sections are used to map IP packets to shortcut paths without using ATM addresses and ATM Forum protocols. The mapping process, however, makes use of the ATM VPI/VCI value as a way associating IP-layer information with a VC.

The characteristics of the IP-to-VC addressing model are

- A single IP address is maintained; no ATM addresses are needed or used.
- A single IP routing protocol is used to advertise IP address prefixes.
- Specialized IP-to-VC mapping protocols are used to associate IP packets with a shortcut path.

The IP-to-VC addressing model is used by all IP switching entities that peer with each other and where ATM Forum protocols are not present or being used to facilitate the establishment of a shortcut path. An example of the use of the IP-to-VC addressing model is IP switches running IFMP and GSMP.

6.4 IP Switching Models

There are two basic models for IP switching, and they differ in their use or nonuse of ATM Forum protocols and in whether the components consist of IP switches or virtual IP switches. The two models are called peer and overlay.

6.4.1 Overlay Model

The overlay model for IP switching consists of an IP layer running on top of a separate ATM layer. In other words, it consists of IP devices with IP addresses running IP routing protocols and ATM devices (IP hosts, IP routers, ATM switches, etc.) with ATM addresses running ATM signaling and routing protocols. The overlay model is probably the simplest to implement because both the IP and ATM elements are available and run for the most part standards-based protocols. But there is some duplication of function in that two address spaces must be maintained and two routing protocols must be supported. An example of the overlay model is MPOA.

The characteristics of the overlay model are that it

- Uses separated addressing.
- Runs separate routing protocols at IP (e.g., OSPF) and ATM (e.g.,

PNNI), which means there are two separate topologies to maintain, and one does not know about the other. For example, IP routers running OSPF have knowledge of the IP network topology but do not see or know about the ATM switches.

- ✔ Requires address resolution between IP and ATM and user network interface (UNI)/PNNI signaling and routing if SVCs are used for either normal or shortcut path establishment.
- ✔ Generally uses virtual IP switches. Another example could be a bunch of IP hosts and routers attached to the perimeter of an ATM backbone.
- ✔ Supports default routed and shortcut paths.
- ✔ Has the same issues for IP over other Non-Broadcast Multi-Access (NBMA) technologies.[5]

An example of the overlay model is illustrated in Fig. 6-3. Separated addressing is implied by the existence of both IP and ATM networking devices. The default routed path follows a path through two router hops, IP control points 1 and 2. A shortcut path that bypasses the router hops may be established between the ingress and egress as shown. Once the ingress determines the ATM address of the egress, it can use standard ATM Forum signaling and routing to establish a shortcut path through ATM switches 1, 3, and 4. Observe that packets placed on the shortcut

Figure 6-3
Overlay model.

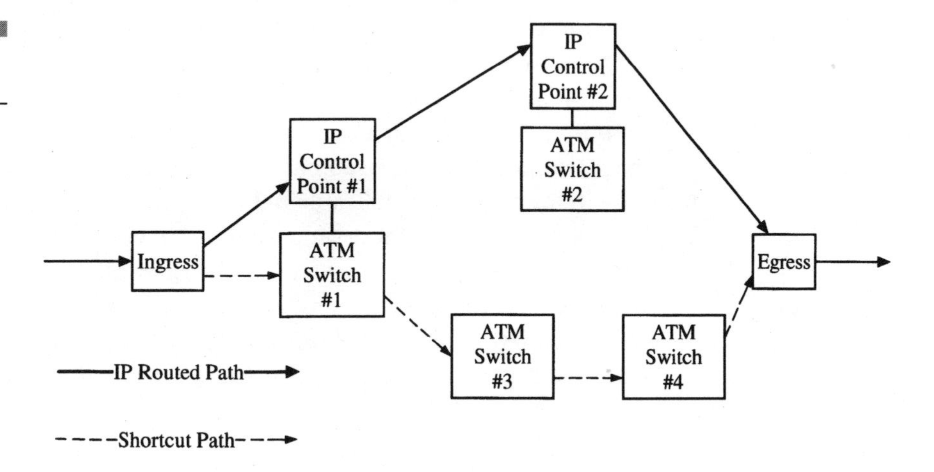

[5]Examples of NBMA networks are ATM, frame relay, and X.25.

path do not pass through the same devices or links as those packets following the routed path. This is because the ATM routing protocol (e.g., PNNI Phase I) selected a path for the ingress-to-egress SVC setup request based on the state of the ATM switch topology, not the IP topology.[6]

6.4.2 Peer Model

The peer model for IP switching specifies that the IP switch components maintain a single IP address space and support a single IP routing protocol. The peer model also implies the existence of a separate control protocol that is used to map IP traffic to shortcut paths. An example of the peer model is a network of IP switches running IFMP and GSMP

The characteristics of the peer model are that

- There is a single IP address space to maintain.
- It has a single IP routing protocol.
- IP switches use special control protocols to map IP traffic to shortcut paths.
- It supports default routed and shortcut paths.

An example of the peer model is illustrated in Fig. 6-4. IP-to-VC addressing is implied because there are no ATM switch components present that are running ATM Forum protocols. Observe that the routed path and the shortcut path pass through the same devices, IP switches 1 and 2 respectively, and that both routed and shortcut path traffic flow over the same links. This is because a single IP routing protocol will compute a best path to a destination based on a single IP network topology. A summary comparison between the peer and the overlay models for IP switching is shown in Table 6-1.

A couple of other important items that will be touched upon in more detail in later chapters are the issues of router adjacency and VC consumption. In the overlay model routers connected to a large ATM (or other NBMA network) network may need to peer with each other to exchange routing protocol updates so as to gain an accurate and timely

[6]This may be a very attractive feature for network providers that need to support a high-capacity backbone that is capable of utilizing all links and nodes in the topology. Note that IP routing protocols do not take into account available network capacity, delay, and other QoS metrics when selecting a path through the network.

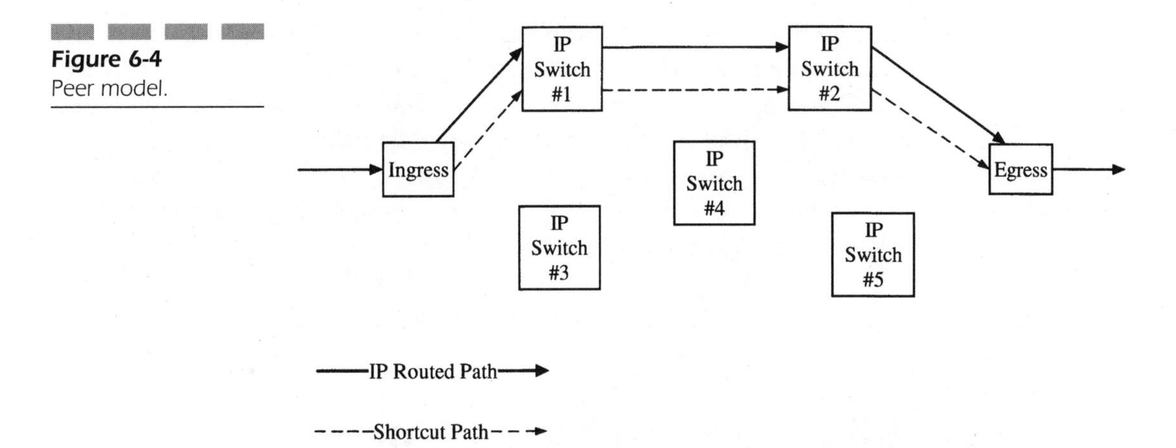

Figure 6-4
Peer model.

representation of the IP network topology. Routers supporting a large number of peer connections will have to process a proportionally large number of routing protocol updates. This is in addition to any data connections the routers may have established with each other. In the peer model, the IP switches must only peer with their neighbors, just like a network of routers would peer with each other.

TABLE 6-1

Overlay vs. Peer Comparison

Attribute	Overlay	Peer
Addressing	Separated (IP and ATM)	Single (IP), uses direct IP-to-VC mapping
Routing protocols	Both IP and ATM required	IP only
ATM forum protocols	Yes	No
Special IP-to-shortcut protocols	No	Yes
Address resolution (IP-to-ATM)	Yes	No
Shortcut path and IP routed path topologies are the same	No	Yes
IP switching components	Virtual IP switches, routers, hosts	IP switches, routers, hosts
Examples	MPOA, Router attached to ATM network	IP switches running IFMP/GSMP, ARIS, etc.

In addition, routers attached to a large ATM network will naturally wish to establish direct VCs with each other because that offers the best performance. A full mesh connecting all routers consumes a large number ($O(N^2)$) of VC resources. Some implementations of the peer model minimize the consumption of VC resources that would otherwise be needed for full-mesh connectivity. One technique introduced by ARIS is called VC merging; it merges multiple upstream VCs into a single downstream VC. This sharply reduces the number of VCs that must be established across the ATM network and maintained by each router.

6.5 IP Switching Types

The criteria for establishing a shortcut path and the classification of traffic that is placed on the shortcut path depends on the type of IP switching solution and protocol. Generally speaking there are two types of IP switching solutions: flow and topology driven. The following sections discuss the basic operation and framework of the two types.

6.5.1 Flow-Driven Solution

The flow-driven model for IP switching acts on a particular IP flow, which is defined as a sequence of packets with the same source IP address, destination IP address, and port numbers. The basic operation of flow-driven IP switching is illustrated in the conceptual diagram in Fig. 6-5 and consists of the following steps:

- ✓ The first N packets of a flow are routed hop-by-hop through one or more IP routing entities (R_1, R_2,…,R_n) to the destination. The IP

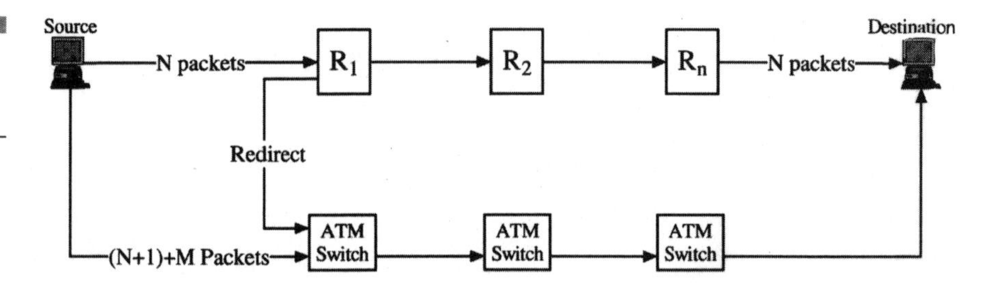

Figure 6-5 Flow-driven IP switching.

routing entities are the routing components of an IP switch or virtual IP switch, or they may be actual routers that are connected to each other by ATM connections.

- ✓ Based on the characteristics of the IP flow (e.g., traffic type, port number, source/destination IP address, arrival rate, etc.), the IP routing entity (either at the edge or middle of the network) triggers the initiation of a redirect[7] process. The redirect process involves the actions required to build a shortcut path and to redirect the IP flow over the shortcut path so that hop-by-hop router processing ($R_1, R_2, \ldots, R_n$) is bypassed. This is accomplished by assigning a new VPI/VCI label to the cells belonging to the IP flow and then having the VPI/VCI connection tables updated within the ATM switch.[8]
- ✓ Once the IP flow is redirected over the shortcut path, the remaining $(N + 1) + M]$ packets of the flow are switched.

Flow-driven IP switching has several interesting properties. First, it may not and should not be necessary to switch all flows. Some flows such as ICMP "pings" and best-effort email may not warrant or derive any benefit from a faster switched path. Second, the redirect process is performed individually for each flow that qualifies to flow over a shortcut path. What qualifies a particular flow as a candidate to be switched depends on implementation-specific policies and heuristics. There is no standard that says, "This flow identified by these header fields will be switched."

And finally observe that the first few packets of the flow are routed through the network, and only after the redirect process has completed does the "tail" of the flow traverse the shortcut path.

The basic functional characteristics of flow-driven IP switching are:

- ✓ Appearance of specific IP flows may trigger the establishment of a shortcut path where a flow is defined as application-to-application or host-to-host traffic. A flow is identified by a sequence of packets that share common IP header information such as source and destination IP addresses and TCP or UDP port numbers.

---

[7]Other names for the redirect process are flow setup and flow mapping.

[8]The illustration shows just R_1 communicating with a switch component. In actuality each IP routing entity associated with a switch component may act independently to redirect the packets through the switch component and update the switch's connection tables.

- Packets can be routed through the network or placed on a shortcut path on a per-flow basis
- The redirect process is performed independently for each flow that is to be placed on a shortcut path. The scope of the redirect process may be limited to a pair of adjacent IP switches, or it could be end to end between the network ingress and egress.
- Each IP switch may perform the redirect process on an individual basis for each flow. In other words the decision to relabel a flow and hence update a switch's connection tables does not depend on what another IP switch does with the same flow.
- Some flows may not qualify to be switched based on the shortcut path decision criteria and policies implemented in the IP switching system.
- A flow-based ingress component may place some flows on a short-cut path and route others.
- If a shortcut path does not exist or it suddenly disappears, packets can be routed to the destination.
- The entire flow is generally not switched. The first few packets are routed so that the IP switches have an opportunity to classify a flow and decide to initiate the redirect process. After the shortcut path is established, the remaining packets of the flow are switched.
- Shortcut paths are typically soft state, which means they will time out unless refreshed.

Operating a flow-driven IP switching solution does incur some overhead. There are the redirect or flow-setup messages that must be exchanged and processed on a per-flow basis with the participating components in the network. Also, the switching elements in this scheme must allocate per-connection resources (e.g., labels, buffers) for each flow that will be switched. The arrival of a large number of switch-qualified flows could introduce some delays in routing normal IP traffic and setting up shortcut paths as well as imposing resource constraints on switch resources.

A flow-driven IP switching solution would probably not work best in the core of a large network where there might be a very large number of individual IP flows. Moving away from the core where the probability of the source and destination sharing the same locality is greater makes this scheme more applicable and deployable. In particular individual host pairs located in a campus or intranet may have one or more applications that can benefit from being switched. As one moves closer to the end user, the volume of IP flows is not as concentrated as it is in the core. This means there are more switch resources available to handle shortcut paths between hosts located on the same side of the core (that is, they do not have to pass through the core to communicate with each other).

Examples of flow-driven IP switching protocols include IFMP and Flow Attribute Notification Protocol (FANP), which will be covered in detail in later chapters.

6.5.2 Topology-Driven Solution

Topology-driven IP switching is based on the IP network topology maintained by routing protocols (e.g., OSPF, BGP) running in the IP routing entity of an IP switch. New labels (VPI/VCI) associated with destination IP prefix(s) reflecting a destination network(s) are generated and distributed to the other IP switches in the routing domain. All traffic destined for a particular destination network follows a switched path based on the new VPI/VCI labels.

The basic operation of topology-driven IP switching is illustrated in the conceptual diagram in Fig. 6-6 and consists of the following steps:

- ✓ IP switches converge on a network topology based on the exchange of routing protocol messages between the IP routing entities (R_1, R_2,..., R_n).
- ✓ New VPI/VCI labels associated with destination IP prefixes in the form of {destination prefix, label} are generated and distributed to the switching components of the IP switches in the routing domain.

Figure 6-6 Topology-driven IP switching.

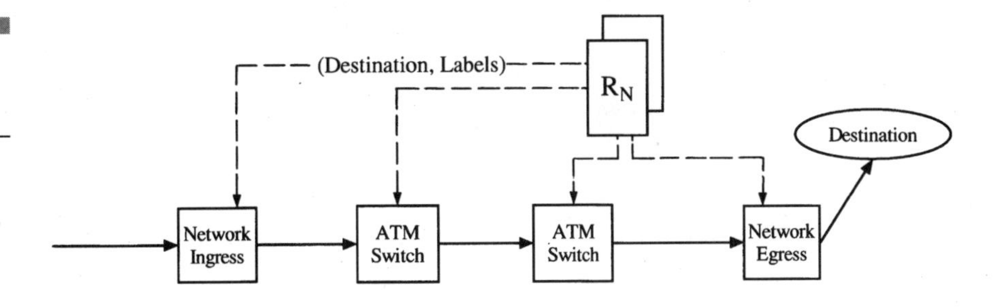

- ✓ The ingress IP switch checks the destination prefix of the packets entering the network. Instead of forwarding packets to the next hop IP address, the ingress IP switch places the packets on a switched path to the network egress. All traffic to that destination flows over a switched path from the network ingress to the network egress.
- ✓ The egress IP switch receives the packets over the switched path and forwards them at layer 3 to the destination.

The topology-driven approach to IP switching offers several performance and scalability enhancements when compared with the flow-driven approach. First, all traffic consisting of one or more flows going to a common destination network is switched. In other words the entire flow is switched, not just the tail as is the case with flow-driven IP switching. Second, the run-time overhead is much lower for topology-driven IP switching because switched paths are built only after a change in topology or after the arrival of control traffic. If the topology is stable, switched paths have been set up and traffic will flow over them. And from the perspective of scalability, the topology-driven approach is superior. This is because the number of switched paths in a routing domain is proportional to the number of routes or the size of the network. In addition, an even higher level of aggregation is possible by creating multipoint-to-point shortcut paths (e.g., using VC merge) that lead to a shared collection of destination networks.

Of course the better scaling and performance properties available to all flows comes with a cost. Topology-driven IP switching builds shortcut paths whether traffic flows over them or not. This means that switch resources are consumed as a result of control traffic (e.g., routing protocol updates) and not data traffic. Another possible issue to consider is that fact that it might be difficult to provide QoS to a flow that is traveling over a topology-driven shortcut. This is because you may have mix of real-time and best-effort IP flows destined for resources on a common destination network flowing over a topology-driven best-effort shortcut path. Of course the best-effort flows won't care, but the real-time flows may experience less than satisfactory performance. And finally, until a network converges, inaccurate routing table updates can cause the formation of transient loops that can lead to a set of labels being generated that form a switched loop. IP routers can limit the damage caused by transient loops by decrementing the TTL field to zero and then discarding the packet. Layer-2 (e.g., ATM) switches may have no such mechanism, so a loop in this case could be problematic.

The basic characteristics of topology-driven IP switching are:

- ✔ Shortcut paths are established based on the existence of destination networks and are established only after a change in topology.
- ✔ All traffic bound for a destination is switched.
- ✔ Ingress IP switches receive a new per-destination label (VPI/VCI). Normal IP forwarding is performed but instead of directing the packet to the next hop IP address, the ingress IP switch appends the new label[9] and places the packet on a switched path.
- ✔ At each intermediate switching component, the packets (or cells) are label-swapped.
- ✔ Egress IP switches receive the packet over the switched path, remove the label, and then forward the packet using normal layer-3 procedures.
- ✔ Packets may be routed to the destination if a shortcut path suddenly disappears
- ✔ It can cause transient loops at layer 2.
- ✔ Topology-driven shortcut paths remain in place until there is a change in topology.

Examples of topology-driven IP switching solutions are tag switching and ARIS. These will be discussed further in later chapters.

6.6 IP Switching Taxonomy

In this chapter the basic concepts of IP switching were introduced. The addressing models, the IP switching models, and the two types, flow and topology, were defined. Much more will follow in subsequent chapters including those solutions that use ATM as a connection vehicle between two hosts on the same network.

In the meantime, the definitions and concepts introduced so far warrant a preliminary taxonomy of IP switching and related technologies, as shown in Fig. 6-7. Down the peer model branch are the two possible

[9] If ATM is used, the ingress IP switch segments the packet into cells and applies the appropriate VPI/VCI label.

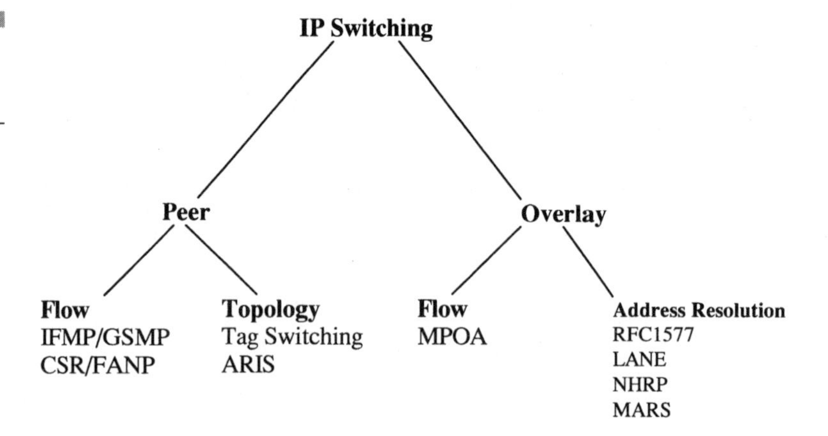

Figure 6-7 Preliminary IP switching taxonomy.

IP switching types, flow and topology. Associated with each type are several specific implementations that will be discussed in more detail in Part 2.

Down the overlay branch, there is a flow-driven solution, MPOA, that will be discussed in a later chapter. Included in this taxonomy is a separate branch called address resolution (AR). AR-based IP switching solutions are those solutions that resolve a data network address (e.g., IP, IP multicast, MAC) with an ATM address for the purpose of using ATM SVCs as a means of communicating between end points. AR-based solutions generally are confined to a local subnet, although one solution, NHRP, attempts to bypass intermediate routing hops. In addition AR-based IP switching must invoke address resolution functions and establish SVCs just to function in a practical manner. This is fundamentally different from IP switching, which exists to increase the forwarding rate over a preexisting routed path.

This preliminary IP switching taxonomy may also serve as a roadmap of topics that will be discussed in the rest of the book.

MORE INFORMATION

Callon, R., "Multilayer Switching including MPLS and Related Approaches," tutorial notes, Next Generation Networks 1997, November 1997, Washington, D.C.

http://infonet.aist-nara.ac.jp/member/nori-d/mlr/.

Metz, C., "MSS and IP Switching," IBM Whitepaper, May 1997.

Petrosky, M., "Short-cut Routing," *The Burton Group Network Strategy Report,* February 1997.

White, P., "ATM Switching and IP Routing Integration: The Next Stage in Internet Evolution," *IEEE Communications,* April 1998.

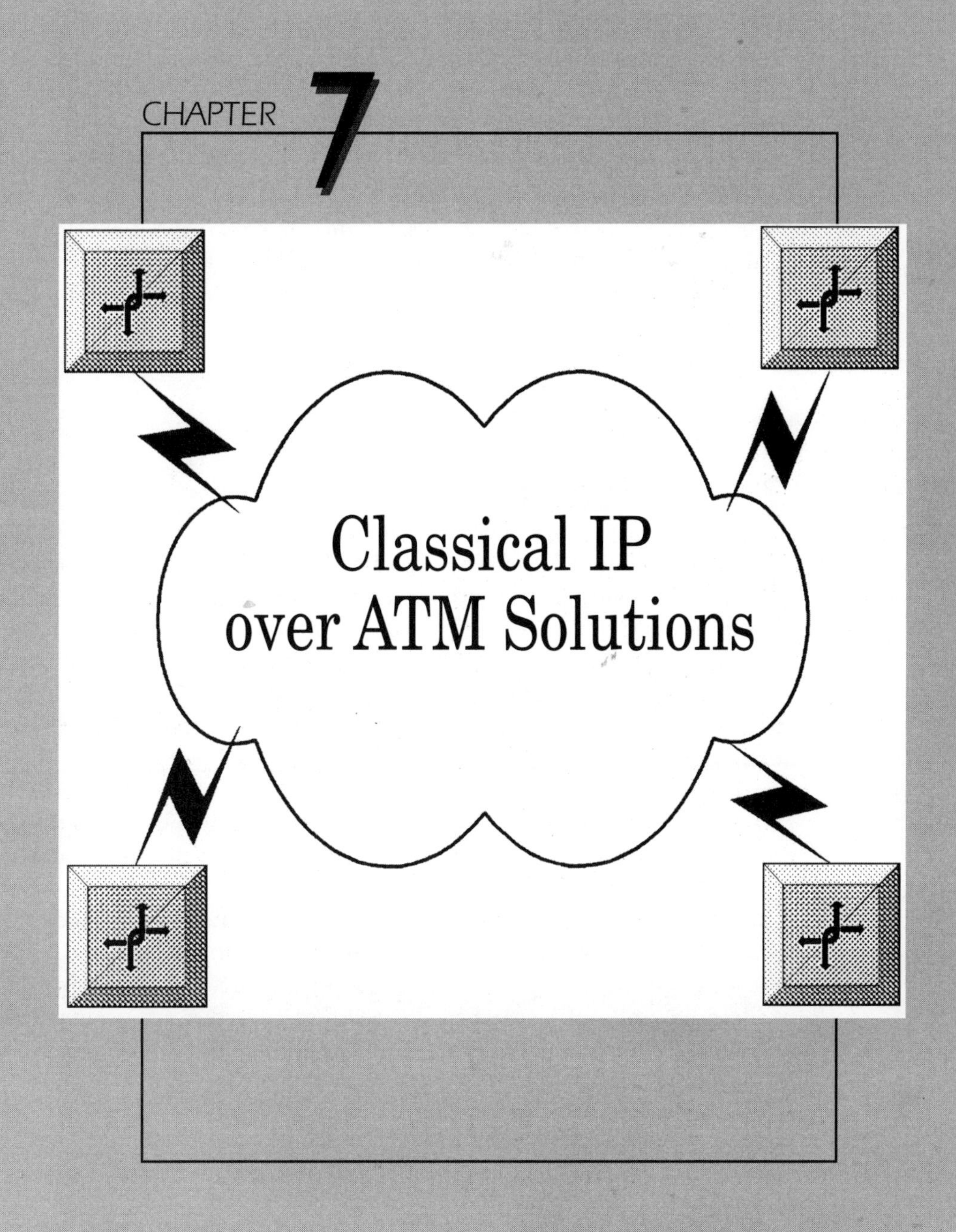
CHAPTER 7
Classical IP
over ATM Solutions

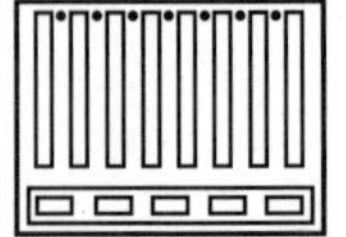

Now that the basics of IP, switching, and IP switching have been covered, it is a good time to step back and review the architecture, protocols, and operation of the classical suite of IP over ATM technologies. Classical IP over ATM is the name usually and correctly attributed to the functions described in RFC1577, Classical IP and ARP over ATM, and then the updated RFC2225. This chapter will apply a more inclusive definition to this term so that not only is RFC1577/2225 covered but also its sister suite of IP over ATM protocols. In addition to RFC1577, the classical suite consists of the NHRP and the multicast address resolution server (MARS). They are classical in the sense that they support traditional IP services and behaviors and are transparent to the applications running above them.

7.1 Architecture

The architecture of the classical IP/ATM[1] technologies incorporates many of the topics previously discussed. IP hosts and routers that share a common address prefix are grouped into subnets attached to a physical ATM infrastructure. IP is the network-layer protocol that provides a connectionless hop-by-hop forwarding service for packets flowing across network boundaries. Routers maintain forwarding tables based on the exchange of routing update messages and forward packets from one network to another. IP devices on the same subnet communicate directly using ATM VCs. The one exception to this standard behavior is the function provided by NHRP that enables IP devices on different subnets to communicate directly over the ATM network.

At the ATM layer, ATM Forum protocols maintain a consistent ATM network topology and support the establishment of PVCs and SVCs between ATM endpoints. IP hosts and routers are assigned IP addresses, and ATM-attached devices and switches are assigned ATM addresses. Because both IP and ATM addresses and topologies are dis-

[1]The terms Classical IP over ATM, Classical IP, CIP, and CLIP are all used to denote the protocols originally defined in RFC1577 and updated in RFC2225. To broaden the definition in this chapter to include RFC1577 (and its successor RFC2225), NHRP, MARS, and Server Cache Synchronization Protocol (SCSP), a small "c" classical IP/ATM will be used to describe this suite of protocols, which all originated from the IETF Internetworking over NBMA (ION) working group, and which all support IP and require an address resolution server, or ARS (exception is SCSP).

tinct, the classical IP over ATM protocols use separated addressing. In fact with the exception of NHRP, there is really no attempt whatsoever to treat ATM as anything more than just a layer-2 transport between communicating IP endpoints on the same network.

There are, however, accommodations that IP must make to communicate over an ATM network. First, IP packets must be adapted at the AAL layer for segmentation into cells at the source and reassembly into packets at the destination. Therefore a standard technique for encapsulating IP packets in AAL frames is needed. Second, ATM uses either preconfigured PVCs or dynamically established SVCs to transport data from a source to a destination. In the latter case, the source must know the ATM address of the destination. Therefore a source IP device that wishes to communicate with a destination IP device over an SVC must, prior to connection setup, resolve the destination IP address to an ATM address. In a LAN environment, the source IP host issues ARP broadcasts that will be delivered to the destination IP host, which in turn sends an ARP reply containing the destination MAC address. ATM is not a broadcast media, so some other technique is necessary for address resolution.

The classical IP/ATM protocols all use an address resolution server (ARS) when the mode of communications is ATM SVCs. The function of an ARS is to maintain a table or cache of IP address(s) and associated ATM address(s)[2] and to respond to queries from source IP devices that wish to resolve an IP address with an ATM address. The ARS can be populated through manual configuration, registration or by passively examining data or control packets that contain IP/ATM address bindings. IP devices will maintain a control VC with the ARS and may set up one or more SVCs with other IP devices for the exchange of data. Of course if PVCs are used, dynamic address resolution is not required and the use of an ARS is not mandated.

The general architecture of the classical IP/ATM protocols is illustrated in Fig. 7-1. The application and TCP/IP sit on top of an AAL and ATM layer in the source and destination, respectively. The source and destination devices (e.g., hosts, routers) are configured with both an IP address and an ATM address. A source IP device will query the ARS for an IP/ATM address binding so that it may establish an SVC with the destination. If PVCs are used, an ARS is not required.

---

[2]These are called IP-ATM address bindings and are denoted {IP.1, ATM.1, IP.2, ATM.2,... IP.n, ATM.n}.

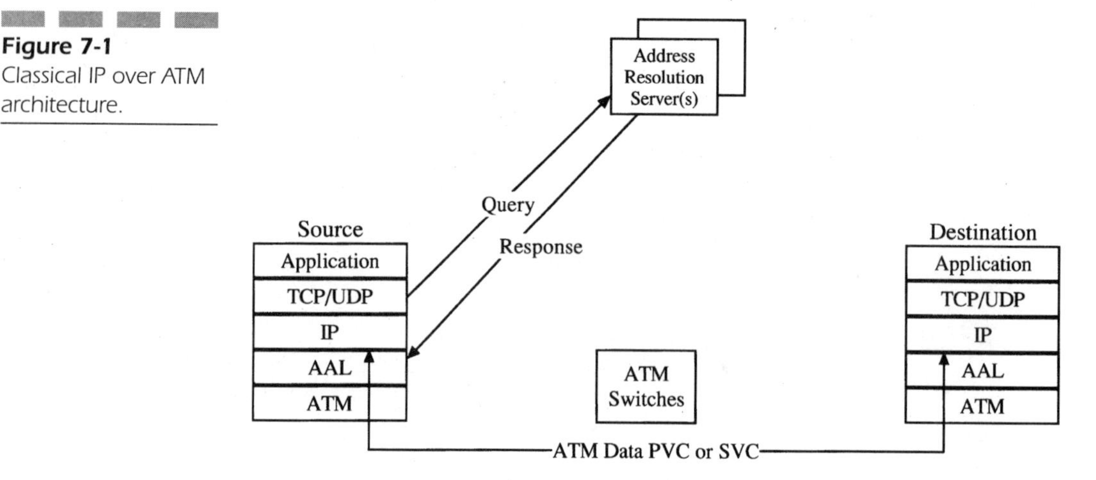

Figure 7-1 Classical IP over ATM architecture.

7.2 RFC1483

IP must be encapsulated in the format of the current data link the packet is traveling over, and this is true for ATM as well. Multiprotocol encapsulation over ATM described in RFC1483 defines a standard technique for encapsulating IP packets and more generally multiprotocol data packets in AAL5 frames for transport over an ATM network. RFC1483 defines two forms of encapsulation: Logical Link Control (LLC)/SNAP and VC multiplexing.

7.2.1 LLC/SNAP Encapsulation

The first technique specified in RFC1483 is called LLC/SNAP encapsulation. It works by appending an 8-byte LLC/SNAP header in front of each AAL5 frame that contains an IP packet or any other network-layer protocol packet for that matter. The LLC/SNAP header identifies whether a frame is routed or bridged and if it is routed, which network layer protocol (e.g., IP, IPX) is being routed. A typical LLC/SNAP-encapsulated AAL5 frame containing an IP packet is illustrated in Fig. 7-2. The OUI and the Ethertype shown comprise the SNAP header. The Ethertype value for IP is x'0800'. Other values for different protocols are possible.

Because the protocol identity is contained in each AAL5 frame, LLC/SNAP encapsulation enables multiple protocols to share the same

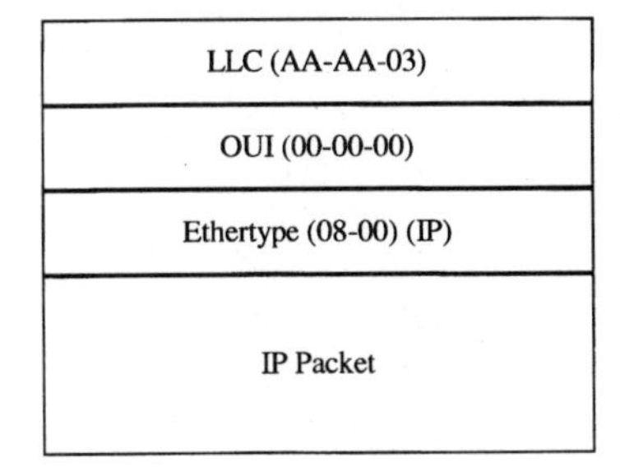

Figure 7-2 RFC1483 LLC/SNAP encapsulation header.

ATM VC. This minimizes the number of VCs required to support multiprotocol traffic over an ATM network. This can be a factor if there is a cost, monetary or otherwise, associated with the number of VCs in the network. In addition, it is not incumbent upon ATM signaling to identify the protocol and to terminate the VC in a protocol-specific entity. If this was the case, it would be up to the ATM system to multiplex different network-layer protocols by establishing a VC for each. Rather it is the presence of the LLC/SNAP header along with an LLC entity present in each VC endpoint that identifies the protocol entity and handles the multiplexing and demultiplexing of data as it is passed up to the application. The trade-off in lower VC consumption and signaling overhead is the additional 8-byte LLC/SNAP header that is appended to each AAL5 frame.

The concept of LLC/SNAP encapsulation for IP over ATM data is illustrated in Fig. 7-3. LLC/SNAP encapsulation is useful if the communicating ATM devices exchange multiprotocol data traffic over PVCs or SVCs. An example of such a scenario would be two routers connected by

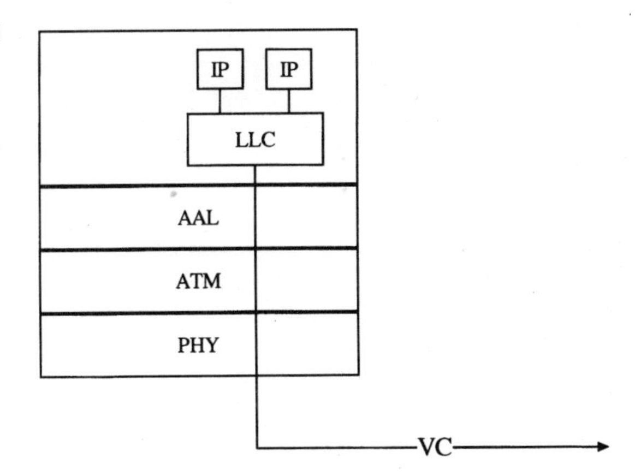

Figure 7-3 RFC1483 LLC/SNAP encapsulation.

a PVC exchanging IP and IPX traffic. LLC/SNAP is the default encapsulation technique used in RFC1577/2225 and in fact in most IP over ATM solutions.

7.2.2 VC Multiplexing

The second technique specified in RFC1483 is called VC multiplexing or null encapsulation. In this mode, no LLC/SNAP header is required. Network-layer protocols flow over separate VCs that are terminated in a network-layer protocol entity present in each ATM endpoint. The identity of the network-layer protocol can be manually configured or dynamically established by ATM signaling during connection setup. Multiplexing and demultiplexing of network-layer protocols is accomplished at the ATM layer implicitly by the existence of per-protocol VCs.

VC multiplexing offers a slight advantage over LLC/SNAP encapsulation in reduced overhead and better performance. This is because the 8-byte LLC/SNAP header is not required for each AAL5 frame, and the VCs are terminated directly in a network-layer protocol entity. This slight advantage, however, may be offset by higher VC consumption and signaling overhead needed to support a multiprotocol ATM environment. If VC resources are plentiful and inexpensive to set up or if IP is the only network-layer protocol, VC multiplexing is the better choice.

The notion of VC multiplexing as described in RFC1483 is illustrated in Fig. 7-4.

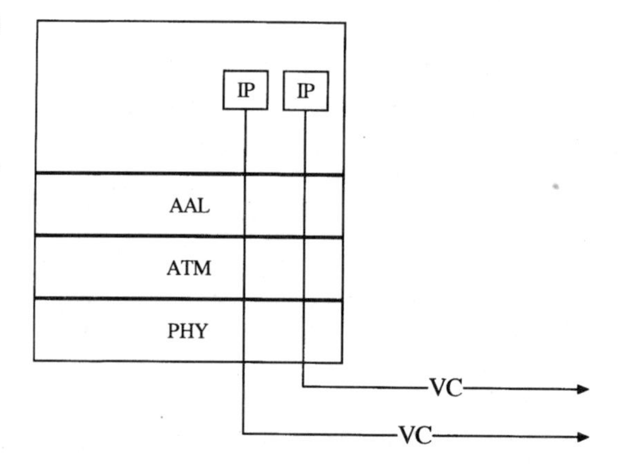

Figure 7-4 RFC1483 VC multiplexing encapsulation.

7.3 Classical IP and ARP over ATM

The functions and procedures collectively termed *classical IP and ARP over ATM* (CLIP) are described in RFC1577 and its successor, the newer RFC2225.[3] CLIP describes mechanisms for establishing point-to-point ATM virtual connections between IP devices located in the same subnet (called an LIS) and provides a service for tracking and distributing IP/ATM address bindings among the members of the LIS. The intent of CLIP is not to emulate the operation of IP and ARP over an existing network topology such as a LAN but rather to enable ATM to serve as a "wire" for the exchange of data between two IP devices on the same subnet. Intersubnet communication remains unchanged and still requires that packets flow through a router.

7.3.1 Components

The components of a network supporting CLIP connectivity consist of the following:

- ✓ *LIS.* This is a collection of hosts and routers connected to a physical ATM network that share the same IP address prefix or subnet number. Any two devices on an LIS should be able to communicate directly over a point-to-point ATM PVC or SVC. Devices on different logical IP subnets must communicate through a router. A router may be attached to one or more ATM LISs and subnets consisting of other media (e.g., Ethernet) as well.
- ✓ *ATMARP.* This defines the address resolution protocol and message formats used between an ATMARP client and server. ATMARP is similar in concept to traditional IP ARP except that it has been modified to carry ATM addresses and work over a connection-oriented ATM network. In addition because there is no native broadcast support with ATM, the ARP address resolution function is performed by a "third party," the ATMARP server.
- ✓ *ATMARP server.* The ATMARP server is the ARS for the members of an LIS. The ATMARP server is responsible for maintaining a

[3]Unless noted, the CLIP functions and procedures discussed in this chapter pertain to those outlined in RFC1577.

table of IP/ATM address bindings {IP.n, ATM.n} of LIS members and for responding to ATMARP queries from LIS clients. A minimum of one ATMARP server is required for each LIS. The ATMARP server function may reside in a workstation, a router, or a standalone device attached to the LIS and is only required if SVCs are supported. An ATMARP server does not forward any data; it only handles control data.

- *ATMARP client.* An ATMARP client function resides in each LIS member that wishes to establish an SVC with another device in the same LIS. The ATMARP client contacts the ATMARP server and exchanges ATMARP messages to resolve IP/ATM address mappings. Another name for the ATMARP client is RFC1577 client.

It should be noted that both the control functions of ATMARP and the data exchange over VCs are confined to a single LIS. There is no attempt on the part of classical IP and ARP over ATM to bypass or shortcut intermediate routing hops. This is true even if there are multiple LISs overlayed on a single physical ATM infrastructure. Classical IP and ARP over ATM impose no change on the traditional end-to-end IP internetworking model.

7.3.2 Address Resolution

The basic operation of a CLIP network consists of address registration, address resolution, SVC establishment, and data exchange.[4] The top portion of Fig. 7-5 shows two hosts and an ATMARP server connected to an ATM switch that together form an LIS. The exchange of messages necessary for Host1 to establish an SVC to Host2 as described in RFC1577 consist of the following:

- CLIP clients supporting RFC1577 first register their IP addresses with the ATMARP server. This is accomplished by an exchange of InvATMARP messages with the ATMARP server. The ATMARP server first issues an InvATMARP request when looking for the IP address of the ATM-attached IP host (it can also be a router on the

[4]Most if not all RFC1577 networks use SVCs amongst LIS members. If the network uses PVCs only, the address resolution process may entail the use of Inverse ATMARP mechanisms to resolve a PVC end point with an IP address.

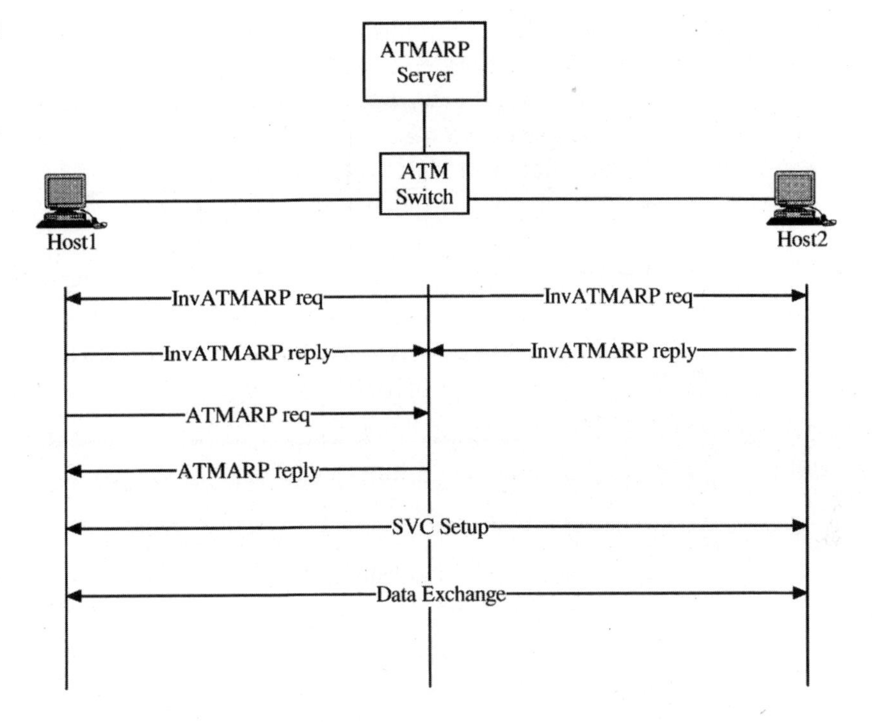

Figure 7-5
RFC1577 address resolution.

LIS). The IP host responds with an InvATMARP reply that contains its IP address. The ATMARP server then stores the IP and ATM addresses of the host in a table for up to 20 min. At the end of 20 min the IP/ATM address binding is aged out unless it is either refreshed by the host or by the ATMARP server over an existing VC. The InvATMARP process takes place over a control VC established between the IP host and the ATMARP server. The control VC is initiated by the IP host that must be configured with the ATM address of the ATMARP server.

- A source IP host on the LIS that wishes to establish an SVC with another device on the same LIS must resolve the destination's IP and ATM addresses. To do so, the source IP host ATMARP client will exchange ATMARP messages with the ATMARP server over the control VC. The ATMARP reply sent to the ATMARP client (Host1 in Fig. 7-5) contains the IP/ATM address binding of the destination (Host2). The basic format of an ATMARP message is illustrated in Fig. 7-6.

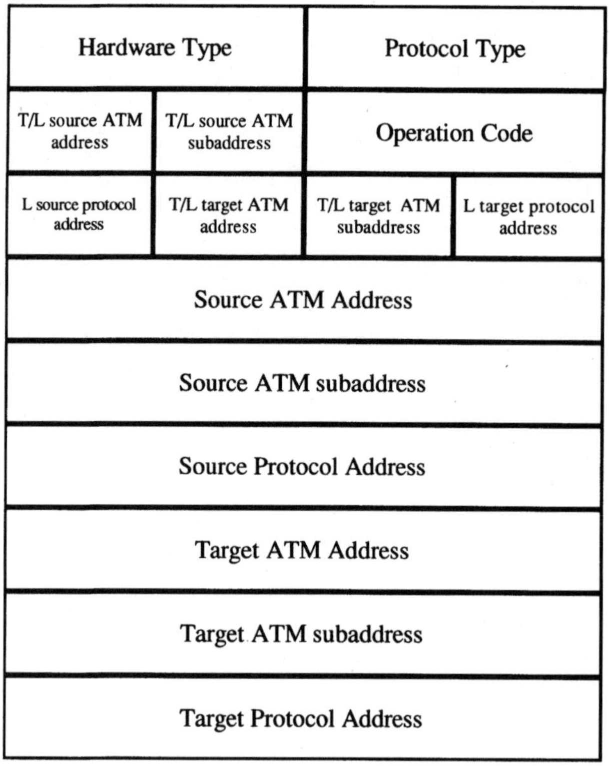

Figure 7-6 ATMARP packet format.

<table>
<tr><td colspan="2">Hardware Type</td><td colspan="2">Protocol Type</td></tr>
<tr><td>T/L source ATM address</td><td>T/L source ATM subaddress</td><td colspan="2">Operation Code</td></tr>
<tr><td>L source protocol address</td><td>T/L target ATM address</td><td>T/L target ATM subaddress</td><td>L target protocol address</td></tr>
<tr><td colspan="4">Source ATM Address</td></tr>
<tr><td colspan="4">Source ATM subaddress</td></tr>
<tr><td colspan="4">Source Protocol Address</td></tr>
<tr><td colspan="4">Target ATM Address</td></tr>
<tr><td colspan="4">Target ATM subaddress</td></tr>
<tr><td colspan="4">Target Protocol Address</td></tr>
</table>

- The source IP host now establishes an SVC to the destination IP host using standard ATM Forum protocols.

ATMARP messages are encapsulated in an LLC/SNAP header and an Ethertype of x'0806', which specifies an ARP process. The important fields to note in the ATMARP message are

- *Operation code.* This defines the type of ATMARP message. The values are
  - ARP_Request = 1
  - ARP_Reply = 2
  - InARP_Request = 8
  - InARP_Reply = 9
  - ARP_NAK = 10

- *Source ATM address.* ATM address of source host.
- *Source protocol address.* IP address of source host.
- *Target ATM address.* ATM address of destination host.
- *Target protocol address.* IP address of destination host.

The CLIP process can be improved in several areas for better efficiency and reliability. One such improvement is described in the new version of classical IP and ARP over ATM, RFC2225. Registration of a host's ATM address with the ATMARP server is now performed by an exchange of ATMARP messages. To register, the IP host transmits an ATMARP request containing its own IP address in the destination IP address field. The ATMARP server stores the source IP and ATM addresses in its cache and returns those values in the ATMARP reply to the IP host. This overcomes some of the unnecessary overhead of the InvATMARP process when it is used for registration. In some cases the InvATMARP process can consume unnecessary resources, and it is not really needed to register an IP host's IP/ATM address binding with the ATMARP server. Consider the case of a multiprotocol router functioning as an ATMARP server, which, by definition, must send InvATMARP requests to all devices on the LIS—even if they do not support IP. Registration of an LIS member's IP and ATM addresses can just as easily be performed using the ATMARP message exchange.

Another area of improvement is ATMARP server availability. If the single ATMARP server on the LIS goes down, a backup is not available. The Server Cache Synchronization Protocol (SCSP) addresses the problem of multiple servers; it is discussed later in this chapter. New CLIP clients can now be configured with a list of ATM addresses representing one or more ATMARP servers serving the LIS. If one goes down, the client may contact the next one in the list.

And finally, CLIP only supports IP unicast communication. Although this will not change, CLIP can be augmented with a MARS function that provides IP multicast and broadcast support over an ATM network.

7.3.3 Classical IP and ARP over ATM Network Example

CLIP is a simple and effective technique for implementing IP over ATM. It is a good solution in which IP hosts and devices can benefit from the high bandwidth and low latency of ATM. In particular, IP unicast applications that operate best over high-capacity, high-bandwidth networks are excellent candidates to run over a CLIP network.

An example of a CLIP network is shown in Fig. 7-7 and illustrates several important features about its operation. First, all members of an LIS can communicate directly over the ATM network. It is not required that the members of an LIS physically attach to the same or nearby collection ATM switches—only that they share the same IP address prefix and the same ATMARP server(s) and can communicate with all other LIS members via ATM. This separation from physical and logical topology gives CLIP a VLAN-like capability without the broadcast support. Second, multiple LISs may be overlayed on the same physical ATM network. However, even though they may share the same ATM network, they can only communicate through a router. In our example, all inter-LIS communications must pass through the router.

Third, CLIP does not alter the "concatenated network" model of IP internetworking. In other words, logical IP subnets are connected to other LISs and other networks by routers. As far as a router is concerned, the LIS is nothing more than another network that it will for-

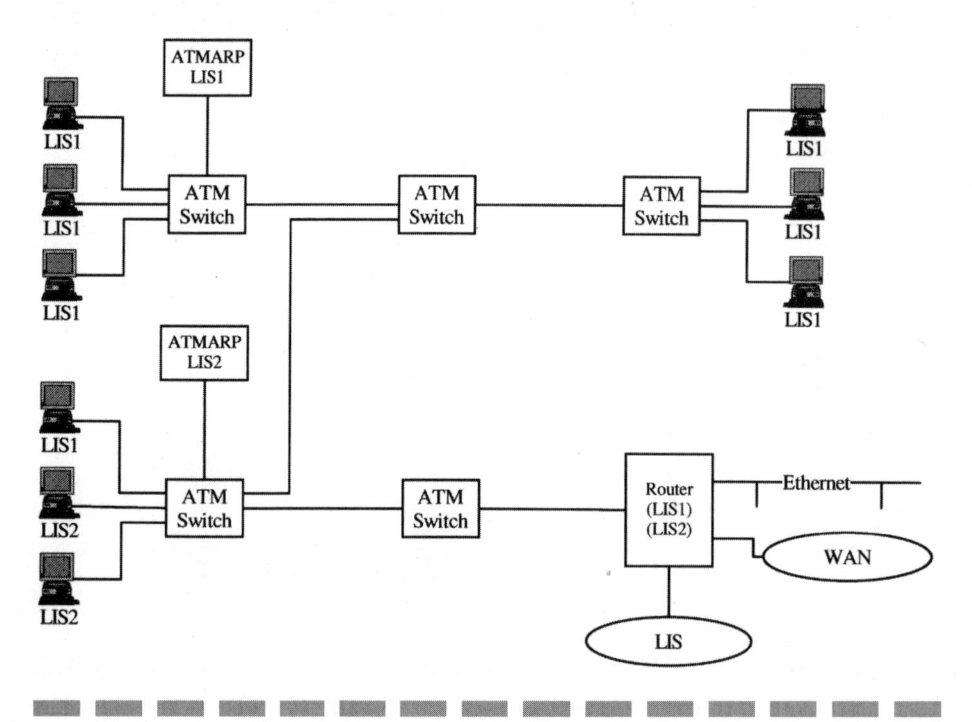

Figure 7-7 Classical IP and ARP over ATM network example.

ward packets to. And finally, CLIP does not support broadcast or multicast. This means that alone it cannot be a direct replacement for traditional LAN communications. This should be a consideration if the host applications require broadcast or multicast support or if one has control functions that make use of broadcast or multicast transmissions (e.g., OSPF). However, if a MARS client and server function is included (perhaps colocated) with the ATMARP client and server functions, respectively, a CLIP-style network could conceivably provide all of the services of a traditional LAN. In theory this is possible, but in practice it is not likely because the ATM Forum LANE protocols have for some time provided unicast and multicast IP over ATM support. And network implementers like to deploy what already works and what most vendors are building and shipping.

7.4 Next-Hop Resolution Protocol

The next protocol to discuss from the suite of classical IP/ATM solutions is the NHRP, which is defined in RFC2332. NHRP extends the notion of the CLIP ATMARP protocol beyond the boundaries of the LIS. NHRP is an intersubnet (LIS) address resolution protocol that enables hosts and routers connected to the same NBMA network to establish point-to-point VCs (shortcut paths) across subnet boundaries.[5] Essentially it enables a source to query the network for an ATM address that can be used build an SVC to a "next-hop," "last-hop," or "egress" entity on that NBMA network that is closer or fewer hops to the actual destination. The objective of NHRP is to bypass intermediate routing hops positioned between hosts and routers connected to the same NBMA network (e.g., ATM) thus achieving higher throughput, lower latency, and a reduced load on network routers and switches.

The basic requirement for NHRP can be best illustrated by following the path of a packet as it is forwarded over several hops through routers that are attached to the same NBMA (or ATM) network. Consider the network shown in Fig. 7-8. Hosts A.1 and C.1 along with routers 1 and 2, respectively, are attached to a single physical ATM network. Observe

[5]NHRP is frequently referred to as the "Next-Hop Routing Protocol." This is incorrect. It is not a routing protocol at all. In fact NHRP requires the use of IP routing as a means of correctly forwarding NHRP messages to the destination.

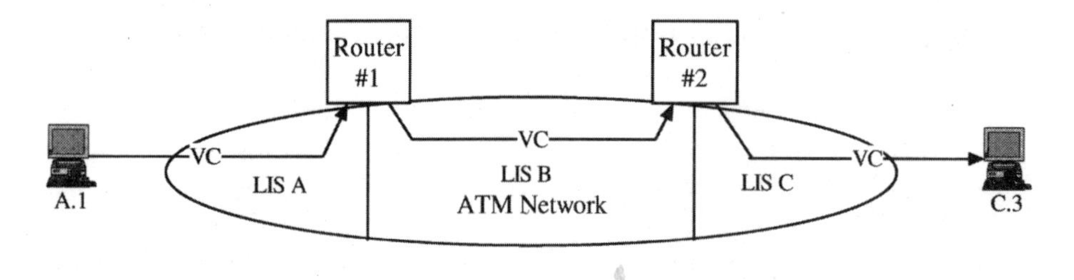

Figure 7-8 Multihop routing through an NBMA network.

that the routers are physically attached to the ATM network by a single ATM interface and thus take on the appearance of a "router lollipop," or "router on a stick." Logical attachment to each subnet is accomplished by one or more LIS-specific VC terminated on the single ATM interface in the router on one side and another host or router in the same LIS on the other side.

Host A.1 wishes to engage in a video over IP session with Host C.1. The two parties could certainly benefit from the high performance, low latency, and QoS that direct ATM communications affords. But according to standard IP routing, all intersubnet communications must pass through routers. Therefore Host A.1 must first establish an SVC with router 1, forward packets targeted for Host C.1 over the VC to router 1, which in turn will perform standard IP forwarding. If it has not already done so, router 1 must build an SVC to router 2 and then forward the packets to router 2, and it must do the same between router 2 and host C.1. All told this requires the establishment of N+1 per-LIS VCs if N is the number of hops and the aggregate delay and latency associated with N number of hops.[6]

Also observe that in the case of the router on a stick configuration, packets (in the form of cells) traverse the ATM switch and UNI physical attachment twice. This is hardly the most efficient path for bandwidth-hungry and delay-sensitive IP traffic. NHRP solves this problem by enabling Host A.1 to communicate directly with Host C.1 over an ATM VC even though they are on separate subnets. It improves the performance of the forwarding path and offloads the intermediate routers from some packet forwarding. NHRP can be considered, then, a relax-

[6]A multi-LIS network overlayed on a single physical NBMA network will always need N+1 LIS-specific VCs to serve as a default routed path between any two hosts. The real performance savings comes in bypassing the N number of hops in the data forwarding path.

ation of the traditional requirement that all intersubnet communications must take place through routers.

7.4.1 Components

The components of a network supporting NHRP consist of the following:

- *NBMA network.* NHRP runs on hosts and routers that share connectivity to the same NBMA network. Overlayed on the single NBMA network may be one or multiple IP subnets. An example of a NBMA network is ATM.
- *Local address group (LAG).*[7] This is a collection of hosts and routers connected to a physical NBMA network that share the same IP address prefix. Unlike an LIS, though, LAG members associated with different IP subnets can forward data traffic through routers or alternatively set up shortcut paths with each other. Stated another way, the "local/remote" forwarding decision is separated from the destination address and coupled with the traffic or QoS characteristics of the flow.[8] This makes sense if one wishes to route low-volume, short-lived, best-effort traffic while reserving the NHRP shortcut paths for high-volume, long-lived, delay-sensitive traffic.
- *NHRP.* This defines the address resolution protocol and message formats used between an NHRP client and server. NHRP is similar to ATMARP except that the ATM addresses are associated with IP host addresses and address prefixes that are on different networks.
- *Next-hop server (NHS).* The NHRP server, known as a NHS, is colocated on a router or route server that provides NHRP service to a collection of NHRP clients. An NHS is responsible for maintaining a table of IP/ATM address bindings that represent destinations on and off the ATM network. A destination in this context

[7]LAG is a term unique to NHRP and probably alien to most people familiar with routing, etc. For the sake of clarity, the term subnet or LIS will be substituted for LAG in the discussion of NHRP but with the understanding that the hosts are capable of originating and terminating intersubnet VCs.

[8]RFC1937, Y. Rekhter and D. Kandlur, "Local/Remote" Forwarding Decision in Switched Data Link Subnetworks.

may be the IP address of an individual host or an address prefix representing a number of hosts. The primary function of the NHS is to provide address resolution for the specific NHRP clients and destinations it serves. If an NHS does not serve a particular host or destination, it must attempt to forward the NHRP request to the NHS that does. The NHS must know the identity (addresses) of the destinations that it is functioning as a server for. A subnet requires at least one NHS.

- *Next-hop client (NHC).* The NHC, or NHRP client function, is located on a host or router. It initiates NHRP requests to an NHRP server and is capable of originating shortcut paths to a destination across an ATM network. An NHC maintains a cache of IP/ATM address mappings derived from NHRP protocol exchanges with the NHS, or it may be constructed from other means. A NHC is also capable of forwarding packets to the default router.

7.4.2 Address Resolution

The basic operation of the NHRP address resolution process is similar to the ATMARP address resolution process for CLIP except that it is performed across subnet boundaries. First, the clients must register their IP and ATM addresses with the server. Next, an NHC that wishes to establish direct ATM connectivity with a destination entity forwards an NHRP request over the routed path toward the destination.[9] The "closest" NHS that serves the destination will respond with an NHRP reply containing the ATM address of the destination host or of the ATM device closest to the destination.[10] The NHRP reply follows the routed path back to the source NHC. The NHC caches the information and may then establish a direct ATM VC or shortcut path that bypasses intermediate routers. The general format of the NHRP mandatory message header

[9]A desire to connect over the NBMA network is probably not a sufficient reason in practice to initiate the NHRP process. Typically it should be kicked off if the traffic flow is long-lived and can benefit from the better performance and lower latency offered by the ATM network. The policies or heuristics needed to make that decision are outside the scope of the NHRP specification but should be available in any NHRP implementation.

[10]ATM is used in the explanation as an example of an NBMA network.

and the client information element (CIE) is illustrated in Fig. 7-9, and notable fields are explained below.

The important fields contained within the mandatory portion of a NHRP message consist of the following:

- *Request ID.* Unique value used to synchronize an NHRP request with its corresponding reply
- *Source ATM address.* ATM address of source NHRP client
- *Destination protocol address.* IP host address or IP address prefix that NHRP is attempting to resolve with a corresponding ATM address

NHRP requests and reply messages may contain one or more CIE sub-messages. The important fields in the CIE are

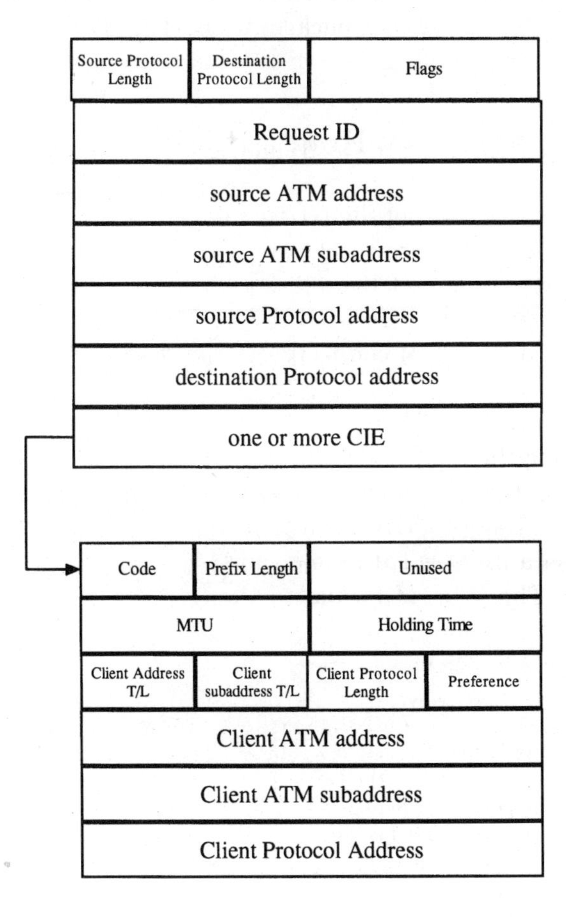

Figure 7-9 NHRP mandatory header and CIE format.

- *Prefix length.* This is used to create an equivalence class of all IP addresses that share the address prefix specified by the prefix length. This enables NHRP to set up a shortcut path that can be shared by multiple destinations contained in the equivalence class.
- *Holding time.* Amount of time that the information contained in the CIE is considered valid. After the holding time expires, any cached information will be discarded.
- *Client ATM address.* ATM address of NHRP client. In an NHRP reply this field will contain the ATM address associated with the destination protocol address.

Figure 7-10 shows a single ATM network consisting of a pair of hosts and routers configured over three separate IP subnets. Router 1 connects LIS1 and LIS2 and Router 2 connects LIS2 and LIS3. Both routers have an NHS and each host has an NHC. The flow of data and NHRP messages required to establish a shortcut path between the two hosts consists of the following:

- NHSs must learn the ATM addresses of the NHCs in their subnets. This can be accomplished in a number of ways including the use of

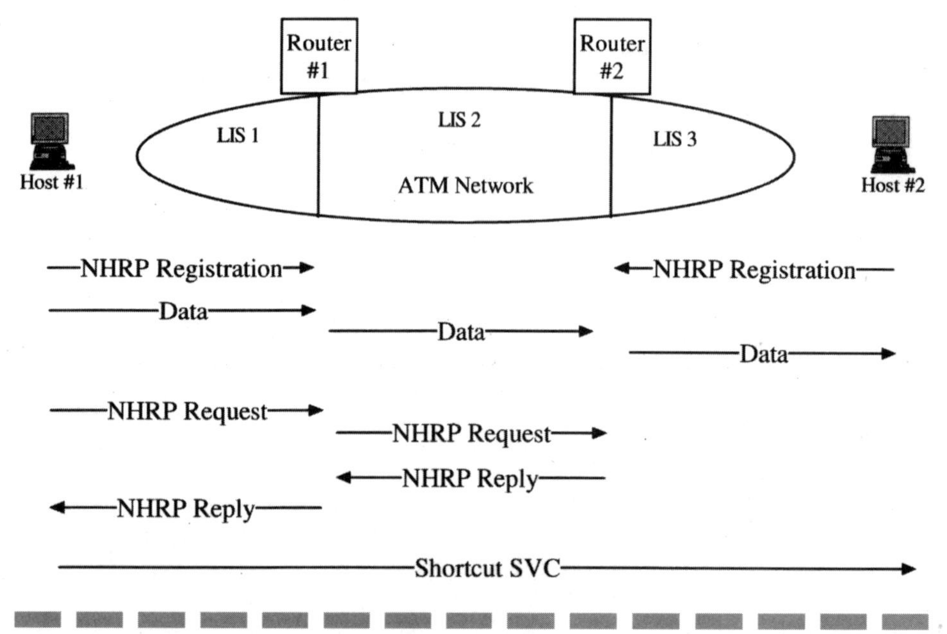

Figure 7-10 NHRP address resolution and data flow.

explicit NHRP registration messages as shown. Another technique is manual configuration. A third technique has NHSs "snoop" NHRP request/reply packets as they are routed to and from points in the network. The first two techniques are used to populate authoritative NHSs. An authoritative NHS is the actual NHRP server of the subnet and is the only NHS that can respond to authoritative NHRP requests. Moreover, it is the last NHS on the ATM network that the NHRP request will flow to.

- ✔ The source host (Host 1) begins sending packets to a remote destination by forwarding them to the default router. The fact that the packets are being forwarded out an ATM interface coupled with a "shortcut setup" policy can trigger the host to issue an NHRP request. An NHRP request contains the source IP address, the source ATM address, and the destination IP address and is routed toward the destination IP host.
- ✔ The NHRP request packet is forwarded over the routed path toward the destination. When it reaches the NHS that serves the destination (router 2), an NHRP reply is generated and sent to the source NHC over the routed path. The NHRP reply contains the source IP address, destination IP address, source ATM address, and information sought by the client, the destination ATM address.
- ✔ After receiving the NHRP reply, the source NHC caches this information and can set up an SVC directly to the destination (Host 2), thus bypassing two intermediate router hops. After the specified holding time expires, the cached information is discarded. If at some point the information becomes invalid, NHRP generates purge messages to remove the invalid data from the cache maintained in the client and the NHS.

The basic function provided by NHRP is fairly straightforward, but there are some issues that should be pointed out. First, NHRP is not a routing protocol. NHRP functions (client and server) coexist in routers that support dynamic IP routing, which is required to forward NHRP and data packets through the network.[11] Second, a common criticism levied against NHRP is the potentially high shortcut setup latency. This

[11]This assumes that static routing is outdated and not used except in extraordinary situations.

can be lessened to some degree by allowing intermediate or transit NHSs to provide nonauthoritative responses to NHRP requests for destinations that they do not explicitly serve. Applying this example to Fig. 7-10, router 1, having previously learned of the IP/ATM address binding associated with Host 2, could respond to the NHRP request generated by Host 1 as long as the nonauthoritative request bit was turned on. One other important issue to raise when discussing NHRP vis-à-vis other techniques discussed later in this book is the fact that, in practice, only point-to-point shortcut paths are supported. NHRP does not currently have a defined technique for supporting point-to-multipoint shortcut paths.

Another consideration with NHRP is the fact that the destinations served by an ATM-attached NHS do not have to be physically attached to the ATM network. In this particular case, the destination ATM address returned in the NHRP reply is that of the egress router or edge device. Observe the configuration in Fig. 7-11. The shortcut path has been created for all traffic destined for network A.1, which is located outside of the NBMA network. The ATM address returned in the NHRP reply by the NHS located on router 3 is that of router 3 itself.

With the addition of extensible TLV fields defined in the protocol, NHRP can be considered a generic address resolution and control protocol for devices attached to and running over an NBMA networks. For example, NHRP contains extensions that enable just about any internetworking information to be communicated through the exchange of NHRP messages. An NHRP client attached to a LAN could use NHRP to resolve the MAC address of a LAN-attached destination host on the other side of an ATM backbone. It is this flexibility and extensibility that prompted the ATM Forum to incorporate NHRP functions into the MPOA specification that will be covered in the next chapter.

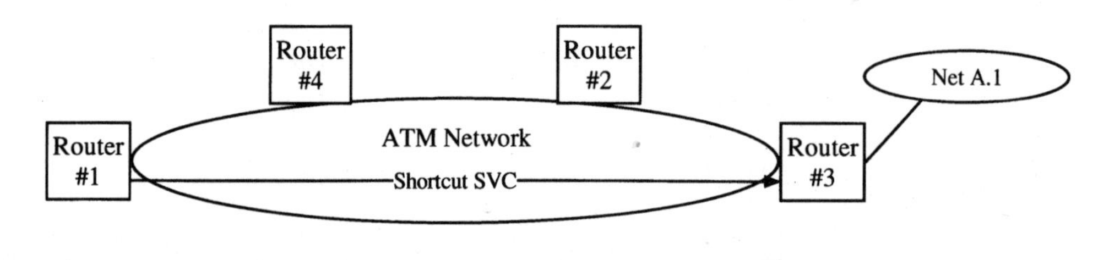

Figure 7-11 NHRP shortcut path to an external NBMA network.

One other significant issue that needs to be raised with respect to NHRP is that in certain topologies persistent routing loops can form. The conditions that must be present for a loop to form in a network running NHRP are

- Traffic is flowing over an NBMA backbone network by traveling over a router-to-router shortcut path. If the shortcut path has a host on either side, a loop cannot form because hosts, by definition, do not forward traffic.
- The existence of a non-NBMA "backdoor" path between the destination and source networks.
- A sudden change in topology unbeknownst to the NHRP shortcut routers.
- A loss of distance or path vector information over the shortcut path that makes it appear that for packets exiting the NBMA network through the egress router that the backdoor path is the shortest path to the destination.

The conditions ripe for the formation of a persistent loop in a network using NHRP are illustrated in Fig. 7-12. RFC1932 provides a description and example of a loop formed in a NHRP environment as well as some possible remedies for avoiding this scenario. One obvious solution is to remove or block any backdoor paths. Another solution would be for the NHRP shortcut routers to exchange routing protocol information or perhaps make some sort of periodic check for route validation. Neither is

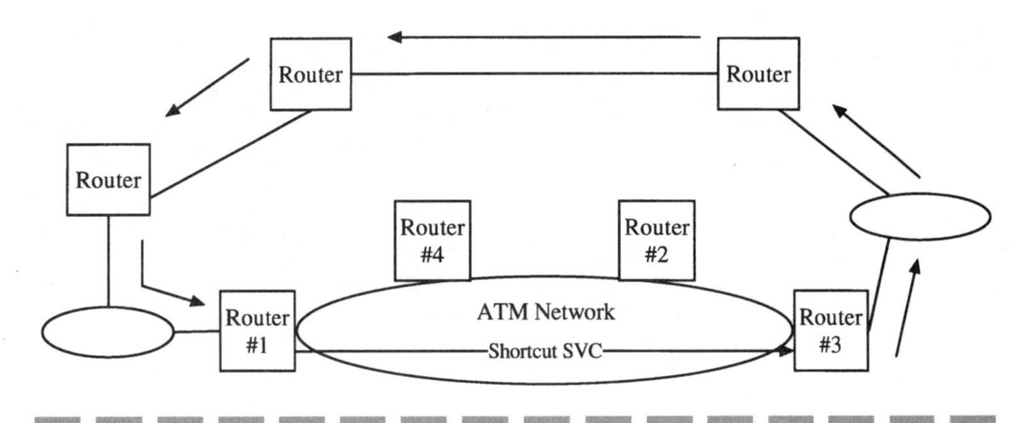

Figure 7-12 Conceptual NHRP loop topology.

very attractive because of the inherent inconsistency between the routed topology and the data path topology in the former and the introduction of excess control traffic and state into the network in the latter. Whatever approach is adopted, it will be a requirement to maintain some sort of shortcut state when the shortcut crosses routing domain boundaries. This is because metrics specific to one protocol running in one domain may not be consistent with those in another domain.

7.4.3 NHRP Network Example

NHRP is used to establish shortcut paths across ATM networks. Two such scenarios are illustrated in Fig. 7-13. In the top diagram a campus ATM network is operating an NHC function on the ATM-attached server and edge device. The NHS function is running on the ATM route server. NHRP can be used to establish a shortcut path for traffic transiting the ATM network. In the other example, low-volume transaction-based client-to-server queries are routed to the server, and the return path

Figure 7-13 NHRP network examples.

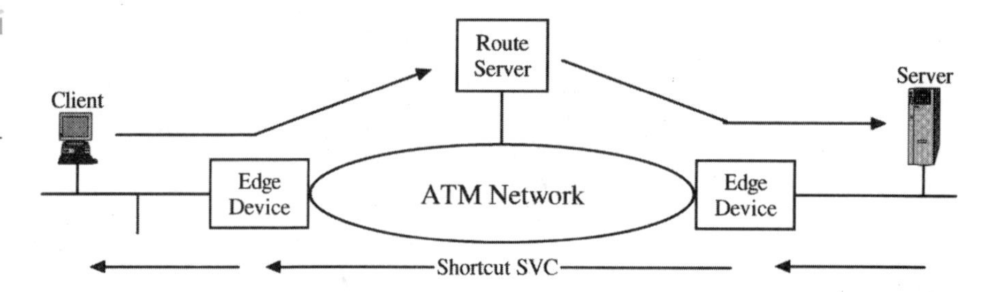

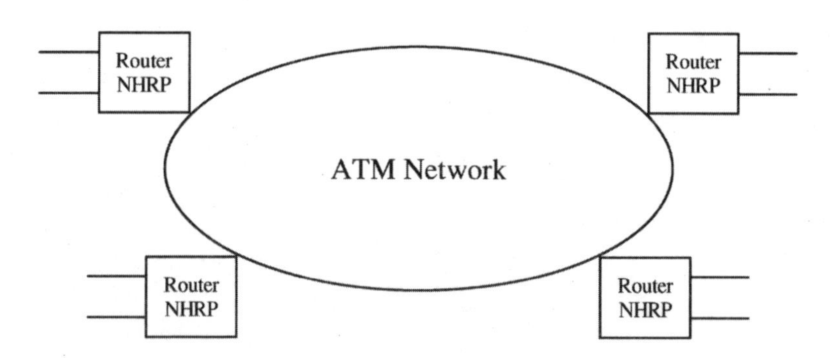

from the server to the client can be an NHRP-initiated shortcut path. This example illustrates the capability of NHRP in the campus. In practice it is likely that MPOA, which is a superset of the NHRP, will run on the clients and servers and provide the same shortcut path services.

In the bottom diagram in Fig. 7-13, the NHRP client and server function is coresident in each of the ATM-attached routers and is used to establish shortcut paths across the switched WAN network for unicast traffic. In this example NHRP provides the mechanisms for establishing shortcut paths across the WAN, but policies for determining which traffic will initiate and flow over the shortcut paths must be instituted.

7.5 Multicast Address Resolution Server

MARS is the third protocol in the suite of classical IP/ATM technologies. MARS extends the ARS concept to include multicast address resolution. Generally speaking MARS provides an address resolution and registration service for the operation of IP multicast over ATM. MARS is defined in RFC2022 and can also be thought of as an IGMP replacement for group registration over an ATM network as well as a means to emulate multicast forwarding behavior over an LIS.

7.5.1 Multicast over ATM

Before discussing MARS, it might be useful to overview the two possible techniques for supporting generic multicastlike transmissions communications over an ATM network. First, ATM does support multicast in the form of point-to-multipoint VCs. This is a unidirectional, single-QoS tree that is rooted at the source (called the root) and branches out to specific nodes called leaves. In UNI 3.1, the root is required to know the ATM address of each leaf it wishes to add to the tree. In UNI 4.0 a leaf node can perform an LIJ to join an existing multicast tree, but the leaf must know a priori a group ID. If the requirements for generic multicast and more specifically IP multicast require support for multipoint-to-multipoint flows, this is only possible in the ATM environment by configuring either a VC mesh or by making use of a multicast server (MCS).

Figure 7-14 illustrates the two methods. A VC mesh consists of a mesh of point-to-multipoint VCs rooted at each host and branching out to all other hosts on the ATM network. The VC mesh offers the best possible performance because there is a direct ATM path from the root to all leaves on the tree and the "multicast forwarding" is done at the cell level over ATM connections.[12] However, the VC mesh configuration consumes the most VC resources. And if there are multiple sources in the group, UNI signaling is required between each source and the ATM network each time a member is added or dropped. This UNI signaling processing grows proportionally with the number of sources and the frequency of membership changes.

The other technique for supporting multicast over ATM is through the use of an MCS. In this scenario, a source(s) establishes a single point-to-point VC with the MCS. Rooted at the MCS is a single point-to-multipoint VC that branches out to all hosts on the network including the sender. Multicast data originating at the sender is sent first to the MCS, which then forwards it to all hosts on the network. Obviously this requires fewer VCs and less signaling overhead. For a particular group,

Figure 7-14 Multicast over ATM techniques.

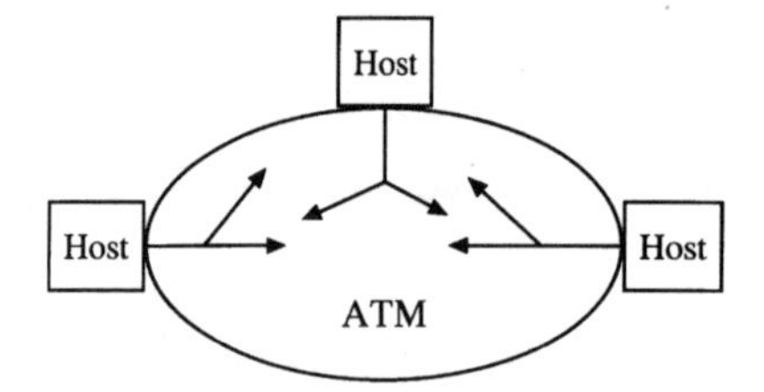

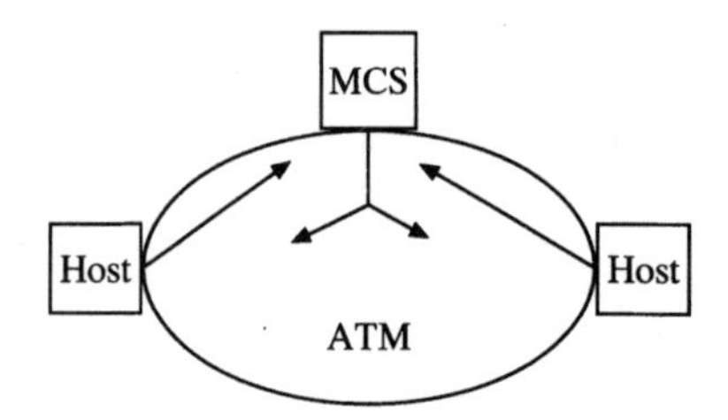

[12]ATM does not have any notion of group addressing at the ATM level. Using the term *multicast forwarding* in the context of ATM means that the inbound cell is copied and label-swapped out one or more outbound ports.

only a single point-to-multipoint VC rooted at the MCS needs to be managed. However, increased latency and delay are associated with this solution because all traffic must first pass through the MCS. This involves not just forwarding the data at the cell level but also reassembling the cells into an AAL5 frame and then segmenting them back into cells for transmission on the outbound point-to-multipoint VC. The senders must also have some way of dealing with their own packets that have been sent back to them from the MCS.

MARS supports both techniques, and it is up to the network administrator to decide which one is used on a per-group basis to forward multicast data over an ATM network.

7.5.2 Components

The components of a network supporting MARS consist of the following:

- *MARS.* MARS is the overall name given to the set of services and protocols developed to support IP Multicast over UNI 3.0/3.1 ATM networks as defined in RFC2022. This includes the protocols and message formats used to register group membership with a MARS server and to distribute group membership information to all members of the cluster and those used by a sender that wishes to resolve IP multicast group addresses with specific ATM addresses.
- *MARS cluster.* A cluster is a group of ATM-attached endpoints that use the same MARS server to register their group membership information with and to receive group membership updates from.
- *MARS server.* The MARS server, or simply the MARS, is the address resolution server for a cluster of ATM-attached devices (e.g., hosts, routers) supporting IP multicast over ATM. It maintains a table of IP multicast Class D group addresses (denoted as IPmc.n) and the corresponding ATM addresses of the group members {IPmc.1, ATM.1, ATM.2,...ATM.n} called the host map. The MARS server also responds to queries from MARS clients that wish to resolve a group address with the one or more corresponding ATM addresses of the group members. The MARS may be colocated with an NHS and/or ATMARP server or may run on a separate device.

- *MARS client.* The MARS client is resident on all members of a cluster supporting IP multicast over ATM. The MARS client is responsible for contacting the MARS server and registering its ATM address and the IP multicast groups that it wishes to become a member of. A MARS client sender may also query a MARS server to resolve an IP group address with the corresponding ATM addresses of the group members. A MARS client sender may add or drop members of a group on a point-to-multipoint VC depending on their membership as tracked and distributed by the MARS. A MARS client may also establish a VC to an MCS to use for forwarding multicast packets to a group through a specific MCS.
- *ClusterControlVC.* The ClusterControlVC is a point-to-multipoint VC that is rooted at the MARS and branches out to all cluster members. It is used by the MARS server to distribute group membership.
- *MCS.* The MCS serves as an intermediate rendezvous point between the MARS senders and the receivers. From a MARS protocol perspective, though, it must function as a proxy cluster endpoint for the sender(s) of a particular group and must behave as a sender when managing the point-to-multipoint VC out to all members of a group. The MCS is responsible for registering its ATM address along with the IP multicast group addresses it will act as an MCS for. The MARS maintains a separate table of IP multicast group addresses and the ATM addresses of the MCS(s) for that particular group {Ipmc.1, MCS.1.} called the server map. The MARS also adds the MCS to a point-to-multipoint ServerControlVC that is used to distribute group membership to the MCS(s). Senders that wish to transmit multicast packets to a group establish a VC to the one or more MCS(s). The MCS then forwards the packet out its group-specific point-to-multipoint VC.

7.5.3 Address Resolution

The basic operation of the MARS protocol is quite similar in concept to the other classical IP/ATM techniques discussed so far. First, a MARS client conveys to the MARS server which IP multicast groups it wishes to join along with its own ATM address. This enables the MARS server to build a host map consisting of IP multicast group addresses and corresponding ATM addresses. At the same time the MARS will add the client as a leaf to its ClusterControlVC. When a potential sender wishes to transmit IP multicast packets to a specific group, it queries the MARS sender for the host map of the multicast group. Upon receiving the information from the MARS server, the sender adds the group members to a source-rooted point-to-multipoint VC and begins transmission.

The basic operation and flow of MARS protocol messages is illustrated in Fig. 7-15 and is explained in more detail below:

- ✔ Cluster members Host2 and Host3 transmit `MARS_JOIN` messages to the MARS server. The information contained in the `MARS_JOIN` message includes the IP multicast group address (Ipmc.x) that the hosts wish to join and their own ATM address (ATM.x).
- ✔ The MARS stores the addresses contained in the `MARS_JOIN` messages and adds the IP hosts to its ClusterControlVC.
- ✔ A cluster member, Host1, that wishes to transmit multicast packets addressed to Ipmc.1 contacts the MARS and transmits a `MARS_REQUEST` message.

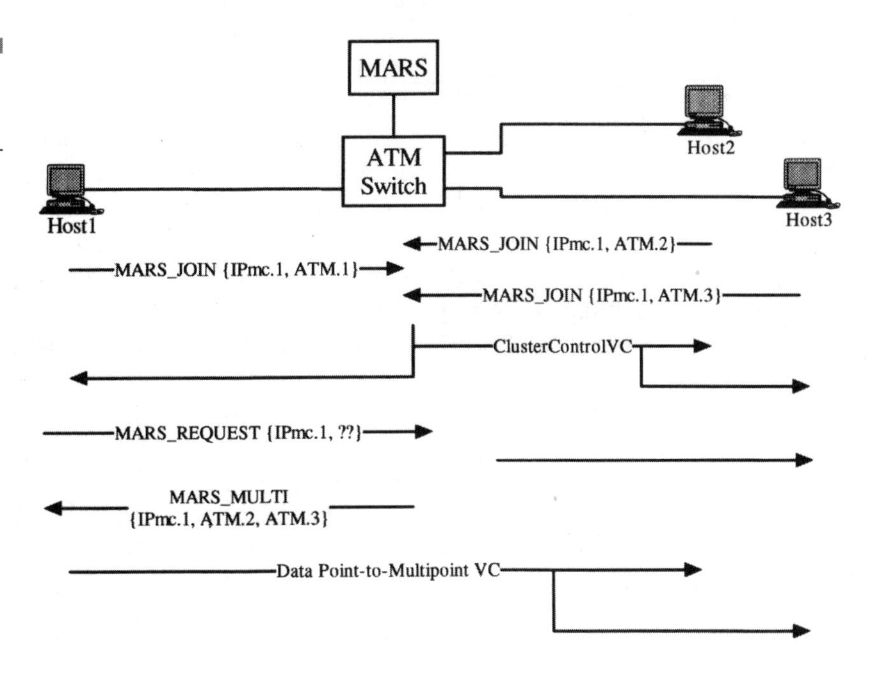

Figure 7-15 MARS address resolution and data flow.

- ✔ The MARS returns a set of ATM addresses corresponding to the members of the group associated with multicast group address Ipmc.1 in a MARS_MULTI message.
- ✔ Host1 builds a point-to-multipoint VC to Host2 and Host3 and begins sending multicast packets.
- ✔ A group member can leave a multicast group by transmitting a MARS_LEAVE message to the MARS. The MARS will then propagate the message out the ClusterControlVC where it will be received by the sender, which will promptly remove the group member from its point-to-multipoint VC. The same procedure, using a MARS_JOIN message, enables a group member to join a group and then have the sender add the new member to an existing group-specific point-to-multipoint VC.

The sender can manage group membership by tracking the MARS_JOIN or MARS_LEAVE messages that are propagated by the MARS and delivered to the sender over the ClusterControl VC. Based on the contents of these messages, the sender can either add members to or drop members from the point-to-multipoint VC.

When an MCS is introduced into the configuration, the fundamental behavior of the endpoints (senders and receivers) remains the same, but the machinery in the middle is slightly modified. There is an MCS that functions as a "reflector" to forward multicast data out to all group members. The MCS must tell the MARS which multicast group it supports and does so by sending a MARS_MSERV message to the MARS. The MARS updates its per-group server map entries in the server map and adds the MCS to a ServerControlVC. The ServerControlVC is established so that the MARS can convey any group membership changes (MARS_JOIN/LEAVE) to the MCS. The MCS uses this information to manage the group-specific point-to-multipoint VC rooted at the MCS. When the MARS receives a MARS_REQUEST message for a particular multicast group, it will return the contents of the server map. The sender will then establish a point-to-point VC to the MCS serving a particular group.

As noted earlier, senders must have some way of handling their own packets that have been reflected back to them from the MCS. This is handled by the use of a unique cluster member identifier (CMI) that is allocated by the MARS when a host joins a cluster. The CMI is placed in the header of each packet that is transmitted by the sender. A sender may then check the CMI of each packet arriving from the MCS and if it matches its own CMI, the packet is dropped. This requires additional AAL frame processing but is nevertheless necessary when an MCS is used.

Given that MARS provides a multicast service to members of a cluster, it is a natural extension to have it support the more general case of broadcast. RFC2226 defines a technique that enables MARS to provide an IP broadcast service to members of an LIS. This may be particularly useful if there is a requirement to support client/server applications that require broadcast functions to locate specific resources on the LIS. Another example is the deployment of control protocols such as RIP that use broadcast transmissions as a means of disseminating control information. An IP broadcast capability would round out the IP over ATM capabilities already defined by RFC1577/2225 for unicast IP flows and RFC2022 for multicast IP flows.

Incorporating IP broadcast support is relatively straightforward using the mechanisms supported by MARS. LIS members that wish to receive broadcast frames send a `MARS_JOIN` message with the IP broadcast address of 255.255.255.255 and their ATM address. When a sender wishes to transmit a broadcast, it obtains the ATM addresses of all other LIS members using a standard `MARS_REQUEST/MULTI` message exchange with the MARS. The fact that multiple ATM addresses can be packed into a single `MARS_MULTI` message reduces the control overhead. Another advantage of using the MARS for broadcast support is the ability to use multicast servers to reduce the amount of VC resources.

Although the discussion so far has centered on multicast communication between IP hosts within a cluster, it is likely that a cluster will also include a router. A router is used to forward unicast traffic to and from different networks and may be used to forward multicast data as well. A router attached to a cluster registers its ATM address and any multicast group it wishes to join or leave using standard MARS protocols messages. In emulating a multicast-enabled router attached to a broadcast LAN such as an Ethernet, the router may opt to join a range of multicast groups, perhaps consisting of the entire Class D address space. In this mode the router is behaving in a promiscuous manner. However, with M transmitting hosts involved with N multicast groups, the router would have to terminate $M * N$ number of VCs and deal with the associated AAL processing overhead. It may be that some multicast traffic does not need to be transmitted outside the LIS, and thus it would be the responsibility of the network administrator to configure the routers to block out certain multicast addresses or ranges of addresses that the router is eligible to join.

Another function that multicast-capable routers attached to LANs perform is the IGMP query process. The objective is to ascertain the existence of multicast group membership on the LAN by transmitting an IGMP query to the all-hosts 224.0.0.1 multicast address. Although simple to do on a broadcast LAN, a typical MARS scenario would require extra MARS_JOIN messages and consume additional VC resources. An alternative process defined in RFC2022 is the use of MARS_GROUPLIST_REQUEST/REPLY messages. These messages provide the router with a list of multicast groups that are active on the cluster. This approach obviously consumes fewer resources and does not require an explicit cluster member initiative.

7.5.4 MARS Network Example

MARS provides an IP multicast service to members of a cluster where a cluster is essentially an LIS. An LIS using MARS for IP multicast is illustrated in Fig. 7-16. The MARS can be coresident with the ATMARP server as shown. In addition a router is also attached to the LIS to provide forwarding of both unicast and multicast frames to and from other networks in the extended internetwork.

Figure 7-16
MARS network example.

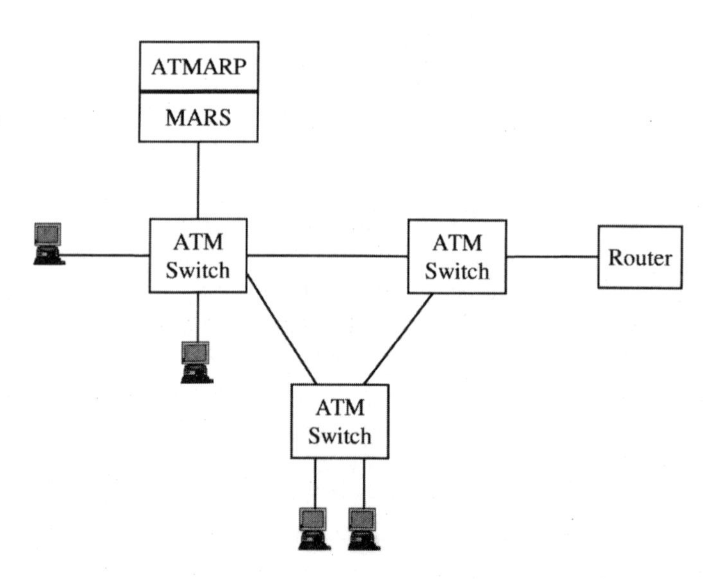

7.6 Server Cache Synchronization Protocol

The existing suite of classical IP/ATM protocols depends on the availability and accessibility to a single server or servers to provide among other things address resolution services. If that single server disappears and is not available to the classical IP/ATM clients, the network ceases to function. Indeed the ATMARP, MARS, and NHS servers are all single points of failure. Therefore to enhance the robustness and flexibility of the classical IP/ATM network solutions, one must devise a way of protecting against the loss of a single ARS server without seriously disrupting the operation of the network. At the same time any failure protection mechanism must attempt to minimize additional overhead placed on the clients, servers, or network itself.

One can envision a possible solution that utilizes a "hot-standby" backup server in case the primary server goes down. In this scenario, the backup server is connected to the network but is configured to operate in hot-standby mode. In other words, although there may be an active connection from the backup server to the network, the backup server is not performing any of the classical IP/ATM ARS functions. At the most it may establish a VC to the primary server as a means of implicitly monitoring the availability of the primary server. So if the VC between the backup and primary goes down, the backup server will know to begin providing basic ARS services to the clients.

This solution is simple to implement and the notion of a hot-standby anything can provide an additional level of comfort to the clients and network operators. However, there are some drawbacks to this solution. First, there is the question of cost. It simply may not be cost effective to populate the network with backup ARS servers that are not doing any useful work other than just standing by. Second, for scalability reasons it might be useful to offload the processing from a single server and distribute it among several in the network. This would require, though, that all servers synchronize their databases,[13] and this is something that the hot-standby approach does not provide. Another concern is that once the clients cut over to the backup server, there will be some delay before

[13]In the context of SCSP, *database* is just another word for cache or table entries. Specifically it is the client information maintained by the server that is made available to other clients in the network through a query/response protocol.

the network returns to its primary server state. In addition there is the question of under what circumstances the clients will return to the primary server once it comes back on line.

To address the limitations inherent in the hot-standby approach as well as provide a general multiserver synchronization and replication solution, the SCSP was developed. SCSP is defined in RFC2334. It enables multiple servers providing a service such as IP/ATM address resolution to synchronize their table or cache information so that all servers have a copy of the same information. If one of the servers disappears, the remaining server(s) can instantly provide the same ARS services. It also contributes to greater scalability by distributing the processing load across multiple servers in the network. In addition, the span of control traffic is limited to the distance between the client and the nearest server.

Rather than define a new cache synchronization/replication technique, SCSP borrows the database synchronization methodology from OSPF. In fact the behavior of SCSP is quite similar to that of OSPF except that there is no need to perform a Dijkstra calculation each time the contents of the database are updated. The only requirements are that the database consisting of client state information (e.g., IP/ATM address bindings) is synchronized among all the participating servers in the group and that database updates are reliably flooded to all servers in the group. Borrowing existing OSPF mechanisms makes sense because it is proven to work in a scalable fashion, and database updates can be quickly and reliably propagated with minimal network overhead.

To best illustrate its operation, the SCSP specification defines three types of servers:

- *Local server (LS).* The LS is the server by which operational and functional definitions are applied. It is only called an LS to provide a frame of reference. Every SCSP-capable server is an LS in its own right, and there is no functional difference between an LS and the other SCSP servers.
- *Directly connected server (DCS).* A DCS is directly connected to an LS by a VC. Every SCSP-capable server is an LS in its own right, and there is no functional difference between a DCS and the other SCSP servers.
- *Remote server (RS).* An RS is not directly connected to an LS but is always one or more SCSP server hops away. There is no functional difference between an RS and the other SCSP servers.

The SCSP specification defines three subprotocols that enable servers to contact each other, perform cache synchronization, and propagate cache updates when required. They are

- *Hello protocol.* The Hello protocol is used by the LS to determine if the neighboring DCS is active and whether the connection between them is unidirectional, bidirectional, or nonfunctional.
- *Cache alignment (CA) protocol.* The CA protocol is used by an LS to synchronize or align its cache with those of its neighbor DCSs.
- *Cache state update (CSU) protocol.* The CSU protocol is used to dynamically update cache entries in a group of SCSP servers. An example of an updated cache entry is a newly registered CLIP client and its associated IP/ATM address binding. The CSU advertisements are reliably flooded to all servers in the group.

An example of three ATMARP servers connected to a single large LIS is illustrated in Fig. 7-17. From the perspective of server 1, which is the LS, server 2 is the DCS and server 3 is the RS. Each time a client registers with server 1 or a table entry is aged out, server 1 transmits a CSU update to server 2, which in turn forward it to server 3. Thus SCSP enables all three ATMARP servers to maintain accurate information about the clients connected to the LIS.

A separate instance of SCSP must operate for each protocol that the server is supporting. It is intended that SCSP can function as a general-purpose server cache synchronization and replication protocol. Therefore it is likely that it will be implemented on servers supporting the classi-

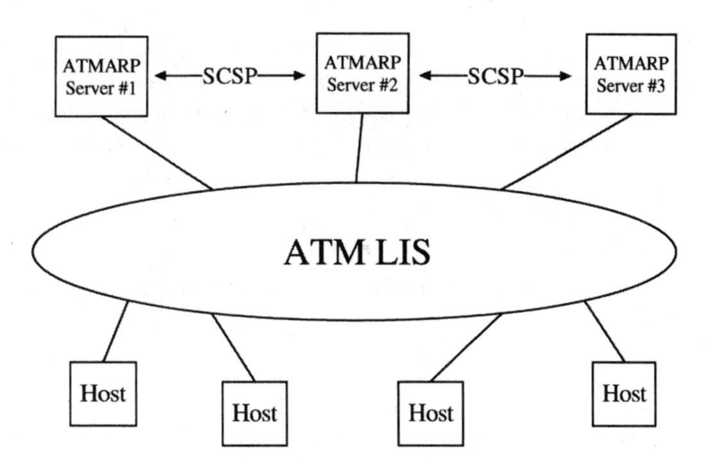

Figure 7-17 SCSP Network example.

cal IP/ATM protocols. In addition the ATM Forum will specify its use in the LAN Emulation Network-to-Network Interface (LNNI) 2.0 specification as a means of providing distributed LAN Emulation Server (LES)/Broadcast/Unknown Server (BUS)/LAN Emulation Configuration Server (LECS) services. And consideration for implementation with other client/server state protocols such as DHCP is a strong possibility.

7.7 IPv6 over ATM

All of the classical IP/ATM protocols discussed so far deal with IPv4 operation over an ATM network. Given that IPv6 is the next-generation version of IP and that future networks will consist of very high-speed broadband transports, it is more than likely that there will be a good deal of IPv6 traffic flowing over ATM networks.[14] Currently it is possible to run IPv6 over ATM using the emulated LAN approach supported by the ATM Forum LAN Emulation specification. Many of the IP switching schemes are architecturally capable of mapping IPv6 flows or address prefixes to an ATM VC-like switched path. But there has not been a specific solution for mapping the IPv6 operation over an ATM network analogous to IPv4 and the classical IP and ARP over ATM function defined in RFC1577/2225.

There are several reasons for this. IPv4 is ubiquitous, and any solution mapping IP over ATM must first address IPv4. Also ATM, and to a much lesser degree IPv6, have yet to be deployed in any meaningful scale that would compel researchers, engineers, and vendors to develop and deliver IPv6 over ATM capabilities. Another issue is that due to continued adjustments and fine tuning, the IPv6 specification is not yet stable. And finally and perhaps most significantly, proper IPv6 host operations depend on the use of multicast as a means to discover and resolve IPv6 to link-layer addressing with other IPv6 hosts in the same network. Therefore the problem of IP multicast operation over ATM had to be solved first.

IPv6 uses multicast transmissions in its neighbor discovery operations. A source IPv6 that wishes to communicate with a destination host on the same subnet first multicasts a neighbor solicitation message addressed to a special destination, solicited-node multicast address. The

[14]Internet2 will most likely deploy IPv6 over ATM and other broadband transports first.

destination host receives a copy of the multicast packet, notes that it was addressed to its specific multicast address, and returns a neighbor advertisement message to the source with its link-layer address.

Preliminary work defining IPv6 operation over ATM was based on the notion that the neighbor discovery mechanism should be left untouched. Or in other words, it should not have to fundamentally change to accommodate operation over a specific data link technology such as ATM. This means though that the machinery required for multicast operation over an ATM network must be present, and in the case of IP multicast over ATM, this means MARS.

Preliminary investigations of various solutions point to a couple of different ways for supporting IPv6 over ATM. One technique is specific to IPv6 endpoints that share the same link. In this case the notion of an IPv6 link has been generalized to a logical link (LL), which is similar in concept to the IPv4 logical IP subnet. IPv6 intra-LL connectivity over an ATM network is accomplished with a MARS server to distribute neighbor discovery messages to other members of the LL. IPv6 inter-LL connectivity over an ATM network is supported by either ICMP redirect messages or possibly router-to-router NHRP as a means of resolving an off-link target address with the corresponding ATM address.

Figure 7-18 illustrates IPv6 over ATM intra-LL connectivity. Host1 and Host2 are connected to an ATM network. The MARS is present as a vehicle to provide IP multicast support and more specifically as a means to forward neighbor discovery messages to hosts on the LL. Before neighbor discovery can be performed though, Host1 and Host2 must register with the MARS server as cluster members. If Host1 wishes to communicate with Host2, Host1 will transmit a neighbor solicitation message up to the MARS. The MARS may either forward the message directly to the intended cluster member if a point-to-point VC to that

Figure 7-18
IPv6 over ATM intra-LL connectivity.

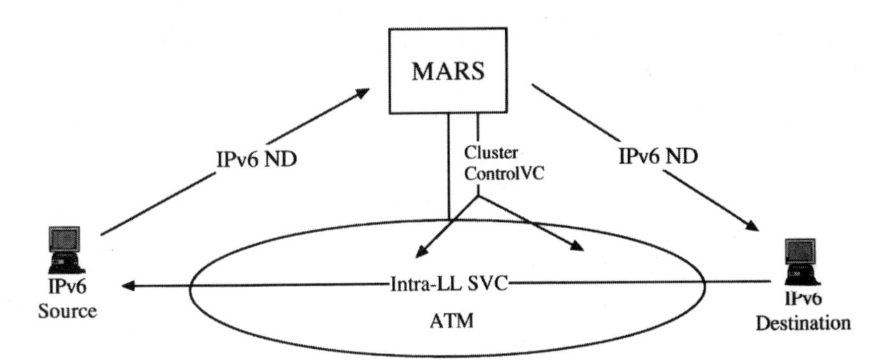

cluster member exists, or it may transmit the message back out its ClusterControlVC. Observe that the MARS in this case has been modified to forward non-MARS control information. When the destination receives the neighbor solicitation, it will extract both the IPv6 and ATM addresses from the message and establish a VC back to the source. ATM connectivity between two IPv6 hosts on the same LL has now been established.

MORE INFORMATION

http://www.ietf.org/html.charters/ion-charter.html

Metz, C., and Sackett, G., *ATM and Multiprotocol Networking,* New York: McGraw-Hill, 1997.

Minoli, D., and Schmidt, A., *Network Layer Switched Services,* Wiley Computer Publishing, 1998.

CHAPTER 8

LANE and MPOA

Classical IP and ARP over ATM initially defined in RFC1577 was the first attempt to map a connectionless network-layer protocol on top of the connection-oriented services provided by ATM. That effort was initiated and performed by the IETF, which naturally addressed the problem only from the IP perspective. But the world of data networking is not and probably will never be a uniprotocol one.[1] There is in addition to IP a large legacy of LAN applications dependent on other networking protocols such as SNA/Advanced Peer-to-Peer Networking (APPN), Netbios, IPX, Appletalk, etc. If ATM is to become a useful transport for data network applications, it must be able to accommodate other protocols including IP.

The ATM Forum tackled that problem and developed the LAN Emulation (LANE) over ATM specification.[2] LANE emulates the functions and behavior of a typical broadcast LAN on top of a physical ATM network. Unmodified LAN applications can operate over an ATM. Legacy LAN workstations and applications can communicate with their ATM-attached counterparts as if they were communicating through a LAN bridge. And just like any old LAN, LANE can support IP as well as other network protocols.

Once the LANE work was completed, the ATM Forum did not stop there. In an effort to utilize the speed and capacity of an ATM network to accelerate the forwarding of routed traffic between networks, the ATM Forum developed the MPOA specification. MPOA uses mechanisms defined in the LANE specification and the IETF's NHRP to establish shortcut paths between devices on different networks, thus bypassing the relatively slow per-packet processing inherent in traditional routers. MPOA is in fact a technique for flow-driven IP switching.

This chapter provides an overview of the components and functions of the LANE and MPOA specifications. Both of course make liberal use of an underlying ATM transport to accelerate the flow of data between IP endpoints.

---

[1]The direction of most networks is of course IP, but in reality there will always be networks that support multiple protocols.

[2]*http://www.atmforum.com.*

8.1 LAN Emulation V1.0

LAN Emulation V1.0 was completed in 1995 and is a product of the LAN Emulation subworking group of the ATM Forum.[3] It defines the components, services, protocols, and functions that enable ATM-attached devices to operate as if they were connected to either a Token Ring or an Ethernet LAN segment. LANE also permits legacy LAN segments to connect to an emulated LAN through an ATM LAN bridge. For applications, LANE creates the illusion that they are running on workstations attached to broadcast LANs. In addition, LANE also enables legacy LANs to use the high bandwidth and low latency of ATM as a backbone transport. Indeed the use of LANE services on an ATM backbone network interconnecting legacy LAN segments is the most common use of ATM in campus networks.

Figure 8-1 illustrates the concept behind LAN Emulation. Observe that as far as the applications and protocols are concerned, there is no difference between a device attached to a legacy LAN and one attached to an emulated LAN supported by LANE.

LANE operates similarly to the Classical IP/ATM technologies discussed earlier in that it defines a query-to-ARS approach for address resolution. However, in the case of LANE, it is MAC-layer addresses, not IP addresses, that are mapped to corresponding ATM addresses and distributed to LANE clients upon request. LANE further supplements the basic ARS approach with services that support the forwarding of broadcast, multicast, and unknown frames as well as services to support auto-

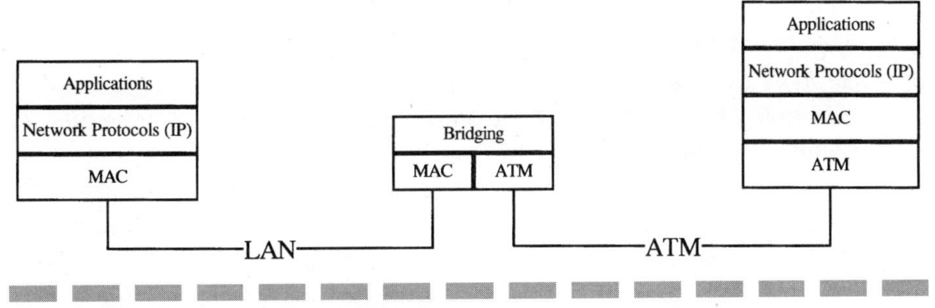

Figure 8-1 ATM Lan emulation architecture.

[3]"LAN Emulation over ATM Version 1.0," ATM Forum, af-lane-0021.000, Jan. 1995.

configuration. There is no question that LANE is a bit more complex than Classical IP, but this additional complexity is required to provide a true emulated LAN service on top of an ATM network.

8.1.1 Requirements

The ultimate goal behind the development and introduction of LANE was to provide a seamless and transparent migration path that LAN applications could use as they moved from the existing legacy LAN infrastructures to newer and more advanced ATM networks. Ensuring this transparency while allowing for a gradual migration to ATM generates a number of specific requirements that the LANE specification must address:

- *Emulation of IEEE 802.3/802.5 LANs.* Most LAN installations and applications run over either Ethernet or Token Ring media. LANE must emulate the functions of one of these two LAN types so that applications and protocols can run unmodified on top of an ATM network. The standard MAC-driver interfaces (e.g., Network Driver Interface Specification (NDIS), Open Driver Interface (ODI), Datalink Protocol Interface (DLPI) and the MAC-layer addressing structure must be maintained and emulated for the application's sake and so that ATM-attached workstations can communicate with their legacy LAN peers through standard bridging techniques. In addition LANE must emulate the connectionless transmission services that normal shared LAN technologies provide. This enables a source to begin transmitting data to a destination without first setting up a connection.
- *Support for broadcast and multicast.* Many LAN applications and protocols use broadcast and/or multicast transmissions for the distribution of control information and to discover other hosts or services on the same LAN segment or network of bridged segments. LANE must therefore provide a service in which a broadcast/multicast packet generated by a source is delivered to all members of the emulated LAN including any ATM-attached LAN bridges.
- *Interconnection with legacy LANs.* LANE emulates the functions of a standard Ethernet or Token Ring. One of those functions is the ability to bridge two or more segments together. A function that

enables a legacy LAN segment to be bridged to an emulated LAN segment supporting LANE must be provided.

- *Use of existing ATM Forum signaling and routing protocols.* Depending on the application, communications on a LAN can be dynamic. For example, a client and server may exchange a large volume of data over a short period of time in the morning and then not talk to each other for the rest of the day. Over a connectionless shared media LAN, this transient interaction consumes nothing more than bandwidth at the time of the transaction. Exchanging data over an ATM network requires the establishment of a VC between the communicating endpoints. Given the dynamic nature of LAN communications and the necessity to perform them over a connection-oriented network, it makes sense to use existing ATM Forum signaling and routing protocols to dynamically establish SVCs between LAN endpoints. Dynamically allocating bandwidth (SVC) when needed conserves resources within the ATM switch fabric and ATM endpoints. Therefore, it is a requirement for LANE to interface between connectionless LAN networking and ATM Forum signaling and routing so that ATM SVCs can be built to exchange LAN traffic and can be released for others to use when the exchange has been completed.
- *Separation between the emulated LAN topology from the physical ATM network.* The topology of the ATM network is separate from that of the emulated LAN provided for by LANE. In fact LANE makes it possible to support multiple emulated LANs attached to the same physical ATM network. There is no association between a LANE workstation and the emulated LAN it is part of nor is there any association between the LANE workstation and the ATM switch it is physically attached to. And finally, because LANE supports a broadcast function, it provides a virtual LAN capability.
- *Autoconfiguration and attachment.* A LAN workstation can typically plug into a port on a hub or switch and become active. LANE should provide the same service.

Developing and engineering relatively complex functions into a common standard took some time and some compromises and trade-offs were made. Nevertheless, LANE is an industry standard that does meet all of the aforementioned basic requirements while leaving some room for implementers to incorporate value-added function.

The elementary function supported by LANE is the emulation of a Token Ring or Ethernet LAN operating over a physical ATM network.[4] LANE-capable devices—workstations, servers, bridges, or routers attached to the same emulated LAN—all "believe" that they are addressing each other and communicating using a standard LAN protocol and MAC addressing scheme and that they can transmit and receive MAC-layer broadcast and multicast frames over a shared media LAN. LANE provides the services that create and maintain this illusion. Therefore connectionless transmissions as well as broadcast and multicast services are supported. Traffic originating from emulated LANs can be bridged or routed to legacy LANs, vice-versa, and between emulated LANs.

With a couple of very minor exceptions there is really nothing that a legacy LAN can do that an emulated LAN cannot do.[5] Because of its linkage to the underlying mechanisms of ATM, an emulated LAN can enhance the operation of a LAN. Broadcast propagation can be reduced, more bandwidth and possibly QoS can be exploited, and the operation of a VLAN is dramatically simplified.

8.1.2 Components

The LAN Emulation specification defines several concepts and components. One is the emulated LAN (ELAN) itself. An ELAN consists of a set of ATM-attached devices that contain one or more client components called LAN Emulation clients (LECs) or LE clients.[6] An ELAN must also provide a set of services and functions that run on a set of LAN Emulation (LE) servers. The LEC is a logical entity that is present in each ATM-attached device that uses LANE services. It essentially sits on top

---

[4]A single instance of LANE can support either a Token Ring or Ethernet segment—not both. Like their legacy counterparts, emulated LANs can be interconnected by routers or bridges. And it is certainly possible to support multiple emulated LANs on top of an ATM network.

[5]An emulated LAN does not emulate every functional nuance of its legacy counterparts. For example, an emulated LAN does not experience collisions such as those that occur on an Ethernet. An emulated LAN does not send out the station management (SMT) or beacon frames found on a real Token Ring.

[6]The terms *LEC* and *LE client* will be used interchangeably. An attempt will be made to avoid using the abbreviation LECs for multiple LE clients because it can be confused with the acronym for the LAN Emulation Configuration Server (LECS).

of the ATM AAL layer and right below the MAC driver. The basic function of the LEC is to link the application and MAC protocol functions above with the ATM services below.

The LE services defined in the LANE specification consist of an LES, a BUS, and the LECS. The LES maintains a table of MAC/ATM address bindings and supports a special address resolution protocol called LE_ARP. The BUS forwards broadcast, multicast, and unknown or unresolved frames to the other members of the ELAN. The LECS provides configuration and ELAN membership information that the LEC retrieves at network initialization. Each ELAN must have one LES/BUS pair, and the LECS is optional.

A conceptual depiction of an ELAN and its components is illustrated in Fig. 8-2. A LEC interfaces with the LE servers using a set of procedures and protocols collectively called the LAN Emulation User-to-Network Interface (LUNI). The four basic functions that are performed between the LEC and the LE servers over the LUNI are network initialization, configuration, address resolution, and data transfer.

The LE servers provide the key services that enable LECs to connect to an ELAN, send and receive broadcast frames, and resolve MAC addresses with ATM addresses so that data exchange may take place over ATM SVCs. Each LE server provides a particular service to the LEC that it is communicating with. An understanding of those services provided by the individual LE servers is given below to help explain the operation of LANE.

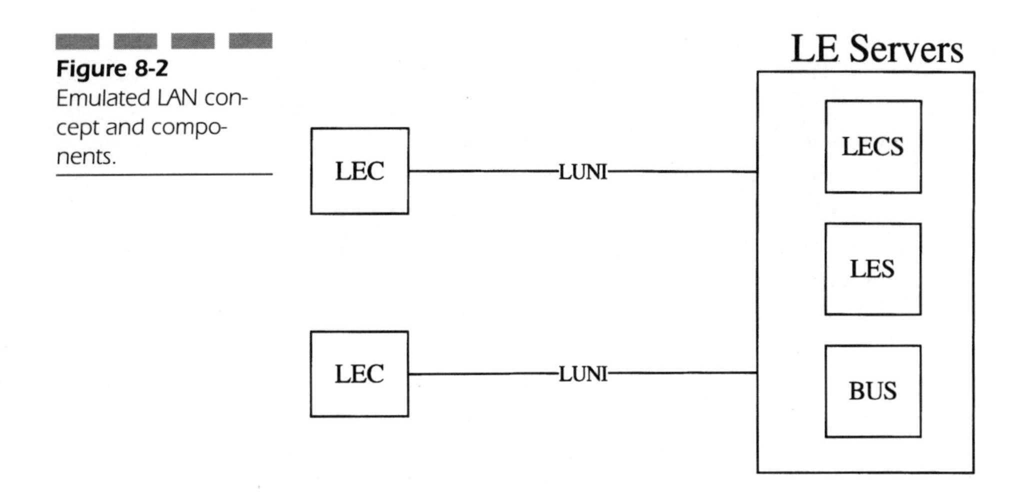

Figure 8-2 Emulated LAN concept and components.

- *LES.* The LES provides the address resolution server function for the ELAN. A LEC registers its MAC (or route descriptor in the case of a Token Ring bridge function) and ATM addresses with the LES. The LES responds to address resolution queries (LE_ARP) from LE clients with the appropriate MAC-ATM address binding.[7] If the LES cannot answer the query with a specific MAC-ATM address binding, it forwards the address resolution request to other LE clients attached to the ELAN. The LES does not forward any data frames. Its sole purpose is to facilitate address resolution for the LE clients that are members of the ELAN.
- *BUS.* The BUS handles broadcast data sent by ELAN members addressed to the broadcast MAC address ('FFFFFFFFFFFF'), multicast data, and initial unicast data in which a target ATM address has not yet been resolved. The BUS provides the service by which LE clients can begin transmitting unicast, broadcast, or multicast frames in a connectionless manner. All broadcast, multicast, and unresolved unicast frames must pass through the BUS. A BUS operates by receiving all cells of a broadcast frame from a source LEC, assembling the cells into a frame, and then segmenting and transmitting them back out to some or all LE clients that are members of the ELAN.
- *LECS.* The LECS provides configuration information upon request to an LE client so that it may join a particular ELAN. The information provided by the LECS consists of items such as the LAN type (Token Ring or Ethernet), maximum frame size, and most importantly, the ATM address of the LES. The LES ATM address enables the LE client to establish contact with a particular LES. Successfully connecting to an LES is an implicit form of stating membership to an ELAN. If an LEC cannot connect to the LES, it cannot join an ELAN. The LECS may also consider some information provided by the requesting LE client (e.g., LEC ATM address, LEC MAC address, etc.) when deciding to either permit or deny membership to a particular ELAN. The LECS therefore is a means by which ELAN (read VLAN) membership policies can be centrally configured and then dynamically distributed to LE clients on demand. The use of the LECS is optional but is generally recom-

---

[7]A MAC-ATM address binding is denoted by {MAC.1, ATM.1}. In the case of a source route vector route descriptor, it is denoted by {RD.1, ATM.1}.

mended. Short of using the LECS, the ATM address of the LES would have to be configured in each LEC.

The LEC is the logical entity contained in data communicating ATM-attached endpoints supporting LANE. An LEC can operate either in a host or an edge device such as an ATM LAN bridge or router. The latter implementation is called a proxy-LEC because it provides LANE forwarding and address resolution services on behalf of all devices behind it. The basic functions of an LEC consist of establishing communications with the LE service components, registering configured MAC and ATM addresses, supporting the `LE_ARP` protocol, and managing direct VC communications with other LE clients in the ELAN.[8]

Figure 8-3 illustrates both a single LEC implementation on a host and a proxy-LEC implementation on an edge device consisting of multiple LEC entities.

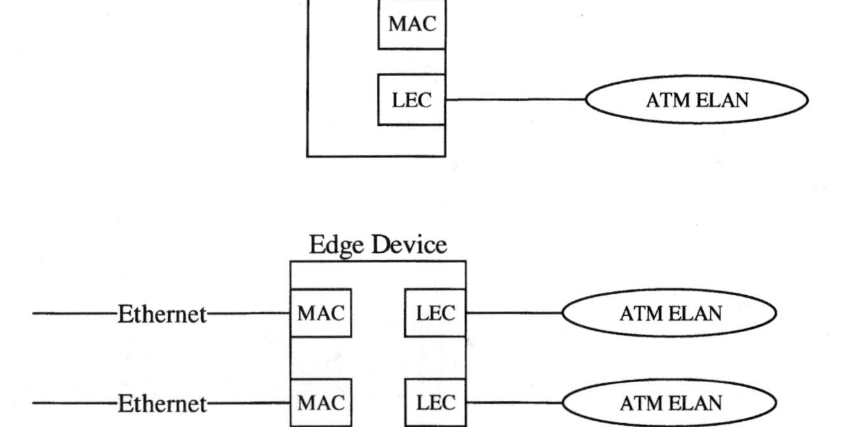

Figure 8-3
LEC and proxy LEC.

[8]A proxy LEC does not register MAC addresses of devices attached to legacy LAN segments. It does, however, provide its own ATM address in response to an address resolution query for one of the MAC addresses attached to a legacy LAN segment.

8.1.3 LANE Connections

There are two basic type of connections that exist within an emulated LAN. The first type consists of several control connections that exist between the LE client and server components. These connections enable the exchange of control information between the LE client and server components so that the operation of a LAN segment can function on top of an ATM network. The second type of LANE connection strictly carries data. Broadcast, multicast, and unknown frames travel between LE clients by way of the BUS, and point-to-point SVCs can be established between two LE clients over the ATM network. The notion of LANE control and data connections in relationship to the LE client and server components is illustrated in Fig. 8-4.

There are three control connections that LANE supports. They are

- *Configuration Direct VCC.* This is an optional point-to-point VC established between the LEC and LECS. Configuration data including the ATM address of the LES is forwarded over this VC. Once the information exchange has completed, this connection can be torn down.
- *Control Direct VCC.* This is a point-to-point VC established between the LEC and the LES. This mandatory connection is used

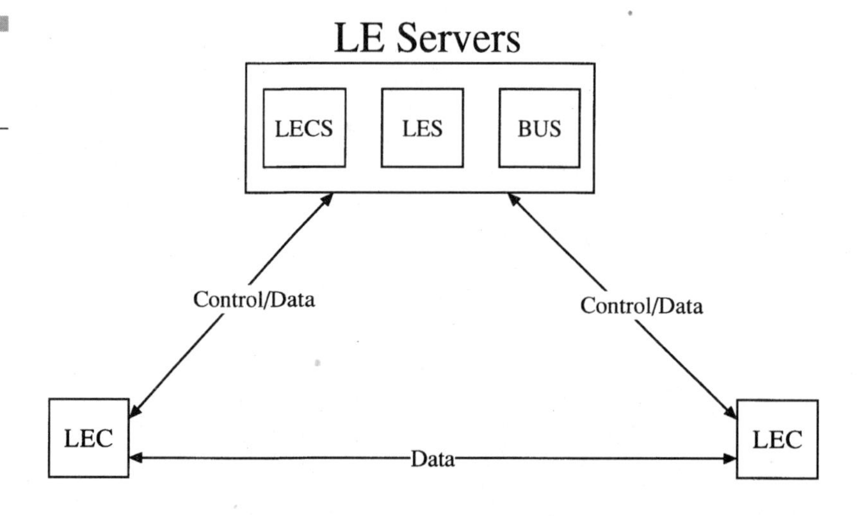

Figure 8-4
LANE control and data connections.

by the LEC for registration and address resolution purposes. This connection must be active at all times.

- *Control distribute VCC.* This is a unidirectional point-to-multipoint VC that is rooted at the LES and branches out to all LE clients attached to the ELAN. This optional VC is used by the LES to distribute address resolution queries and responses to LE clients.

The three basic control connections are illustrated in Fig. 8-5.

LANE supports three data connections. They are

- *Multicast send VCC.* This is a point-to-point VC established between the LEC and the BUS that is used by a source LEC to transmit and possibly receive broadcast, multicast, and unknown frames.
- *Multicast forward VCC.* This is a unidirectional point-to-multipoint VCC that is rooted at the BUS and branches out to all LE clients attached to the ELAN.
- *Data direct VCC.* This is a point-to-point VCC established between two LE clients on an ELAN.

The three basic data connections supported by LANE are illustrated in Fig. 8-6.

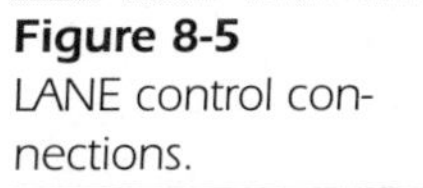

Figure 8-5
LANE control connections.

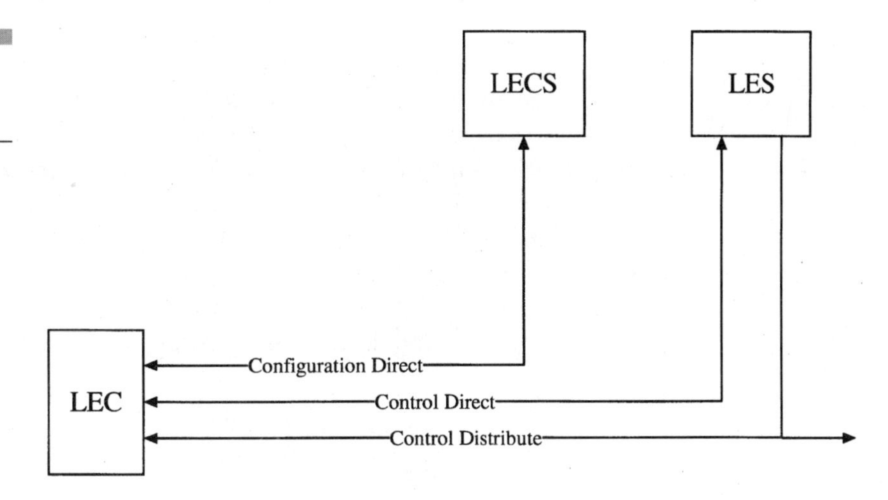

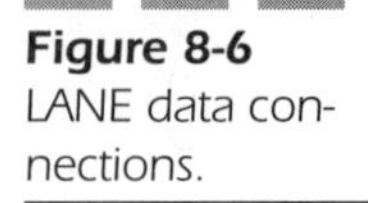

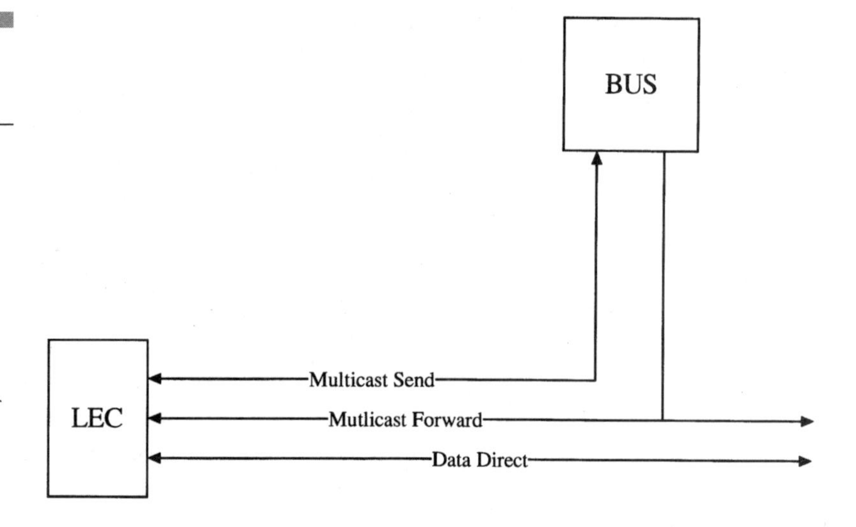

Figure 8-6 LANE data connections.

8.1.4 LANE Address Resolution and Data Flow

The address resolution and subsequent data flow exchange over a newly established SVC between ATM-attached endpoints is relatively straightforward and similar in concept to the ARS-based solutions discussed in the last chapter. LE clients register their MAC/ATM address information with the LES. The LES responds to queries from LE clients with MAC/ATM address bindings so that LE clients can talk to each other over ATM connections. But to fully emulate the operation of a LAN and to do so in a scalable and efficient manner requires some additional steps. Autoconfiguration and initialization through the LECS enables LE clients to bootstrap themselves onto an ELAN. The BUS provides a means by which a LAN broadcast channel can be set up so that connectionless broadcast frames can be propagated to ELAN members. And if the LES cannot resolve a MAC-ATM address query, it will forward the query to other LE clients that can respond.

The basic flows for initializing two LE clients and then performing address resolution and data direct VCC establishment are illustrated in Fig. 8-7. The steps are as follows:

- ✔ The first step is network initialization. This involves manual configuration, or it can be performed automatically by connecting to and interacting with the LECS. Both LEC1 and LEC2 in this

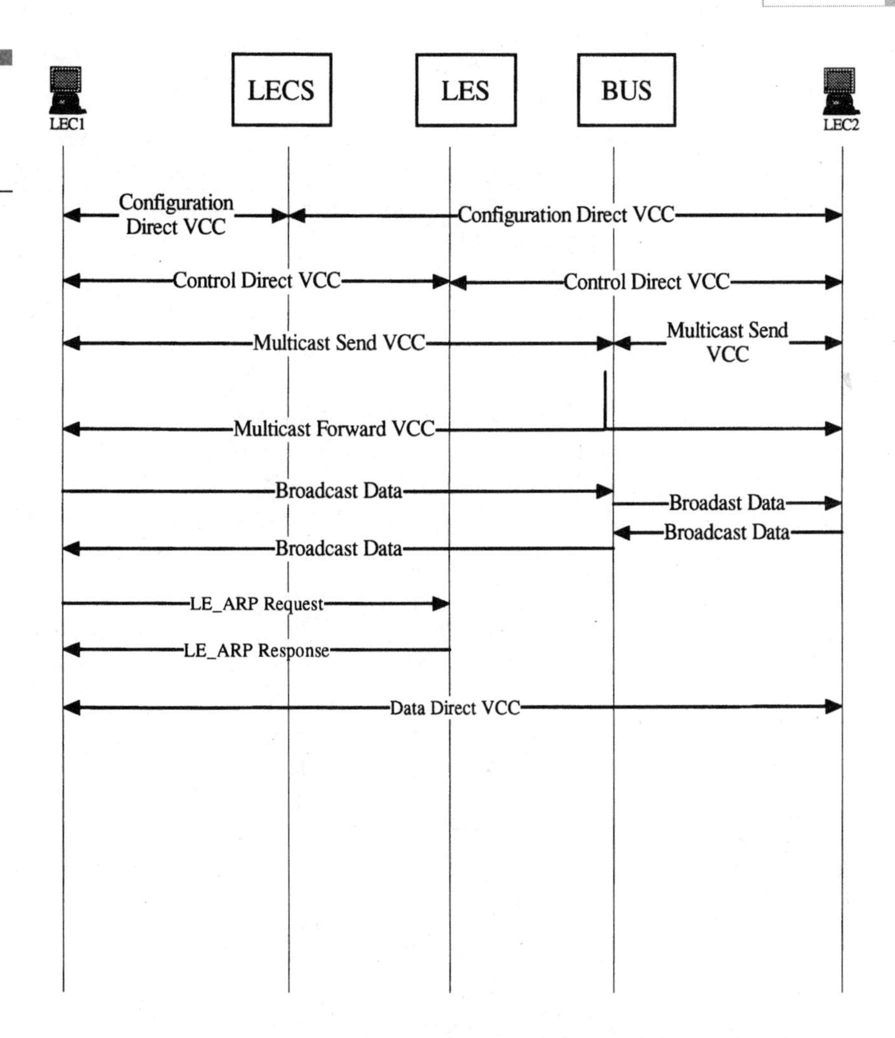

Figure 8-7 LAN Emulation address resolution and data flows.

example have been configured to establish a configuration direct VCC with the LECS. To achieve this, the LE clients must somehow determine the ATM address of the LECS. This can be accomplished in one of several ways: using a preconfigured LECS ATM address, retrieving an LECS ATM address from an ATM switch using an SNMP Interim Local Management Interface (ILMI) Get operation, or by using the well-known LECS ATM address.[9] The LECS provides LEC1 and LEC2 with the ATM address of the LES along with

[9]The well-known LECS ATM address is X"47007900000000000000000000-00A03E000001-00."

LAN type, maximum frame size, ELAN name, etc., using an exchange of LE_CONFIGURE messages.

- The next couple of steps for the LE clients involve joining the ELAN and then registering relevant MAC and ATM addressing information with the LES. LEC1 and LEC2 first establish a control direct VCC with the LES. The LE clients then exchange LE_JOIN messages with the LES, which contain information such as the client's ATM address, LAN type, a proxy-LEC indicator, and so on. Once the LEC1 and LEC2 have successfully joined the ELAN, they exchange LE_REGISTER messages with the LES over the control direct VC to register all of the MAC destinations they know about.[10] In addition the LE clients transmit an LE_ARP request for the MAC broadcast address of ('FFFFFFFFFFFF') to obtain the ATM address of the BUS. The LES then provides the ATM address of the BUS.
- Following the joining and registration steps, a broadcast channel for the LE clients must be established. The multicast send and multicast forward VCC, respectively, serve this purpose. LEC1 and LEC2 establish a multicast send VCC with the BUS and are then added as leaves on the point-to-multipoint multicast forward VCC. The ELAN is now operational and the transmission of real data can commence.
- The first actual data transmitted in this example is a series of broadcast packets that originate from LEC1 (IP ARP broadcast). The broadcast frames are forwarded up the multicast send VCC to the BUS and then back out from the BUS to all LE clients over the multicast forward VCC. A source LEC discards any of its own frames that may have been sent back from the BUS by examining the LECID value in each frame. If the LECID value in the frame matches the LECID value of the LEC itself, the frame is discarded.
- LEC2 returns its MAC address in the IP ARP reply to LEC1. Again the frames flow over the LANE broadcast channel provided by the multicast send and forward VCCs, respectively.
- LEC1 knows the MAC address of the LEC2 but not its corresponding ATM address. To resolve the ATM address of LEC2, LEC1

---

[10]A proxy LEC will not register the MAC addresses on legacy LAN segments that it is bridged to.

transmits an LE_ARP request over the control direct VCC to the LES. The LES responds with an LE_ARP reply containing the MAC-ATM address binding for the target, LEC2. If the LES does not contain a registered MAC/ATM address binding for LEC2, it forwards the LE_ARP request to the other LE clients and proxy LE clients attached to the ELAN. The appropriate LEC should respond to the forwarded LE_ARP request by sending an LE_ARP reply back through the LES, which in turn relays it to the source LEC. The source LEC stores the MAC-ATM address binding in an internal LEC cache.

✔ LEC1 uses normal ATM Forum signaling and routing protocols to establish a data direct VCC to LEC2.

Subsequent data frames that arrive at an LEC will first check the LEC internal cache for a data direct VCC. If one is not available, the LEC will query the LES. While awaiting a response, the LEC may transmit the data frames via the BUS. Once the LES responds to the query, the LEC establishes a data direct VCC. Connections that are idle for some period of time are released and the resources freed up for new connections.

At first glance, the whole process can appear and probably is complex. By stepping back one can see that the additional complexity was engineered into LANE for good reasons. First, LANE would not be a very attractive solution if each individual LEC with multiple ATM addresses and parameters had to be manually configured. To make the process of configuring and initializing an LEC as simple as possible, the LECS was introduced. Second, LAN protocols make very liberal use of broadcasts. A LAN Emulation broadcast mechanism was mandatory and so the BUS was included. And finally, additional control mechanisms and steps were included so that if the LES was not able to directly answer an LE_ARP request, it would propagate it to other LE clients on the ELAN and subsequently relay the response it received to the source LEC. In effect, the LES is not only an address resolution server but also an address resolution broker. Therefore the LES does not have to perform the unscalable and almost impossible task of keeping track of every single MAC address and corresponding ATM address that exists both on the ELAN itself and those legacy LAN segments that it is bridged to.

The LANE specification contains some areas that creative implementers can exploit for better performance and control of network resources. Broadcast management is one such area. In a flat bridged network, a large number of broadcasts can lead to bridge and segment congestion. Other than the time-consuming and error-prone task of installing MAC address filters in the bridges or changing the application and network-layer protocol altogether, there is really no way to limit broadcast propagation. Of course one could migrate to a routed network, but that introduces a whole new set of possible performance bottlenecks and network complexity. In LANE, all broadcasts must flow through the BUS. The BUS could be programmed to examine specific broadcast frames and either forward them out the multicast forward VCC to all LE clients, forward out a multicast send VCC to a specific LE client, or perhaps drop it altogether.

For example, when an IP ARP broadcast arrives at a BUS, it is typically flooded out the multicast forward VCC to all LEC and proxy-LEC components on the ELAN. Every LEC sees the broadcast including workstations on the LAN segment(s) attached to a proxy LEC. An intelligent BUS could scan all IP ARP requests and replies to learn the location of the IP addresses on the ELAN and store those values in a broadcast cache. When a subsequent IP ARP request arrives, and if the destination IP address matches an address in the broadcast cache, the intelligent BUS can direct the IP ARP request to the specific interested LEC over an existing multicast send VCC. The intelligent BUS in this case has dynamically reduced the broadcast overhead for *N* LE clients down to a single LEC.

Another area that implementers can exploit in LANE is ELAN membership policies. Recall that during the initialization phase an LE client establishes a connection to the LECS and exchanges `LE_CONFIGURE` messages. Ostensibly, the LEC is receiving from the LECS ELAN-specific parameters, including the ATM address of the LES. However, it is possible to extend this configuration process to institute membership policies such that an LEC will only be granted permission to join the ELAN based on some policy configured in the LECS. An example of such a policy would be to deny ELAN membership for a certain client that attempted to join during off-hours or whose MAC address falls outside a certain specified address range. It really is up to the implementer to decide how the initial configuration process can be utilized. The point here is that the LECS is a single point of control for administering ELAN membership policies. This is typically not the case with the standard VLAN membership techniques on LAN switches where VLAN membership information must be configured and distributed to all switches in the network.

There are some issues to point out about LANE V1.0. First, the whole LES/BUS/LECS complex is a single point of failure. If the LE services disappears, the ELAN becomes inoperable. There are vendor-specific techniques for providing a standby LES/BUS server and switching over when the primary LES/BUS goes away. A possible drawback to this solution is that the entire network must reinitialize with the backup LES/BUS, and this could introduce significant network and signaling overhead during the frantic postcrash network restart period.

Another often-sited limitation of LANE is the fact that it hides the underlying ATM mechanisms from the LAN. LAN applications cannot exploit some of ATM's more powerful functions such as QoS. Indeed the LANE V1.0 specifies that data direct VCCs *should* use Class 0 (best-effort) for QoS. In all honesty this argument rings hollow. One needs to remember that the principal objective of LANE is to emulate the functions and behaviors of either a Token Ring or Ethernet segment. Natively these LAN technologies offer no QoS mechanisms whatsoever. In fact it has only been quite recently that the notion of QoS (RSVP/IntServ) has been introduced into the IP world—and that effort still requires explicit interactions with the underlying data link layers (e.g., Ethernet and ATM). It really would not have made sense to engineer a QoS mechanism from the "bottom up" when it is really the job of the application from the "top down" to determine and request special services from the network. It should also be pointed out that an LEC can establish and receive a data direct VCC with QoS if it is enabled to do so and that the LECS or LES could be a logical place to store and retrieve QoS parameters for QoS-capable LE clients. The case made here is that LANE should not be held responsible because it does not explicitly define mechanisms for supporting QoS. When LAN applications become capable of requesting QoS from the network, they will need a QoS-capable data link such as ATM to work properly. Until then, one can exploit ATM QoS in a LANE environment, but it does require nondefault configuration of the participating LE clients.

8.1.5 LANE Network Examples

When LANE was being developed, it was envisaged that over time ATM would replace legacy LAN technologies such as Ethernet. Certainly the increased bandwidth and the ability to deliver ATM to the desktop was an alluring possibility. The 25-MB ATM was in particular viewed as a relatively inexpensive way to introduce ATM to the typical desktop. Application servers, edge devices with ATM uplinks, and perhaps routers could attach to the ELAN via higher speeds such as OC3. The ability to attach devices at different speeds to the same ELAN is an often overlooked feature of LANE. Interconnecting ATM switches with one or more OC3 or even OC12 links would provide a flexible and scalable backbone. In addition, specialized niche applications such as MPEG2 over ATM that require native ATM services could be deployed and utilized over the same network infrastructure. This scenario is illustrated in Fig. 8-8.

A funny thing happened to ATM on the way to its christening as *the* next-generation network technology: Ethernet. It turned out that Ethernet could provide comparable bandwidth at lower prices without requiring a suite of new and alien protocols and components. And if there were no applications that needed or could utilize QoS, it did not make sense to install ATM. If bandwidth was the issue, Ethernet was simpler, less expensive, and less risky to implement.

For the most part that is sound reasoning for why ATM is not as successful as Ethernet at the desktop. However, one needs to provide a scalable and flexible backbone for interconnecting all of those Ethernet workgroups together, and surprisingly ATM has quietly emerged as an

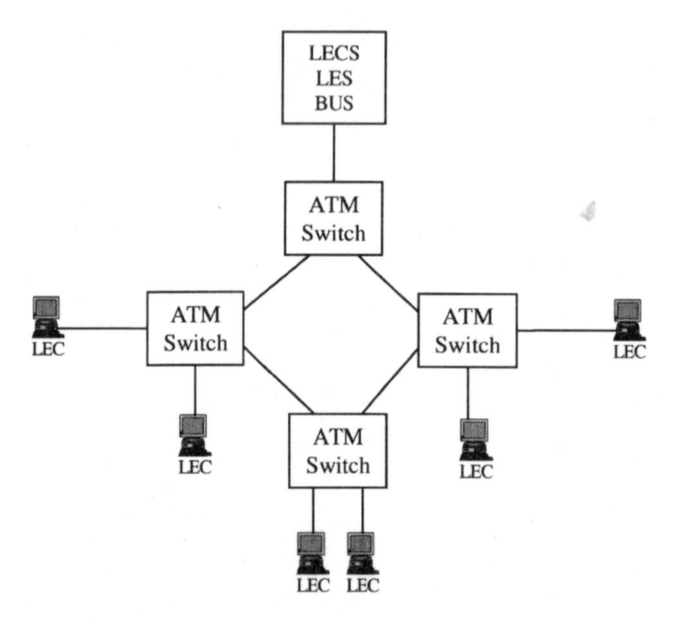

Figure 8-8 Standard LANE network.

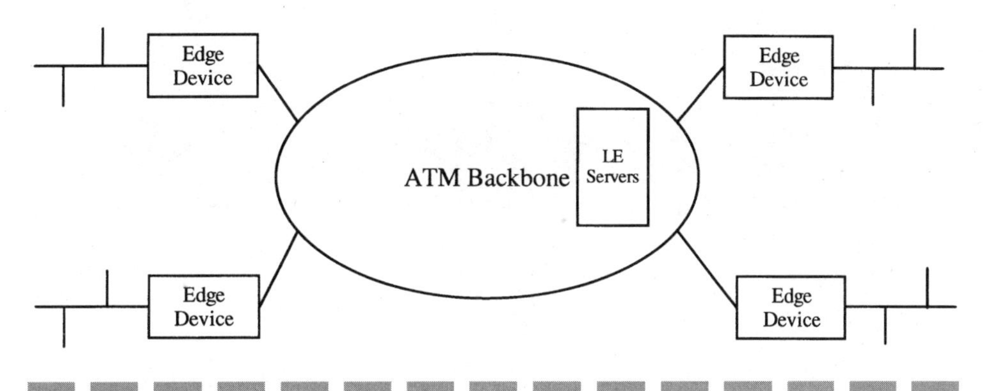

Figure 8-9 Ethernet workgroup backbone using LANE.

excellent backbone solution. As illustrated in Fig. 8-9, Ethernet workgroups can be attached into the ATM backbone via edge devices with ATM uplinks or perhaps by Fast Ethernet uplinks that terminate directly in the ATM switch. Both provide high-performance access into the ATM backbone. The backbone runs LANE and is simply used as a high-speed transport of interworkgroup traffic. If more bandwidth is needed, one can upgrade or add more ATM trunks or deploy more edge devices.

The illustration also shows single ELANs, which like traditional LANs can support one or more IP subnets. To forward traffic between subnets a router is still needed even if multiple ELANs are overlayed onto a single ATM network. For a subnet that comprises multiple ELANs, standard LAN bridging techniques are used to forward traffic between LE clients on different ELANs.

8.1.6 LANE versus Classical IP Comparison

Classical IP over ATM defined in RFC1577 and LANE V1.0 were the initial efforts supporting data networking over an ATM network. Table 8-1 compares the two solutions.

8.2 LANE V2.0

LANE V2.0 is a new version of the LAN Emulation specification developed by the ATM Forum. It provides the same basic services for emulating IEEE 802.3 and 802.5 LANs on top of an ATM network. It offers a

TABLE 8-1 Classical IP vs. LANE Comparison

Attribute	Classical IP (RFC1577)	LAN Emulation V1.0
Protocol supported	IPv4 only	IP, IPX, Appletalk, APPN, etc.
ARS	ATMARP	LES/BUS
Autoconfiguration	No	Yes
Broadcast/multicast	No	Yes
Devices	Hosts, routers	Host, routers, bridges
Extensible	No	Yes
Encapsulation	RFC1483	LANE
Intersubnet communications	Router	Router
Environments	High-performance IP subnets	LANs, LAN backbone

number of new and enhanced features that allow LE clients to make better use of the underlying mechanisms of ATM. The robustness and scalability of the LE service components are significantly improved with the addition of SCSP as a means of synchronizing LES/BUS cache information among multiple LES/BUS pairs. This enables multiple LES/BUS entities to provide address resolution and broadcast services to the members of an ELAN and to eliminate the single point of failure limitation inherent in LANE V1.0.

Because the new functions and enhancements individually address the behaviors and functions of the LE clients and servers, the specification was broken into two separate documents. LUNI 2.0 addresses those functions specific to the behavior and functions of the LE clients.[11] It supports new capabilities such as LLC multiplexing for better VC utilization, support for ABR flow control and QoS, enhanced support for multicast, and support for MPOA. LNNI 2.0 addresses the interserver aspects of LANE 2.0. In addition to SCSP, a new server component called the Special Multicast Server (SMS) is defined.

[11]"LAN Emulation over ATM Version 2—LUNI Specification," ATM Forum, af-lane-0084.000, July 1997.

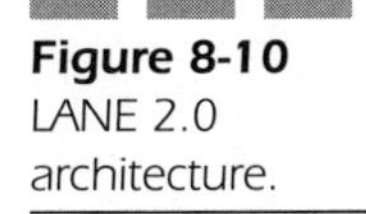

Figure 8-10
LANE 2.0 architecture.

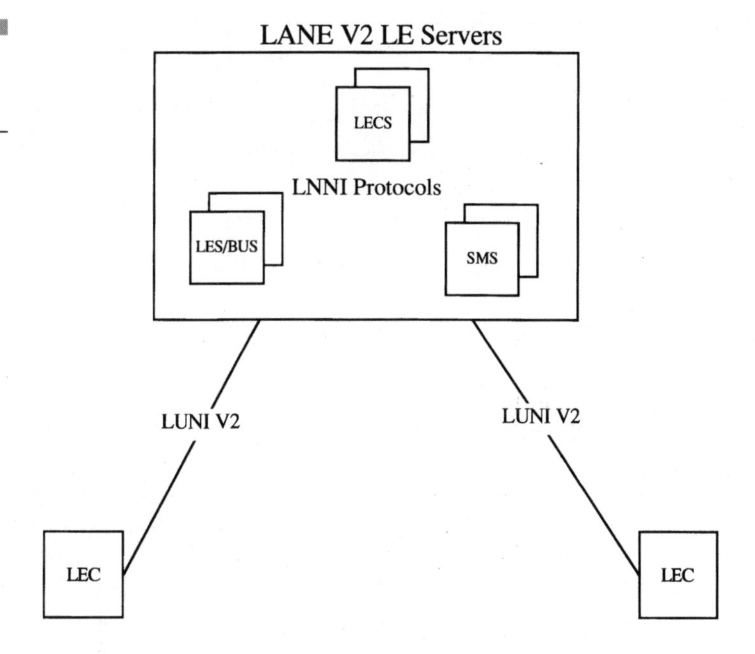

8.2.1 Architecture

The architecture of the LANE 2.0 specification is illustrated in Fig. 8-10. LE clients access the LE server components using the LUNI 2.0 protocols. However, in the case of LANE 2.0 there are now possibly multiple LES/BUS entities that the LE clients can utilize. LE clients may also establish multicast send VCCs to an SMS and become leaves on a multicast forward VCC rooted at an SMS. The scope of the LNNI 2.0 protocols is confined to interserver communications and services.

8.2.2 New Functions

The following new functions and enhancements are defined in the LANE V2.0 specification:

- *LLC multiplexed data direct VCCs.* In LANE V1.0, a data direct VCC was dedicated for a specific traffic flow between two LE clients on the same ELAN. In LANE V2.0, multiple traffic flows from one or more ELANs, including non-LANE traffic flows, can be multiplexed onto a single shared VCC using LLC multiplexed

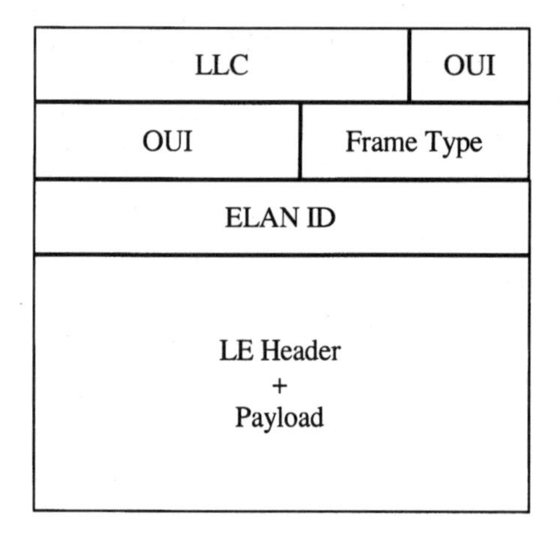

Figure 8-11 LANE V2.0 LLC multiplex LE data frame format.

VCCs. An additional 12-byte header, illustrated in Fig. 8-11, is appended to the normal LE data frame. It consists of a 6-byte LLC/OUI header, a 2-byte frame type (indicating an LLC multiplexed IEEE 802.3 or 802.5 data frame), and a 4-byte ELAN identifier. LLC multiplexed VCCs enables fewer VC resources to be consumed by allowing multiple LE clients to share the same data direct VCC. This could be especially useful for cutting down on the number of VCCs established between multi-LEC edge devices across an ATM backbone.

- *ABR support.* LANE V2.0 enables LE clients to establish ABR connections using UNI 4.0 signaling protocols.
- *QoS support.* LANE 2.0 LE clients can establish data direct VCCs with specified ATM qualities of service. LE clients register which QoS services they are capable of supporting using a special TLV during the registration process. Source LE clients are then informed during the LE-ARP process that a target LE client is capable of supporting QoS.
- *Selective multicast.* In LANE V1.0 all multicast traffic was forwarded through the BUS. LANE V2.0 can off-load multicast processing from the BUS (which also must forward broadcast and unknown frames) by enabling LE clients to establish separate multicast send VCCs to other LE clients or even an SMS. At registration time, an LE client must indicate that it is capable of selective multicast, and it must register which multicast group addresses it would like to receive. This is so the BUS will not send frames

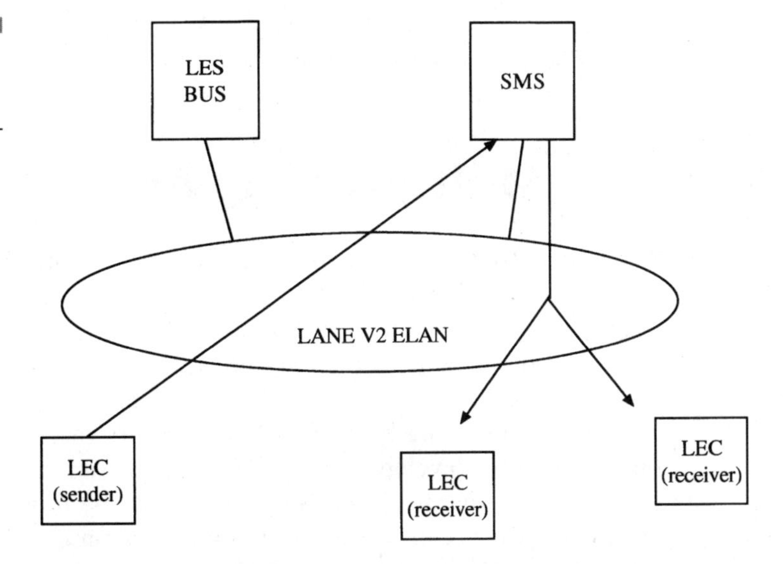

Figure 8-12 LANE V2.0 selective multicast.

addressed to those registered multicast addresses to the LE client. Another reason is that the LES can inform the group-specific SMS that there is an LE client that should be added as a leaf to the SMS multicast forward VCC. LE clients that want to transmit to the group address can then use `LE_ARP` for the individual multicast group addresses so that they may establish multicast send VCCs to the respective SMS. A typical LANE V2.0 multicast scenario is illustrated in Fig. 8-12.

- ✔ *MPOA support.* LANE V2.0 provides a set of TLVs that MPOA relies upon for autoconfiguration, device discovery, and default routing. MPOA clients may retrieve configuration parameters from the LECS. And MPOA-specific TLVs appended to normal LANE registration and ARP control processes can assist MPOA clients and services in discovering each other.[12]
- ✔ *SCSP support.* In LANE V1.0, only one active LES/BUS pair was allowed. LNNI 2.0 incorporates SCSP as a means of enabling mul-

[12]MPOA requires only a minimal version of LANE 2.0 to operate. A minimal version of the LE client supports the set of configuration variables and TLVs identified in the LANE V2.0 specification, associates TLVs in `LE_ARP` responses with entries in its `LE_ARP` cache, and provides a higher-layer interface to the TLVs associated with the `LE_ARP` entries. An MPOA client is not required to support LLC multiplexed VCCs, QoS, selective multicast, etc.

tiple LES/BUS pairs to offer a single LANE service to the clients of an ELAN. This enhances scalability and robustness.

8.3 Multiprotocol over ATM

LAN Emulation supports the operation of an IP-based (other protocols as well) LAN on top of an ATM network. However, a router is still required to cross subnet boundaries. The next logical extension for LANE would enable LE clients on different subnets and presumably ELANs to communicate directly over the ATM network and bypass intermediate routing hops. The ATM Forum proceeded with work along these lines and developed the MPOA specification.[13]

MPOA incorporates and indeed requires the functions of LANE and NHRP to support intra- and intersubnet communications over an ATM network. An MPOA client uses LANE to communicate with other clients on the same subnet and "NHRP-like" protocols to resolve intersubnet addresses and establish shortcut paths to clients on other subnets. Whereas LANE emulates LAN function over ATM with no performance boost other than what ATM bandwidth can provide, MPOA is one of the first attempts to exploit the high bandwidth and low latency of ATM and apply it to traffic flows whose performance is gated by conventional router processing. Indeed the objective of MPOA is to redirect the intersubnet forwarding path away from routers and onto the ATM switch fabric. Perhaps another way of stating the relationship between LANE and MPOA is that LANE provides a layer-2 service on top of ATM, and MPOA provides a layer-3 internetworking service on top of ATM.

8.3.1 Requirements

The primary objective of MPOA is to enable shortcut paths between clients on different subnets. That said, there were a number of requirements and factors that drove the development and introduction of MPOA; they are summarized in the following:

---

[13]"Multi-Protocol over ATM Specification V1.0," ATM Forum, af-mpoa-0087.000, July 1997.

- *High-performance intersubnet communications.* Many campuses and intranets consist of multiple subnets. Most data traffic has to traverse one or more routing hops that introduce latency and delay. An MPOA solution must provide a discernible improvement in performance for traffic crossing subnet boundaries.
- *Leverage existing standards.* The ATM Forum invested a good deal of time and effort in the development of the LANE specification—and it works. It makes sense to leverage this effort along with complimentary work such as the IETF's NHRP to deliver a solution whose mechanisms are well understood. It also shortens the overall work effort and time to market and provides a visible migration path for initial LANE adopters.
- *Simplify the cost and complexity of the routing process.* In a typical multinetwork environment, multiple routers each run a particular routing protocol (e.g., OSPF) to perform path calculation, which can contribute to additional network complexity, operational involvement, and possibly cost. An MPOA solution should attempt to reduce or minimize the instances where path calculation is performed. Doing so might reduce the overall complexity of routing in general and lead to the implementation of lower-cost, higher-speed forwarding devices.
- *Leverage ATM technology.* Besides its well-known technical strengths, ATM also enjoys the backing and continued investment from the networking industry at large. MPOA should certainly attempt to exploit ATM technology where possible. At the same time the MPOA solution must allow components attached to non-ATM LAN infrastructures to benefit from the performance and capacity enhancements enabled by ATM switching.

MPOA provides the mechanisms and protocols needed to support traditional layer-3 internetworking over an ATM network. It takes into consideration the aforementioned requirements and factors in providing routing services by instituting the notion of a virtual router. The MPOA virtual router is illustrated in Fig. 8-13. It consists of a route server or "CPU" that runs traditional routing protocols and performs path calculations just like the central processing unit (CPU) in any old standalone router. The "adapters" of the virtual router are client forwarding components resident in workstations or edge devices that receive packets on the input side and transmit them to the output side based on the appro-

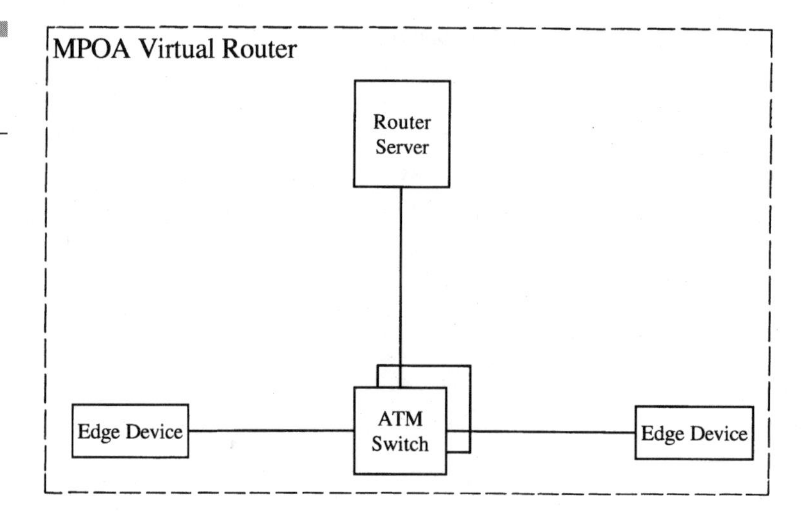

Figure 8-13 MPOA virtual router model.

priate forwarding table entry. To expedite the forwarding process, the client components may cache forwarding table information received from the route server for only those flows where the client is in the forwarding path. And finally the "backplane" or switching fabric of the virtual router is an ATM network.

The benefits of the virtual router model can be viewed from an internal and external system perspective. Internally, the use of ATM switching provides a scalable and high-speed transport between the client forwarding components. Path computations are only performed on the route server and only relevant forwarding information is downloaded and stored in the client forwarders upon request. This has the effect of separating and distributing the route computation and packet forwarding processes over multiple components in the network. It is also quite simple to scale performance and capacity upward by simply adding client forwarding components at the edge or by adding ATM switching capacity in the core. Externally the virtual router looks and behaves like any other router. It runs routing protocols and peers with other routers to exchange routing protocol updates. It can attach to and offer routing services for workstations and devices running on ATM and non-ATM networks. And because it is running just a single instance of a routing protocol, the complexity of the overall routing system is reduced.

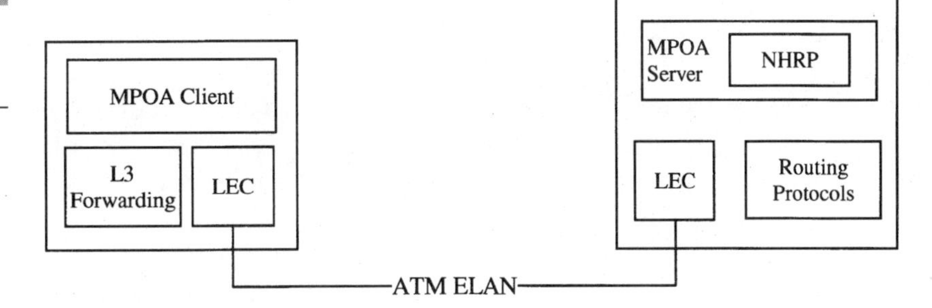

Figure 8-14 MPOA client and server components.

8.3.2 Components

The virtual router is the model or architecture that MPOA is built around. Specifically the MPOA specification defines two important components that are illustrated in Fig. 8-14 and described below:

- *MPOA client (MPC).* The MPC can reside in a workstation, edge device,[14] router, or route server. The primary function of an MPC is to serve as a source and destination for shortcut paths. The MPC contains an LE client so that it may connect to an ELAN and communicate with the MPOA server and MPCs within the same subnet using LANE protocols. It also contains a layer-3 forwarding component, but it does not run a routing protocol. An MPC will query the MPOA server for the information it needs to set up shortcut paths to other MPCs. A further distinction of MPC functions can be made by examining those performed during ingress and egress processing, respectively. An ingress MPC handles packets entering the MPOA system, uses bridging to default forward them to the MPOA server using LANE protocols, and possibly sets up a shortcut path to an egress MPC. The egress MPC terminates shortcut paths and manages a connection-specific cache needed to process frames as they exit the MPOA system.
- *MPOA server (MPS).* The MPS operates in either a route server or router. It runs standard routing protocols (e.g., OSPF), contains one or more LE clients to interface to one or more ELANs, and

[14]An edge device forwards LAN traffic onto an ATM network and vice versa. In LANE the edge device only performed layer-2 bridging. In MPOA the edge device is capable of both layer-2 bridging and layer-3 forwarding.

runs a version of the NHS as defined in NHRP but with additional MPOA extensions. Like its client counterparts, an MPS can operate in ingress and egress modes. The ingress MPS responds to queries from ingress MPCs by downloading forwarding information so that it may set up shortcut paths. The information that is downloaded consists of a network-layer address-to-ATM address binding. The ingress MPS may also serve as a default gateway for ATM-attached MPC hosts or legacy LAN devices on the other side of an MPC edge device. The ingress MPS is also capable of converting an MPOA query into an NHRP format so that it may be forwarded to another MPS. The egress MPS can receive NHRP queries and convert them to an MPOA format. At the same time it can communicate with the egress MPC to impose cache information specific to the pending shortcut connection.

8.3.3 MPOA Control and Data Flows

Like any networking system, MPOA components establish relationships with each other to exchange control information and actual data. MPOA does not specify a formal set of control and data connections as LANE does, but rather it classifies its information flows as either control or data. These are illustrated in Fig. 8-15.

MPCs uses standard LANE control mechanisms to retrieve configuration information from the LECS. A control flow exists between the ingress MPC and MPS for address resolution and between the egress

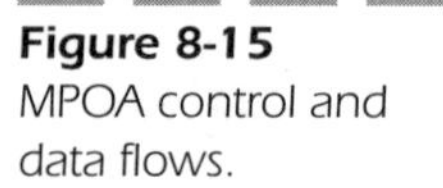

Figure 8-15 MPOA control and data flows.

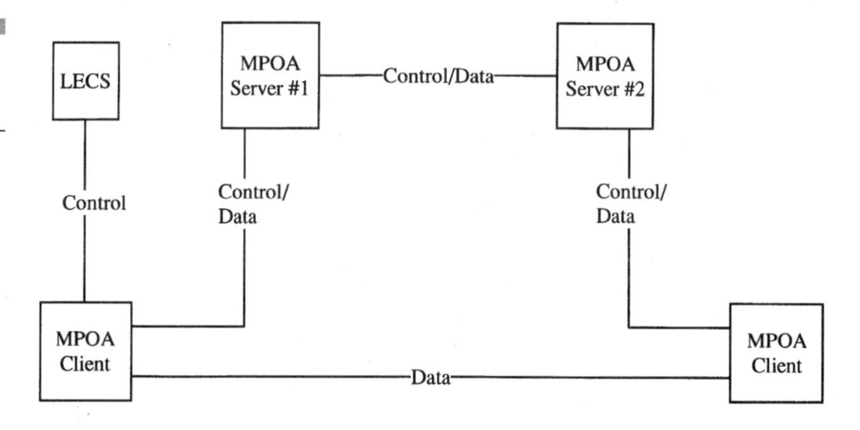

MPS and egress MPC for cache management. There are no specific inter-MPS protocols defined other than standard routing and NHRP.

A default hop-by-hop path is provided by the MPSs for routed traffic and is used by short-lived and/or low-volume data flows or by any data flow prior to the establishment of shortcut path. For data flows that can benefit from ATM switching, a shortcut path can be established between MPCs.

8.3.4 Address Resolution and Data Flow

The address resolution and subsequent shortcut setup processes of MPOA are not unlike mechanisms already discussed. But there are some specific differences relative to MPOA. MPOA utilizes standard LANE configuration mechanisms to retrieve MPOA parameters from the LECS. MPOA also utilizes LANE to discover other MPOA devices in the same subnet. Both are accomplished by appending special MPOA TLV values to the LANE control messages. Once the system is operational, ingress MPCs monitor individual packets entering the MPOA system as it forwards them to the ingress MPS using standard bridging techniques and LANE protocols. The packets then proceed on a hop-by-hop routed path to the destination. If a flow is detected, the ingress MPC forwards an MPOA resolution request addressed to the actual layer-3 destination.[15] The MPOA request eventually arrives at the serving or egress MPS, which instead of responding with an IP/ATM address binding, first exchanges cache management messages with the egress MPC. The cache management message exchange provides the egress MPC and the MPS with the information necessary to complete the establishment of the shortcut path and process packets exiting the MPOA system. The egress MPS then formulates a response and forwards it to the ingress MPC, which can then set up a shortcut path to the egress MPC.

[15]An the context of MPOA a flow is defined as a stream packets originating from the same source and addressed to the same destination that have arrived at the ingress MPC or MPS within a specific time interval. For example if 10 frames with the same layer-3 source and destination addresses, respectively, arrive within a 1-s interval, the ingress MPC considers this to be a flow and initiates the MPOA target resolution process. Prior to flow detection and address resolution, all packets are bridged to the ingress MPS using LANE protocols.

MPOA is a little unique from its close cousins, LANE and NHRP, because it performs flow detection at the ingress to determine when and if a shortcut request should be initiated. Additionally MPOA employs a cache imposition protocol exchange between the egress MPS and egress MPC that is used to update the egress cache information prior to shortcut establishment.

The basic operation of MPOA is outlined below:

1. *Configuration.* MPOA devices retrieve configuration information from the LECS.

2. *Discovery.* MPOA devices discover the presence of each other on the same subnet by the presence of MPOA TLVs in various LANE control messages. For example an MPC must know the MAC and ATM addresses of the MPS so that it can send MPOA requests.

3. *Ingress MPC to ingress MPS processing.* Once a flow is detected that can benefit from a shortcut path, the ingress MPC generates and then forwards an MPOA resolution request to the ingress MPS. The ingress MPS can initiate egress processing if it is the serving MPS for the destination. Otherwise it will convert the request to an NHRP format, replace the source address with its own, and then forward the message along the routed path to the destination. It uses its own address rather then the ingress MPC address as the source so that it can guarantee that it will see the response when it returns.[16]

4. *Egress MPS to egress MPC processing.* When the NHRP request arrives at the serving MPS, it triggers a cache imposition protocol exchange between the egress MPS and egress MPC. The cache imposition protocol has several purposes. First, it verifies that there are sufficient resources at the egress MPC to receive the shortcut path. Second, it is used by the egress MPS to impose a cache state on the MPC. The cache state consists of the ingress MPC ATM address, the destination network layer address, and the data-link layer (DLL) header that the egress MPS would use if it was the actual device that the frame was exiting the MPOA system through. Observe from Fig. 8-16 that the egress MPC (MPOA client #2) is where the packet exits the MPOA system. The appropriate DLL encapsulation for the media that the packet will be placed on is required and must be installed at the egress MPC. The DLL information is discovered by the egress MPS using the

[16]An ingress MPC may not even have an IP address.

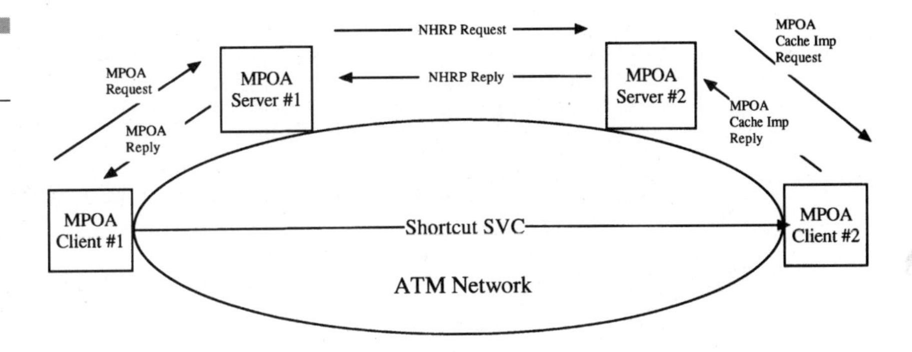

Figure 8-16
MPOA flows.

LANE discovery mechanisms discussed earlier. If the egress MPC has sufficient resources to support a shortcut path, it responds with a cache imposition reply message that contains the ATM address of the egress MPC and creates a cache entry as shown in Fig. 8-17.

The egress MPC may also include a tag in the cache imposition reply. The tag differentiates traffic flows that share the same network-layer destination address and the same source and destination ATM addresses but have a different DLL. A tag may also in some cases speed up cache lookups at the egress. The tag value is included in each frame sent by the ingress MPC. If the tag value is used, an egress cache entry with a tag illustrated in Fig. 8-18 is created at the egress MPC.

After the egress MPS receives the cache imposition reply from the egress MPC, it converts it to an NHRP reply and forwards it over the routed path to the ingress MPS.

5. *Ingress MPS to ingress MPC processing.* When the ingress MPS receives the NHRP reply it converts it to an MPOA resolution reply, and sends it to the ingress MPC. An ingress cache entry as illustrated in Fig.

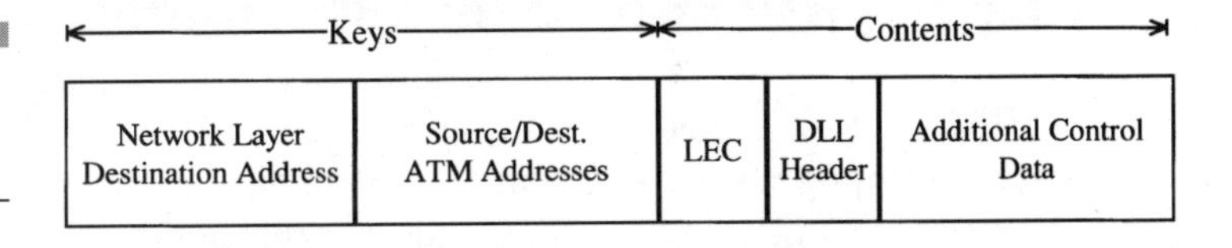

Figure 8-17
Egress MPC cache entry.

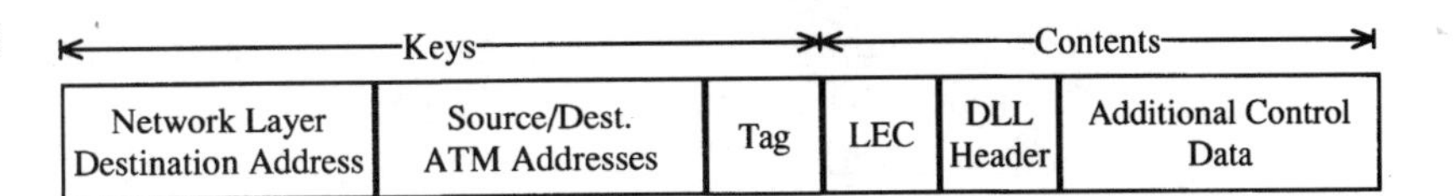

Figure 8-18
Egress MPC cache entry with tag.

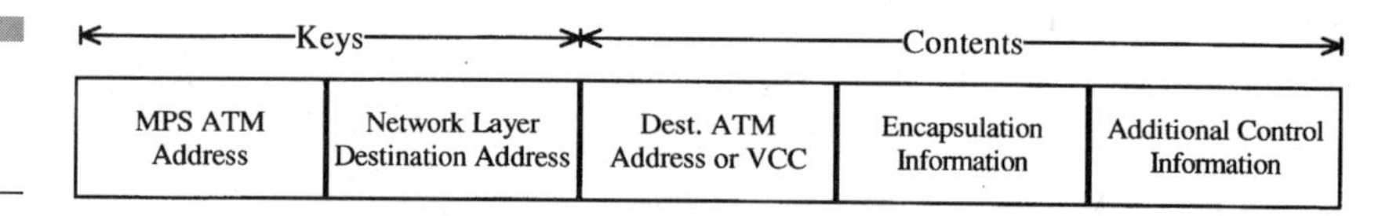

Figure 8-19 Ingress MPC cache entry.

8-19 is created, and the ingress MPC can then establish a shortcut path to the egress MPC.

For a more general picture of the flows required to perform address resolution and shortcut path setup refer back to Fig. 8-16. The endgame is of course the establishment of a unicast shortcut path between MPCs on different subnets. Several additional items should be pointed out about MPOA operations in general. First, all MPOA traffic is encapsulated in LLC/SNAP headers as defined in RFC1483. This enables MPOA flows to share the same VC with other LLC/SNAP encapsulated traffic as well as with LANE V2.0 LLC multiplexed traffic. Second, MPOA provides a vehicle by which an ingress MPC can establish a shortcut path with QoS. A special MPOA service category TLV can be communicated from the egress MPC to the ingress MPC during the address resolution process. This TLV tells the ingress MPC what QoS the egress MPC is capable of supporting.

Another area to mention is flow detection. It does not necessarily have to be the ingress MPC that is responsible for flow detection. Because all routed traffic is initially forwarded through the ingress MPS, it is quite feasible to install a specific flow setup policy in the ingress MPS. When the ingress MPS detects the presence of a certain flow based on the preconfigured policy, it can send an MPOA trigger message to the ingress MPC, instructing it to initiate the MPOA address resolution process.

8.3.5 MPOA Network Examples

MPOA is best suited for campus networks, in particular as the backbone for a campus network. From the technical perspective, MPOA is a flow-based solution that does not scale for the middle of large core networks where there could be potentially hundreds or thousands of individual flows to manage. Another reason for its use in campus networks reflects the reality that most workgroups and end stations sit on shared and switched LAN networks, specifically Ethernet. It would be quite economical and simple to plug switched LAN networks into an MPOA-

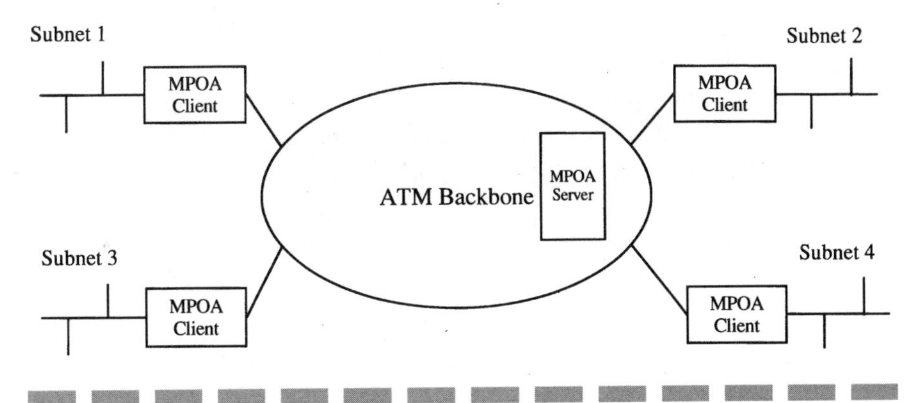

Figure 8-20 MPOA campus backbone.

enabled ATM backbone to provide scalable, high-performance network-layer routing between workgroups in the campus. An MPC edge device is the component that plugs the switched LAN workgroup into the ATM backbone. To scale the performance and capacity of this routing system, one simply needs to add more MPC edge devices or increase the bandwidth and trunks connecting the ATM backbone switches.

An MPOA-enabled ATM campus backbone is illustrated in Fig. 8-20. As far as MPOA operation over a WAN is concerned, it certainly is possible given sufficient signaling and VC capacity. Ingress and egress MPSs can send and receive, respectively, NHRP requests and replies, so it is possible to have two MPSs communicate through a network of NHRP-based routers. As long as the number of flows does not place inordinate demands on the system, MPOA is quite capable of providing both best-effort hop-by-hop routing and high-performance shortcut switching services.

It has been suggested that MPOA could be slightly modified to scale better over a WAN while still providing shortcut routing over an ATM infrastructure.[17] Clearly the flow-based approach of generic MPOA may not scale in a large network. It would operate in a more scalable fashion if the mechanisms were topology based. The fundamental modifications needed to achieve this are

- ✔ Remove the ingress MPC bridging function.

[17]Siemens and Newbridge have suggested this idea and call it Carrier Systems Interconnect (CSI).

- Download the entire forwarding table to the ingress MPC. The topological information needed to build or complete the establishment of a shortcut path is in place prior to the arrival of specific data flows. This lends a topology-driven flavor to this approach.
- Still have the route server run the routing protocol(s) and perform path computations. It must also monitor the status of the forwarding components at the edge.
- Enable the ingress MPC to perform classification of flows, packet filtering, and standard IP forwarding. This would enable nondefault services to be distributed to the edge of the network. For example, a classifier could identify all packets of a particular flow that should be placed on a separate shortcut path with ATM QoS.

MORE INFORMATION

Alexander, C., "MPOA Basics," IBM Technical Report TR292320, Sept. 1997.

http://www.mpoa.com.

"MPOA Multiprotocol over ATM," *ATG's Communications & Networking Technology Guide Series,* 1997, available from *http://www.techguide.com.*

Part 2

IP Switching Protocols

CHAPTER 9
IFMP and GSMP
(Ipsilon)

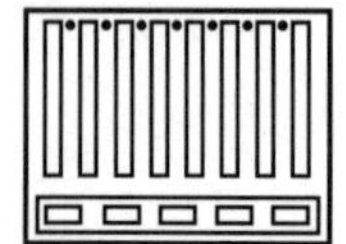

Ipsilon first coined the term *IP switching* in 1996 when they introduced (and defined) a new class of internetworking devices known as IP switches. At the same time they defined a revolutionary new way for improving IP routing performance using ATM technology. It was called IP switching, and they described a technique in which individual flows could be dynamically redirected through forwarding engines comprised of ATM switches. To explain how individual IP flows could be classified and redirected through an ATM switching fabric they first defined a new device called an IP switch. An IP switch is essentially a router attached to an ATM switch that has been stripped of all ATM Forum protocols. They also developed and published, as informational RFCs, two protocols, IFMP and GSMP. The Ipsilon Flow Management Protocol (IFMP) as defined in RFC1953 enables two adjacent IP switches to classify and then relabel the cells of a flow with a new VPI/VCI value. The General Switch Management Protocol (GSMP), outlined in RFC1987 enables the router entity of the IP switch to control the resources and connection table inside the ATM switch. Operating in concert over a network of contiguous IP switches running normal IP routing protocols, IFMP and GSMP provide an additional capability to classify and determine if some flows should be switched rather than routed and then to redirect those flows over a layer-2 switched path consisting of ATM cell switches.

9.1 Architecture

The architecture of the Ipsilon IP switching model is based on several assumptions about the price and performance of routing versus switching, the characterization of Internet (or IP) traffic flows, and the desired behaviors and state maintained in IP forwarding devices. First, the Internet and IP traffic volumes are growing by orders of magnitude. More bandwidth and capacity will be needed to keep up with this growth. This is a fairly safe assumption and one that cannot be argued. Second, traditional routers and shared bus architectures coupled with software-based routing mechanisms are just too slow and cannot keep up with the growth of IP traffic. The introduction of hardware-based gigabit routers, although helpful, may not alleviate the problem because of their high price tag and core network locality. The assumption is that as a high-performance forwarding engine capable of delivering QoS, traditional routers are insufficient. An ATM switch on the other hand is designed to forward cells at gigabit rates with a minimum of effort that

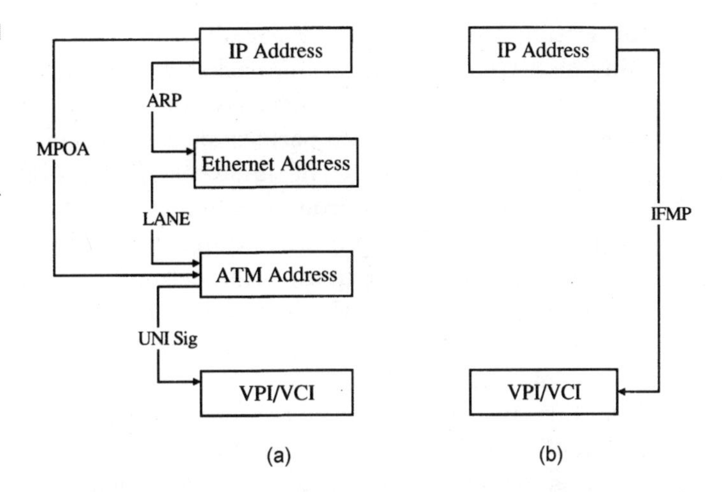

Figure 9-1 Address resolution steps for *(a)* ATM forum versus *(b)* IP switching solutions.

is confined to hardware. As a forwarding engine, ATM switches offer superior capacity and performance at a cost lower than similar function provided in routers.[1]

Another assumption made with respect to the Ipsilon model is that ATM Forum signaling and routing protocols are too heavyweight and complex. To support IP over ATM using the traditional approaches such as Classical IP over ATM, LANE, and MPOA, one needs to support the overlay model that mandates two address spaces, two topologies, and two routing protocols. Factor in ATM signaling and the per-VC state maintained in ATM switches on an end-to-end basis, and it is not hard to see why one would think of ATM as being too complex. A comparison of the address resolution steps required to map an IP flow over a VC between the ATM Forum and IP switching models is illustrated in Fig. 9-1.[2]

The final two assumptions are in line with the philosophy of IP in general. First, all IP traffic can be characterized by a series of flows, where a flow is defined by a stream of packets traveling from a particular source host/application to a particular destination host/application. Of these flows some can be characterized as being long-lived, or in other words, the communicating parties may exchange multiple packets over a specific interval in time. Other flows are short-lived, where the com-

[1]At the time this was a sound argument. But advances in silicon-based routing are rapidly closing the price/performance delta between routing and switching.

[2]As discussed in Chap. 6, the traditional IP over ATM solutions of LANE and MPOA are examples of the overlay model with separate addressing. The Ipsilon IP switching solution is an example of the peer model with IP-VC addressing.

municating hosts exchange a packet or two before moving on to something else. Given the duration, packet volume, and perhaps bandwidth and delay constraints of the long-lived flow, it is logical to assume that there will be advantages in performance and productivity if the network provides a faster pipe. By the same token short-lived flows should and can continue to be forwarded over existing default hop-by-hop paths because there would really be nothing to gain by provisioning a faster pipe. In fact the effort to build the faster pipe just to forward a packet or two is probably more work than to just forward the packet(s) over the default route. Examples of common long-lived and short-lived IP flows are shown in Table 9-1.

And finally, the most basic and perhaps elementary assumption of all is that IP switching should attempt to preserve the connectionless and stateless behavior of IP networking. A normal IP router maintains no state about the individual traffic flows passing through it, and there are no dependencies on the behavior of its upstream or downstream neighbors. A router makes local and independent decisions about where a packet should be forwarded. This same approach is applied to IP switching. An individual IP switch makes a local decision about which flows should be switched. Per-flow connection state in each IP switch is managed locally, not on an end-to-end basis as is the case with traditional ATM. And although maintaining flow-specific state in an IP forwarding device may run counter to the stateless, connectionless philosophy of IP, it nevertheless is "soft," meaning that it will time out unless refreshed. Local control of soft flow state in each IP switch allows IP switching to closely approximate the dynamics and robustness of pure IP routing while achieving some of the bandwidth and performance benefits of connection-oriented cell switching. Indeed, an IP switch can be viewed as an ATM switch that has been mutated into an IP router.

TABLE 9-1

Typical IP Flows

Long-Lived	Short-Lived
File transfer	DNS
NFS	NTP
HTTP 1.1 persistent Connections	email
Telnet, TN3270	ICMP
Multimedia downloads	SNMP

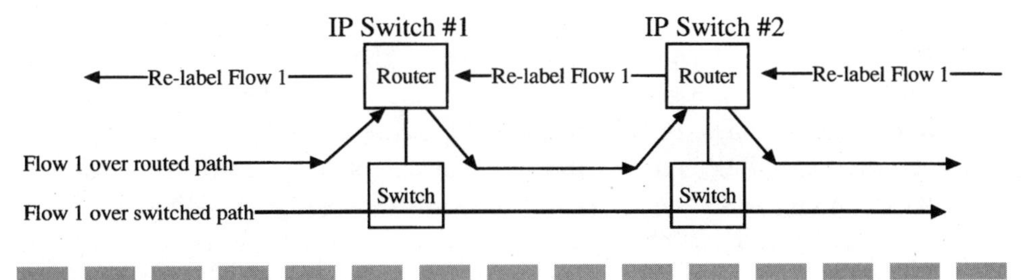

Figure 9-2 IP switching functional model.

The basic architecture of the Ipsilon IP switching model is best explained by examining the functions of two adjacent IP switches as depicted in Fig. 9-2. Observe that the cells of flow 1 are routed hop-by-hop through two IP switches, IP switch 1 and IP switch 2. The default channel established between IP switches for a flow that is being forwarded hop by hop is VPI/VCI = 0/15.[3] The layer-3 processing performed at each hop involves the standard routing table lookup, TTL decrement, and header checksum calculation. This is no different from the layer-3 processing performed on any standard router. Present within the router engine of each IP switch is a flow classifier, which determines if future packets belonging to the flow should be forwarded through the faster ATM switch rather than the slower router component. This determination is purely a local decision made by each IP switch that is based on a configured policy. After a decision is reached to switch the remaining packets of the flow, an IP switch will send a message to its upstream neighbor, instructing it to relabel the cells of the specific flow with a new VPI/VCI value. Presumably each IP switch along the path that flow 1 is traveling will arrive at a similar conclusion and instruct its upstream neighbor to relabel the cells of the flow. Once a flow has been relabeled with a nondefault VPI/VCI on the input and output ports, it is a simple job for the router component of the IP switch to "splice" the incoming and outgoing VCs together to form a shortcut path through the device. Put another way, the collective result of the local flow classification and relabeling procedures performed by each IP switch is an ingress-to-egress shortcut path that bypasses intermediate router hops.

[3]The default channel of VPI/VCI = 0/15 is established when adjacent IP switches are initialized. All default traffic and control messages are forwarded over this VC. This VC exists between every pair of IP switches.

Another way of looking at the IP switching architecture is to examine the origin of the basic building blocks and then see what actually makes up an IP switch. First, an IP switch runs normal IP routing protocols and performs basic IP forwarding. This basic "IP software" component is common to all IP routers and any host capable of forwarding packets between subnets. The ATM switch is just a basic run-of-the-mill hardware-based cell switching fabric that has traditionally been controlled by the protocols and procedures defined in ATM Forum specifications. However, in this case the ATM Forum software has been removed. The third component, independently developed by Ipsilon, is a set of flow classification, relabeling, and switch control procedures. Thus an IP switch and more generally the IP switching model consists of IP software integrated into an ATM switch with an additional component (IFMP and GSMP) to dynamically map flows to VCs and control the switch resources. The basic building blocks of an IP switch are illustrated in Fig. 9-3.

9.2 Components

The basic functional and protocol components of the Ipsilon IP switching solution consist of the following:

- *Flow.* A flow is a sequence of packets from a particular source to a particular destination (unicast or multicast) that are associated with a particular path and the service (default or nondefault) that

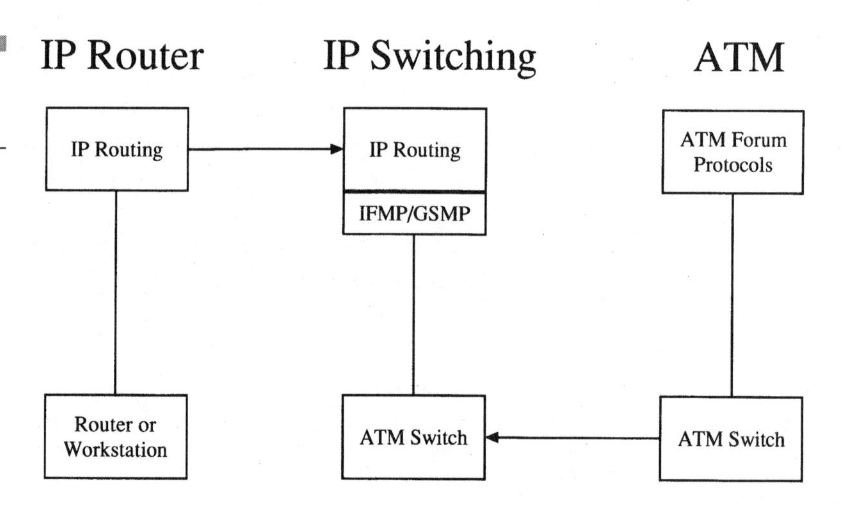

Figure 9-3 IP switching building blocks.

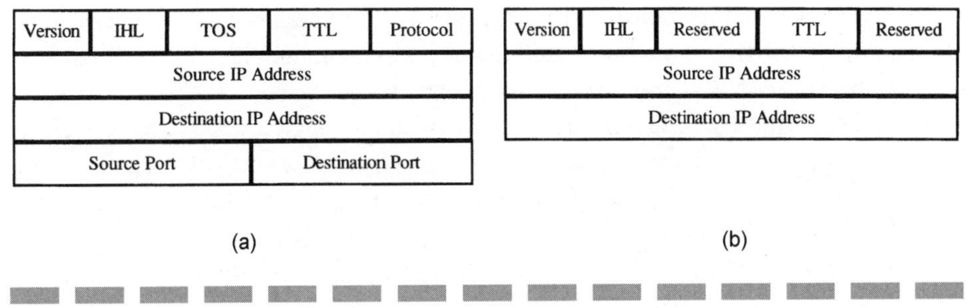

Figure 9-4 Flow identifiers. (*a*) Type 1. (*b*) Type 2.

these packets will receive by each forwarding device over that path. The packets of a flow typically share the same source address, destination address, and perhaps TCP/UDP port numbers.

- *Flow type.* Ipsilon IP switching processes flows at the host-to-host or application-to-application level.[4] A Flow Type 1 is defined as the packets flowing between two processes or applications. The fields in the IP packet necessary to distinguish a Flow Type 1 consist of the source and destination IP addresses along with the respective source and destination port numbers. A Flow Type 2 is defined as the packets flowing between two hosts, and the source and destination IP address fields in the packet header are sufficient to identify this particular flow type. Note that a pair of hosts can support a single Flow Type 2 and multiple Flow Type 1s.
- *Flow identifier.* A set of IP and transport layer fields that identify a particular flow type. The flow identifiers used for Flow Type 1 and Flow Type 2 are shown in Fig. 9-4.
- *Flow classifier.* The component resident in the routing entity of each IP switch that determines if a particular flow should be routed or switched. Flow classification is a local policy decision that is made by each IP switch, although the desire for optimal network performance should dictate consistent policies across the network. A flow classifier works by inspecting certain fields within the header of each packet and then consulting a local policy database. For example, the flow classifier may extract a certain source/desti-

[4]Ipsilon IP switching implements a flow-driven approach to IP switching. It is possible that with the addition of another flow type identifying address prefixes it could support topology-driven IP switching as well.

nation IP address pair and then learn from the local policy database that the packets of this flow should be switched. Or the flow classifier may trigger the redirection process based on the number of packets arriving within a certain time interval.

- *IP switch.* An IP switch formally consists of an IP switch controller attached to an ATM switch.[5] The IP switch controller runs normal IP routing protocols and performs standard IP forwarding. It also contains a flow classifier and exchanges IFMP protocol messages with its upstream and downstream neighbors. The IP switch controller contains a GSMP master control point that communicates with a slave GSMP agent in the ATM switch. A basic IP switch is illustrated in Fig. 9-5.
- *IP switch ingress/egress.* An IP switching system must have an ingress and egress to allow traffic to enter and exit the network. An Ipsilon IP switch ingress and egress is defined as the endpoints of a flow that has been mapped to a layer-2 ATM VC. An ingress/egress function can operate on a router, edge device, or an IP switch and is capable of transmitting and receiving IFMP messages.

Figure 9-5
IP switch.

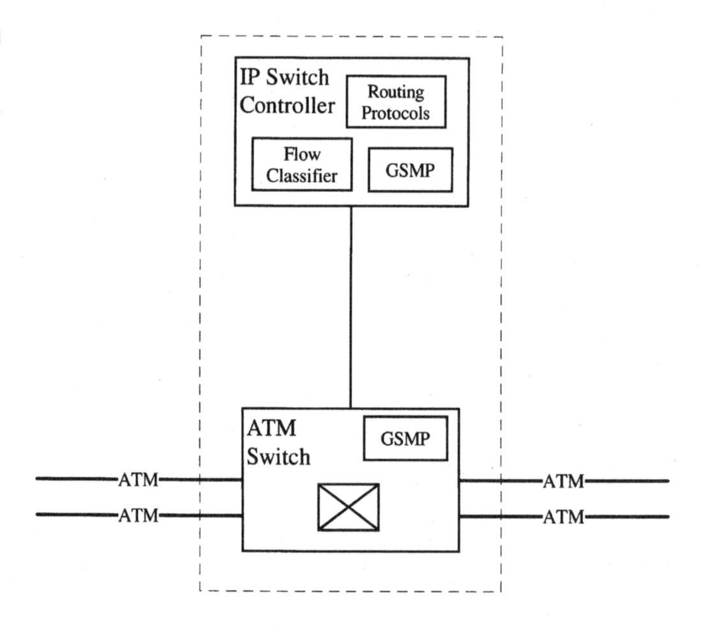

[5] *IP switch controller* is the Ipsilon name for the router or router component that is attached to an ATM switch to form an IP switch.

- ✓ *IFMP.* IFMP is used between adjacent IP switches to relabel the cells of a particular flow. IFMP protocol messages flow from an IP switch to its upstream neighbor. IFMP functions include an adjacency protocol to discover a peer IP switch on a link and a redirection protocol to manage the labels (VPI/VCI) assigned to specific flow.
- ✓ *GSMP.* GSMP is used by the IP switch controller to manage the resources of the ATM switch. It enables the IP switch controller to establish and release connections across the ATM switch.

9.3 Ipsilon Flow Management Protocol

Ipsilon Flow Management Protocol (IFMP) serves as the flow-label distribution protocol in a network of IP switches. It operates between adjacent IP switch controllers (or IP switch ingress/egresses) across external ATM data links. Specifically it is used by a downstream IP switch (receiving end) to inform the upstream switch (transmitting end) that it should associate a particular flow with a specific VPI/VCI value for a specified period of time. The VPI/VCI value is chosen by the downstream IP switch. The flow is identified by the contents of a flow identifier. The downstream IP switch must refresh the flow state within a certain period of time or it will be discarded.

IFMP only operates between adjacent IP switches and is not dependent on the behavior or functions of other IP switches. In other words the receipt of an IFMP redirect message from a downstream IP switch does not trigger the local IP switch to send a similar message to its upstream neighbor. The primary objective of IFMP is to distribute new VPI/VCI labels associated with specific flows so that the forwarding function can be accelerated and possibly switched, thus contributing to improved overall throughput on a per-flow basis.

9.3.1 IFMP Redirection Protocol

IFMP defines two subprotocols as part of the exchange of messages between adjacent IP switches. The first subprotocol, the IFMP adjacency protocol, is invoked when two IP switches initiate communications over an external ATM data link. It enables an IFMP-capable device (IP switch, host, router, etc.) to discover the identity of a peer device at the other end of the link. It is also used to synchronize state across the link, detect peer identity changes, and exchange a list of IP addresses assigned to the link. The adjacency protocol must complete state synchronization (ESTAB state) across the link before subsequent IFMP messages can be processed by the receiving node.[6]

Distribution and management of flow/label binding information are handled by the IFMP redirection protocol. A downstream (receiving) IP switch is responsible for initiating the redirection process, which consists of allocating a new label (VPI/VCI) on the receiving port and sending that information along with a flow identifier and a cache lifetime value in an IFMP message upstream to the sending IP switch. IFMP does not require an explicit acknowledgment. The receipt of subsequent data with either default (VPI/VCI = 0/15) or nondefault labels will tell the receiving node if the message was processed successfully.

The IFMP redirection protocol is receiver-initiated and was designed that way for several reasons. First, it is not good network protocol practice to transmit information in a nondefault manner unless the sender is reasonably (if not 100 percent) sure that the receiver can receive and process it accordingly. Unless the receiving IP switch allocates the exact nondefault label (VPI/VCI) on the input port that the sending IP switch decides to use on its outbound port (extremely low probability), cells will be thrown away and a period of error recovery and label synchronization will follow. The only way the sending IP switch can know for sure that a particular label has been allocated by the receiving IP switch is by receiving an explicit message containing the label to use. Therefore it is a single IFMP message sent by the receiving IP switch that informs the sending IP switch that a nondefault label has been allocated. Second, nondefault communications using cell-based label swapping cannot succeed unless a label has been allocated on the receiving port. IFMP allocates this resource first when initiating the redirection process.

An IFMP message sender encapsulates IFMP information in an IPv4 header with the destination IP address of the peer IP switch at the other end of the external link. The IP address of the peer IP switch is obtained from the adjacency protocol. The five message types supported by the IFMP Redirection Protocol consist of the following:

---

[6]The state tables for the IFMP adjacency protocol are located at the beginning of RFC1953.

- *Redirect.* The redirect message instructs an adjacent upstream IP switch to attach a specific label to a specific flow for a specified period of time. The format of a redirect message element (contained inside of an IP and IFMP header) is illustrated in Fig. 9-6 and explained below:
  - *Flow type.* Specifies the flow type (1 or 2) of the flow identifier contained in the flow identifier field.
  - *Flow ID length.* Length of flow identifier field in number of 32-bit words.
  - *Lifetime.* Length of time in seconds that the flow identifier and label association is maintained within the IP switch that received the redirect message.
  - *Label.* Contains a 32-bit label value. For an ATM data link, this value will contain a VPI and VCI value as illustrated in Fig. 9-7.
  - *Flow identifier.* Identifies the flow that will be associated with the label.
- *Reclaim.* The reclaim message instructs an adjacent upstream IP switch to unbind one or more flows from their associated labels and to release the labels for allocation to future flow identifiers. This may become necessary if there is a change in network topology or if the downstream IP switch becomes label constrained. The unbound flow will be returned to a default forwarding state, and the labels will be released for future use. Each reclaim message element contains a flow type, flow id length, label, and flow identifier.

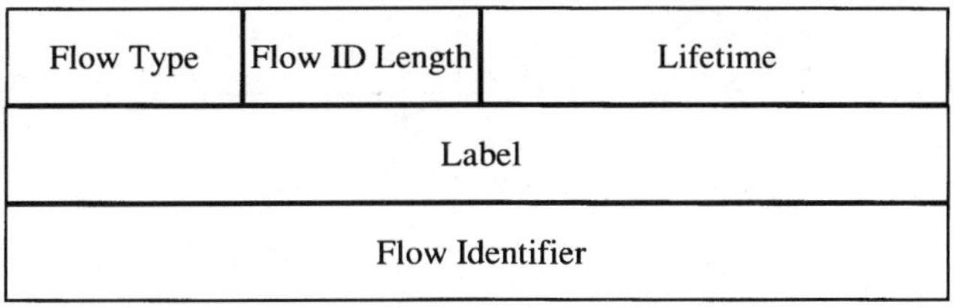

Figure 9-6 IFMP redirect message element.

Flow Type	Flow ID Length	Lifetime
Label		
Flow Identifier		

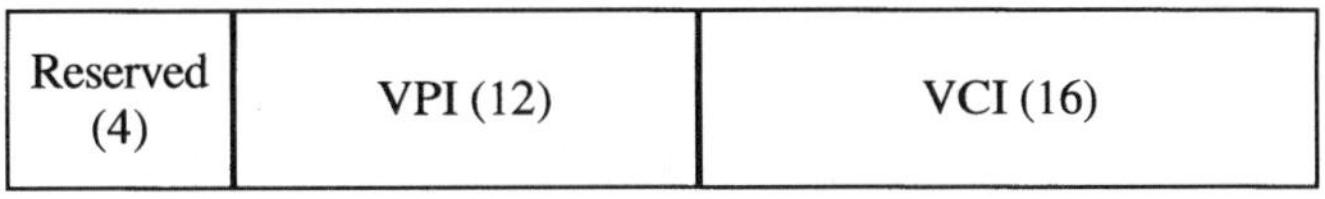

Figure 9-7 IFMP labels for ATM data links.

Reserved (4)	VPI (12)	VCI (16)

- *Reclaim ACK.* The reclaim ACK message prompts an adjacent IP switch to acknowledge the successful release of one or more labels per an earlier RECLAIM message.
- *Label range.* The LABEL RANGE message is sent by an upstream IP switch to a downstream IP switch in response to a REDIRECT message that contained a label that the upstream (sending) IP switch cannot allocate. The LABEL RANGE message contains a minimum and maximum label value. A label that falls within that range is supported by the upstream IP switch.
- *Error.* The ERROR message is sent in response to an IFMP REDIRECT message and informs the downstream IP switch that the sender cannot successfully process the REDIRECT message because of an error condition. An example of an error condition is an unknown flow type.

9.3.2 Flow Label Encapsulations

IP switching supports several different encapsulations for transporting IP packets over ATM data links as specified in RFC1954. The encapsulation used depends on the flow type information conveyed by the downstream IP switch in an IFMP redirect message. The default encapsulation for IP packets traveling over the default channel is shown in Fig. 9-8. It is based on the LLC/SNAP encapsulation technique defined in RFC1483.

IP packets that have been relabeled as a result of the IFMP redirection process are encapsulated based on their respective flow type. Flow

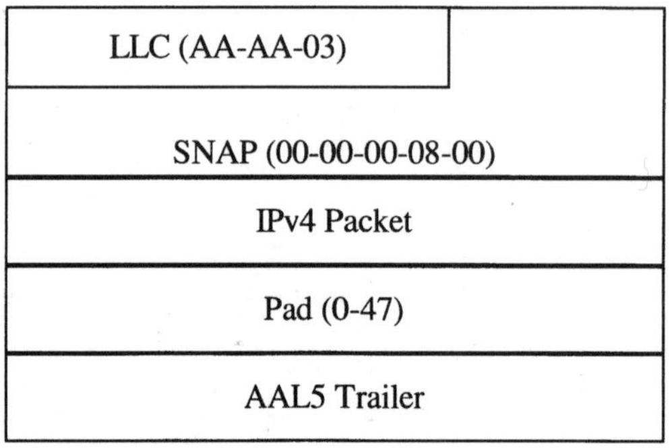

Figure 9-8 Default encapsulation.

LLC (AA-AA-03)
SNAP (00-00-00-08-00)
IPv4 Packet
Pad (0-47)
AAL5 Trailer

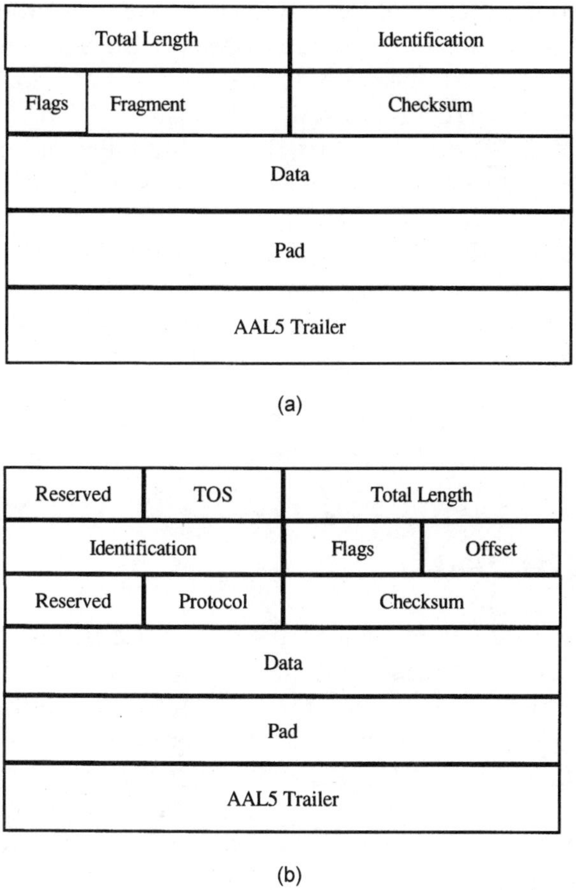

Figure 9-9 Flow type *(a)* 1 and *(b)* Flow type 2 encapsulations.

Total Length		Identification
Flags	Fragment	Checksum
Data		
Pad		
AAL5 Trailer		

(a)

Reserved	TOS	Total Length	
Identification		Flags	Offset
Reserved	Protocol	Checksum	
Data			
Pad			
AAL5 Trailer			

(b)

Type 1 and Flow Type 2 encapsulations are illustrated in Fig. 9-9. In both cases the payload is placed inside a standard AAL5 PDU using the VC multiplexing or null encapsulation as defined in RFC1483. No LLC/SNAP header precedes the payload. In Flow Type 1 encapsulations, the following fields in the IP header are not transmitted: version, internet header length (IHL), type of service (TOS), time to live (TTL), protocol, source IP address, and destination IP address. In addition, the four octets following the IP header that correspond to the source and destination TCP or UDP port values are not transmitted either. In the Flow Type 2 encapsulations, the following fields in the IP header are not transmitted: version, IHL, TTL, source IP address, and destination IP address. In both cases, the IP fields that are not transmitted are stored in the downstream IP switch that issued the redirect message and associated it with the nondefault label. This enables the IP switch that initiated the redirect process to reconstruct the entire packet if necessary. It also provides a measure of security in that it would be difficult for a potential intruder to establish a switched flow connection using one flow-to-label mapping and then having done so, attempt to switch to another flow (different source/destination IP address pair) associated with the same label. The objective on the part of the intruder would be to use the IP switching process to escape detection at layer 3 and perhaps gain unauthorized access to computing resources or data. But the downstream or terminating IP switch is able to reconstruct the original packet from the fields that were cached when the original IFMP message was sent upstream.

9.4 General Switch Management Protocol

General Switch Management Protocol (GSMP) is a simple, general-purpose switch management and control protocol that is designed to manage switch resources in an IP switch.[7] GSMP is the second of two solution-specific control protocols that are part of the Ipsilon IP switching solution. Whereas IFMP operates between adjacent IP switch controllers over an external ATM data link, GSMP operates internally inside of an IP switch between the IP switch controller and the ATM switch. GSMP enables an IP switch controller to establish and release connections across a switch, add and delete leaves on a point-to-multipoint connection, manage switch ports, and request configuration information. ATM switches may also provide unsolicited event information to the IP switch controller using GSMP.

The components that exchange GSMP protocol messages to manage switch resources are the IP switch controller and the ATM switch of an IP switch. There is a master-slave relationship between the two with the IP switch controller acting as the master and the slave being the GSMP agent in the ATM switch. GSMP protocol messages flow over a standard control VC (VPI/VCI = 0/15) established between the IP switch controller and the ATM switch. LLC/SNAP encapsulation as defined in

[7]It is so simple that it has been implemented on a number of different ATM switches and used by other IP switching solutions to manage ATM switch resources.

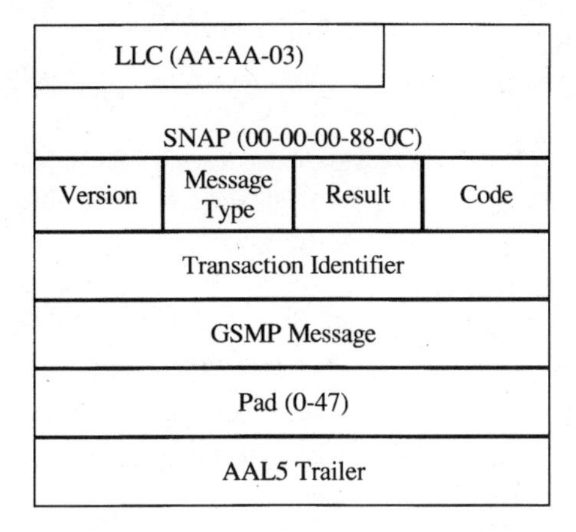

Figure 9-10 GSMP message format.

RFC1483 is used to encapsulate the header and body of a GSMP message as illustrated in Fig. 9-10. LLC/SNAP encapsulation was chosen so that other protocols, perhaps SNMP, could be multiplexed over the same VC. Like its sister protocol IFMP, GSMP defines an adjacency protocol that is used to synchronize state across the link and identify the two communicating parties at each end of the link.[8]

The message type field indicates the type of GSMP message that is contained in the payload. The result field is used to specify that a response is required to a particular request if the request is successful. Given that GSMP is a request/response protocol, a transaction identifier is contained in the GSMP header to associate a particular request with the corresponding response.

9.4.1 Message Types

GSMP supports five message types: connection management, port management, statistics, configuration, and events. The function and purpose of each of these message types is explained in the following:

- ✓ *Connection management.* Connection management messages are used by the IP switch controller to establish, delete, modify, and verify connections that traverse the ATM switch component.[9] These

[8]The GSMP adjacency subprotocol and the state tables are defined at the end of RFC1987.

[9]In the context of GSMP, a connection is an association between an input port, VPI and VCI and output port, VPI, and VCI within the same ATM switch.

TABLE 9-2

Connection Management Message Types

Connection Management Message Type	Description
Add Branch	Establish a connection or add additional branch to a point-to-multipoint connection
Delete Branch	Delete a branch of a point-to-multipoint connection or in the case of single/last branch, delete the connection
Delete Tree	Delete entire connection
Verify Tree	Verify number of branches on a connection
Delete All	Delete all connections on the switch's input port
Move Branch	Move a single branch from its current output port/VPI/VCI association to a new output port/VPI/VCI

are the most common form of GSMP messages because connection state is updated in the switch based on the arrival of IP flows and the subsequent triggering of IFMP messages. Table 9-2 describes the six connection management message types.

Figure 9-11 attempts to better illustrate the role of GSMP in building a connection for a specific IP flow. After sending an IFMP redirect message upstream and receiving the cells of the flow over the allocated nondefault VC (VCI = 27), the IP switch controller

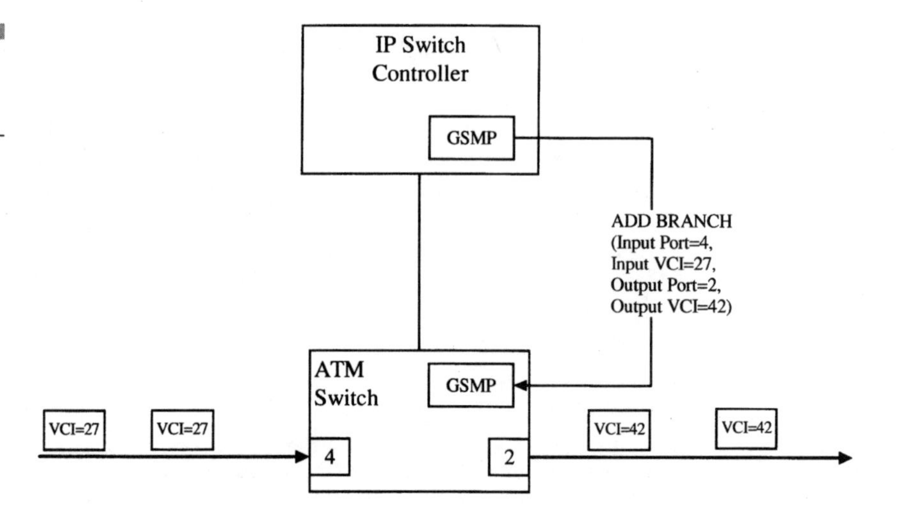

Figure 9-11 GSMP Add Branch example.

checks to see if the flow has been relabeled in the downstream direction. In this example a nondefault VC (VCI = 42) for the flow in the outbound or downstream direction has already been established. So the IP switch controller then issues a GSMP Add Branch command to the ATM switch to build an "intraswitch" connection that associates input port 4 and input VCI = 27 with output 2 and output VCI = 42. The result is that cells belonging to the flow will be cell switched through the ATM switch component and will bypass the IP switch controller.

One interesting point to raise about GSMP connection management is that there is no distinction between managing unicast and multicast connections. The same Add Branch command is used to establish a unicast connection or add a new branch to an existing point-to-multipoint connection. In addition, most connection management messages (exception being the Move Branch command) take up just one ATM cell, so overhead is minimal and processing is fast.

- *Port management.* Port management messages are used to activate (bring up) a switch port, take it down (take down), perform loopback tests, and reset it. When a switch port is activated, all connections that arrive at the specified input port are deleted, and the VPI/VCI label space controlled by GSMP on the port is only available for IP switching connections. A new port session number is generated. The port session number is a 32-bit random number that is generated by the switch each time a port is activated. All GSMP requests made against the port must include a matching port session number or the request will be ignored. The port session number enables the IP switch controller to synchronize with the current active state of the port and protect the port from processing outdated commands. After the switch port is brought up, the port status is active.
- *Statistics.* The statistics messages enable the IP switch controller to request and receive various statistical counters associated with the ports and connections. Information available to IP switch controller includes:
  - Input and output cell counts
  - Input and output frame counts
  - Input and output cell discard counts
  - Input and output frame discard counts

- Input HEC count
- Input invalid VPI/VCI count

- *Configuration.* The configuration messages enable the IP switch controller to discover the capabilities and functions of the ATM switch component. In addition to switch name, switch type, and firmware version number, configuration messages enable the IP switch controller to discover information specific to a port such as VPI/VCI label ranges supported, the maximum cell rate, status, interface type, and whether the port is capable of priorities.[10]
- *Event messages.* Event messages allow the ATM switch to inform the IP switch controller of certain events such as port up and port down.

9.4.2 GSMP v2

GSMP v1.1 defined in RFC1987 covered the basic aspects of management of a generic ATM switch. It enables an IP switch controller to quickly and easily manage connection and port state and to receive some information about port, switch, and connection status. It did not address nor was there support specified for several areas that are common to ATM technology and switches such as

- Virtual path switching
- Usage Parameter Control (UPC) traffic policing parameters
- Per-VC or per-class scheduling
- Enable/disable VC functionality such as Early Packet Discard

GSMP v2 is a new version of GSMP defined in RFC2297. Besides several enhanced message types, the most important enhancement is support for QoS. The new QoS messages allow an IP switch controller to group virtual path connections and virtual channel connections into QoS classes and to allocate resources on a per-class or per-VC basis. Rather than represent itself as a simple cross-connect switch fabric, an ATM switch can offer QoS functions as depicted in the abstract sense in Fig. 9-12.

[10]GSMP v1.1 as defined in RFC1987 only assumes simple strict priority output queues of which multiple priority queues may be defined for a single port.

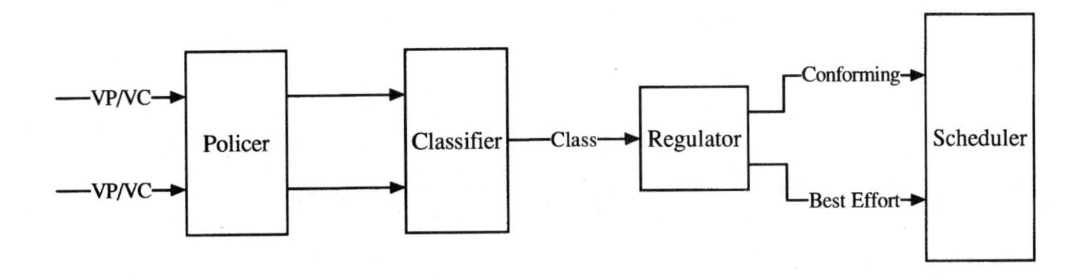

Figure 9-12 GSMP-QoS abstract switch model.

It consists of four primary components: classifier, policer, regulator, and scheduler. The classifier groups multiple connections (VP or VC) into a single class so that QoS can be applied to the aggregate group. The policer enforces a traffic contract on individual classes or connections by possibly tagging or discarding cells. The regulator can perform either a policing or shaping function. The policing function of the regulator evaluates cells to determine if they conform to the rate specified by the regulator parameters. Cells that do not conform may either be tagged, discarded, or subjected to differentiated scheduling in which the comforming cells receive QoS and the excess cells are treated as best effort. The shaping function of the regulator paces cells out, on a per-class or per-connection basis, at the rate specified by the regulator parameters. The scheduler is located on the output port and schedules cells for transmission based on the QoS requirements of the class or connection.

In addition to QoS support, the enhancements specified in GSMP v2 include

- ✔ QoS messages
- ✔ Virtual path switching
- ✔ Message to extract all connection state from a switch port
- ✔ Message to reconfigure VPI and VCI ranges on a switch port
- ✔ Message to delete a list of connections
- ✔ More specific error message definitions
- ✔ Window flow control to avoid overflow to receive buffer
- ✔ Definition of loss of adjacency
- ✔ Enhancement of adjacency protocol
- ✔ Specification of set of mandatory messages

✓ Extension to multiaccess link, Ethernet encapsulation
✓ Version negotiation

9.5 IP Switching Example

Building on the description and explanation of the IP switching components and protocols given so far, it might be useful to look at an example of how an IP switch maps a flow to an ATM connection from the perspective of a single IP switch.

✓ *Default forwarding.* Figure 9-13*a* depicts a single IP switch with upstream and downstream nodes. The default channel of VPI/VCI = 0/15 is used to forward all traffic between adjacent IP switches. When cells arrive at the IP switch controller, they are assembled into a packet, processed at layer 3, segmented back into cells, and transmitted over the outbound default channel to the downstream node.

✓ *Flow classification.* In Fig. 9-13*b*, the IP switch controller has detected an IP flow that is suitable for switching based on a configured policy. The IP switch controller transmits an IFMP redirect

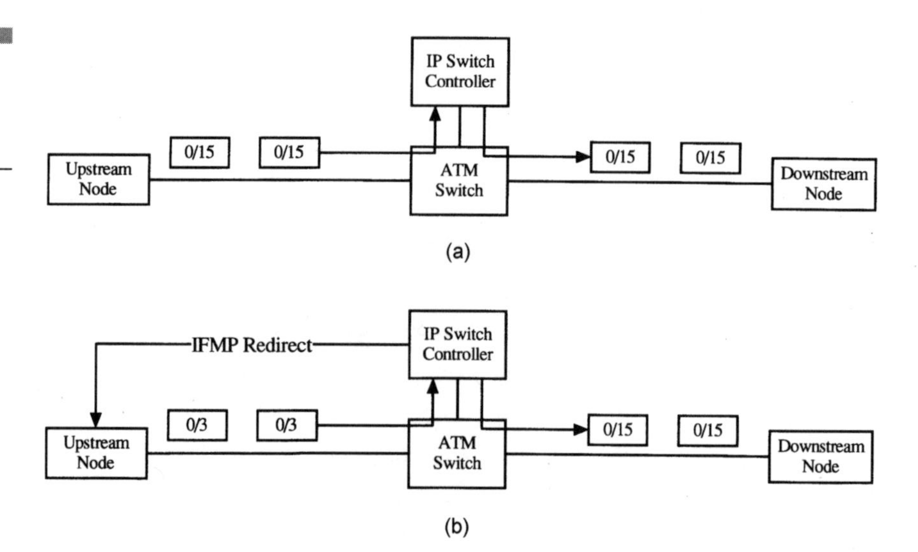

Figure 9-13 (*a*) Default forwarding and (*b*) flow classiication.

message to the upstream node that contains among other things a flow identifier and a new label of VPI/VCI = 0/3. The upstream node receives the redirect message and begins transmitting the cells of the flow on VPI/VCI = 0/3.

- *Downstream node flow classification.* In Fig. 9-14*a*, the downstream node also decides that the flow can be switched. It transmits an IFMP redirect message that includes a new label of VPI/VCI = 0/7 to the IP switch controller. The IP switch in turn begins transmitting the cells of the flow on the new VPI/VCI value of 0/7. At this point, the flow is still being forwarded by the IP switch controller at layer 3, but the cells of the flow have been relabeled on both the input and output ports of the IP switch.
- *VC concatenation.* Once the IP switch controller detects that the flow has been relabeled on both the input port and the output port, it issues a GSMP ADD BRANCH command to concatenate the two together. The connection is now complete and the cells of the flow are forwarded at ATM switch fabric speeds through the IP switch.

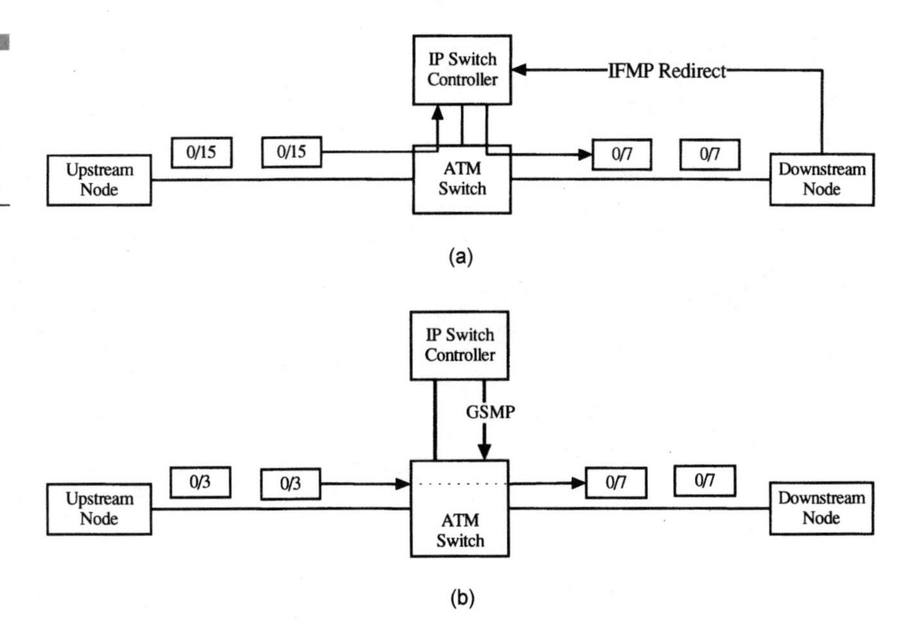

Figure 9-14 (*a*) Downstream node flow classification and (*b*) VC concatenation.

9.6 IP Switching and Multicast

Depending on the application, the profile of IP multicast traffic can be one where high bandwidth and low latency is essential to the operation of the application. An example is a multiparty video conference or a one-to-many video lecture where the video stream must flow downstream over the delivery tree with minimal packet loss, sufficient bandwidth, and low latencies. It is a relatively straightforward operation for IP switching to support IP multicast in a manner similar to its support for unicast traffic. IFMP redirect messages still flow upstream from the receiver to map a flow to a new VC. Indeed the only slight differences are that the destination address in the flow identifier is a multicast group address and at various branch points in the network, GSMP will execute more than one ADD BRANCH operation depending on the number of downstream branches in the delivery tree. A conceptual depiction of multicast support for IP switching is illustrated in Fig. 9-15.

9.7 IP Switching Network Examples

From a connectivity perspective, IP switching can operate in a variety of different network scenarios. It forwards IP traffic using standard hop-by-hop processing and so is compatible with standard IP routers in that respect. However, the fine-grained approach toward accelerating IP traffic that this solution takes requires sufficient per-flow switch resources

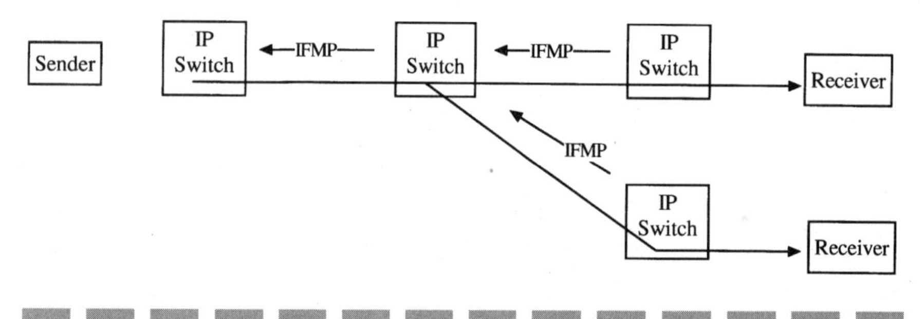

Figure 9-15 IP multicast support.

and that the ingress and egress edge components are IFMP-capable. One scenario that comes to mind would be a campus or small enterprise IP network as depicted in Fig. 9-16. In this environment Ethernet-based workgroups are attached to the IP switching backbone through an IFMP-capable edge device. High-speed servers could be attached directly to the backbone using an IFMP adapter, and the core consists of IP switches. Because of the locality of the communicating parties, the volume of IP flows that are candidates to be switched and that in fact would be switched probably would not exceed the maximum capacity of the switched resources in the network. Total end-to-end throughput would depend on the forwarding capacity of the ingress/egress edge devices. This particular scenario offers the same support as MPOA except that IFMP is the control protocol.

Another possible arena where IP switching might be useful is at the edge of a network. IP switching mechanisms could be employed to serve as a flow discriminator or traffic shaper to identify, label, and service specific individual flows for faster or slower access into and out of an ISP backbone. Consider the two examples illustrated in Fig. 9-17. In the first case, an IP switch is the front end of a Web server farm, and the classifier has been programmed to identify and label certain flows that should be given higher priority. This would enable high-priority flows to traverse a high-performance switched path into the ISP backbone versus others that should be given just basic best-effort hop-by-hop service. The second case, an interesting one, would use an IFMP-capable resi-

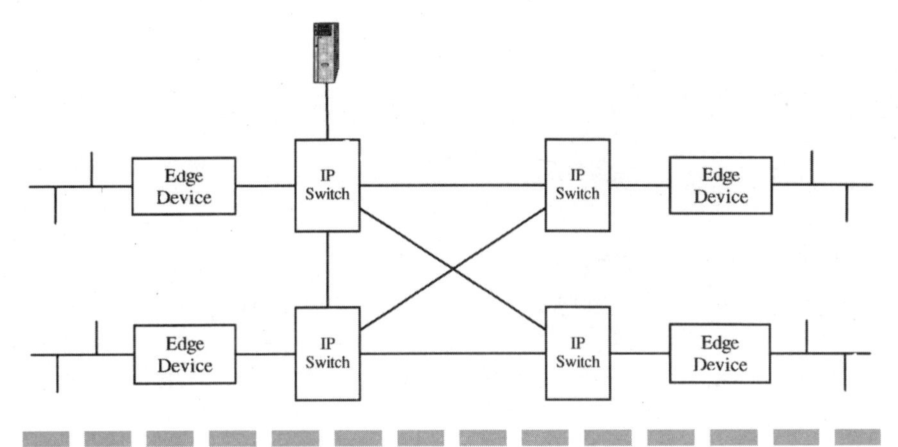

Figure 9-16 IP switching in an enterprise.

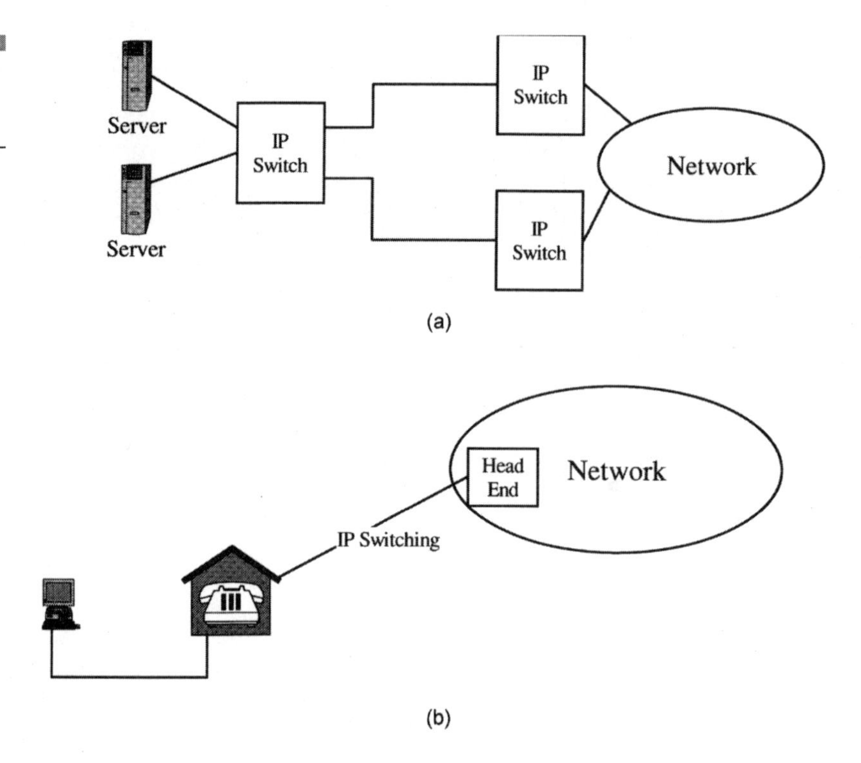

Figure 9-17 IP switching edge services.

dential broadband modem (cable or xDSL) to transmit IFMP redirect messages upstream to the head end with the result being faster service for specific IP flows entering the home.

MORE INFORMATION

http://www.ipsilon.com/~pn/.

McKeown, N., and Lin, S., "A Simulation Study of IP Switching," ACM/Sigcomm Proceedings, 1997.

Newman et al., "IP Switching and Gigabit Routers," IEEE Communications, vol. 35, no. 1, Jan. 1997.

Newman et al., "IP Switching: ATM Under IP," IEEE/ACM Transactions on Networking, April 1998.

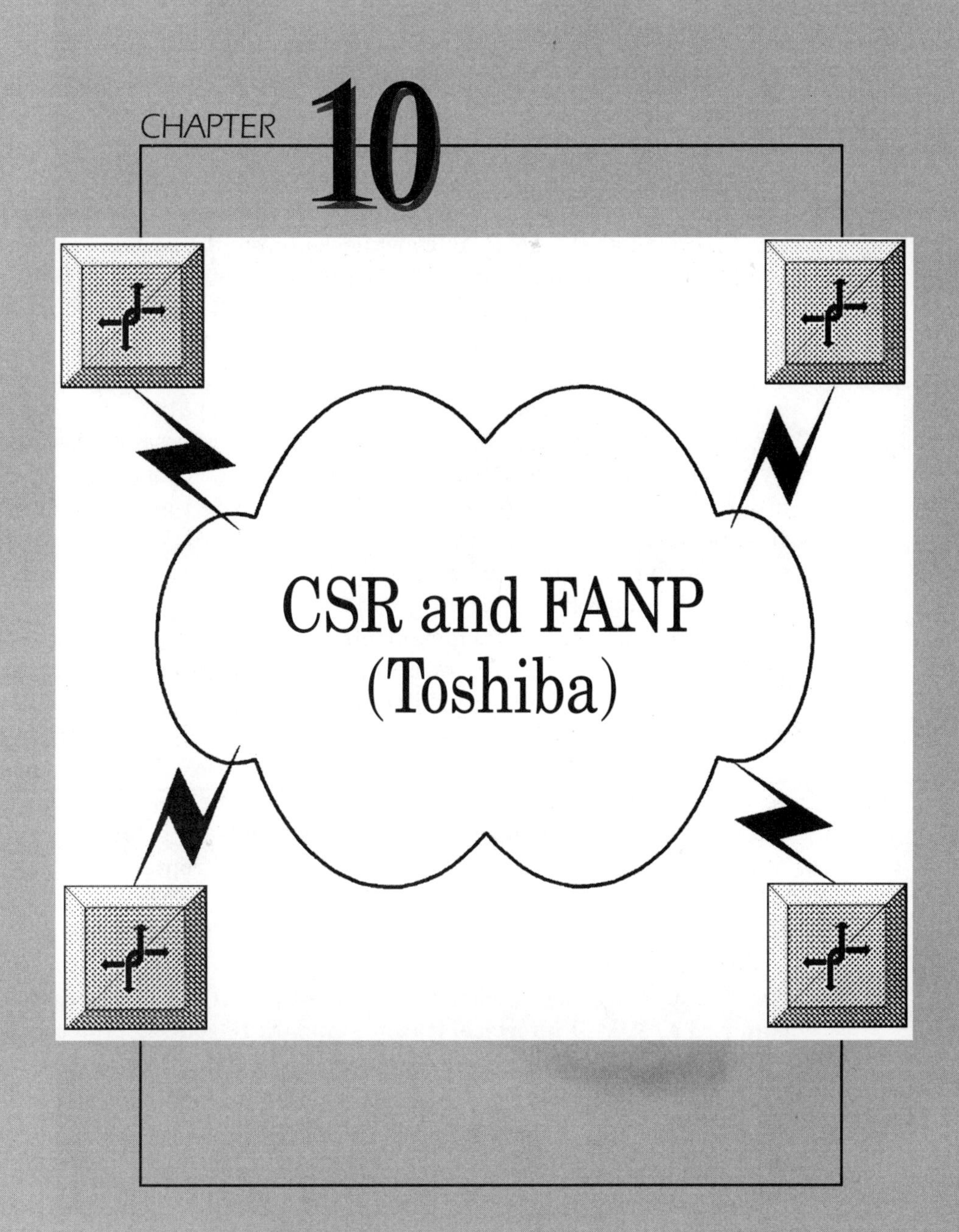
CHAPTER 10
CSR and FANP
(Toshiba)

The integration of IP and ATM technology is not an easy problem to solve. The connectionless, dynamic, and robust properties of IP must somehow be married to the hard-state connection-oriented nature of ATM. The lure of high bandwidth, low latency, and QoS that ATM offers is just too enticing for IP to pass up. Early efforts to define IP over ATM solutions confined IP's use of ATM bandwidth to a single subnet (Classical IP) or emulated LAN (LANE) while retaining the scalable properties and limited throughput of router-based internetworking. NHRP and MPOA relaxed this model somewhat and permitted ATM-established shortcut paths to bypass intermediate routers and thus expose IP traffic to much higher bandwidth. But the price to pay for the higher bandwidth and lower latency that the ATM shortcut path provides is the overhead of the address resolution process and the possible discontinuity between the routed (IP) and forwarding (ATM) topologies. The revolutionary approach of Ipsilon IP switching solved those problems but required that all end systems and intermediate routers (IP switches) scrap standards-based ATM Forum protocols in favor of the new IFMP and GSMP protocols.

While these early development efforts were underway within the IETF, ATM Forum, and elsewhere, a group of prominent Japanese researchers was working on a different solution to the problem of IP and ATM integration. As outlined in a series of Internet drafts and white papers that first appeared in late 1994 and early 1995, this solution suggested that an ATM network should not be treated as one large cloud but rather as a series of concatenated ATM data links (or subnets) interconnected by routers.[1] However, unlike traditional IP routers that perform per-packet layer-3 processing, the routers in this architecture are capable of forwarding at layer 3 packet by packet or at layer 2 cell by cell.

Central to this architecture is the notion of a cell switch router (CSR), which is a device that is connected to multiple ATM subnets through a standard ATM UNI interface. It runs standard IP routing protocols (e.g., OSPF, PIM) and performs conventional IP packet forwarding between subnets. It is also capable of building an internal CSR shortcut path by switching cells at layer 2 between two different ATM subnets. This is accomplished by concatenating or splicing together an incoming VC from one ATM subnet to an outgoing VC on another subnet.

---

[1]In this architecture, an ATM data link represents one IP subnet. An example would be the Logical IP Subnet as defined in RFC1577/2225. The term *ATM subnet* will be used in the remainder of the chapter to mean a single IP subnet overlayed on an ATM datalink.

The other element in this solution is the Flow Attribute Notification Protocol (FANP). FANP is used between a pair of CSRs to map IP traffic with different granularities to a nondefault VC. Once the CSR detects that a particular IP flow has been mapped to a nondefault VC in both upstream and downstream (with respect to data flow direction) subnets, it can splice them together, creating, in effect, an intra-CSR shortcut path. An ingress-to-egress shortcut path for an IP flow is created when contiguous CSRs along a routed path have concatenated inbound and outbound nondefault VCs.

The CSR/FANP solution is similar in many respects to Ipsilon IP switching. Both solutions are flow-driven, provide default and shortcut routing, and utilize a special peer-based flow-driven control protocol that relabels the cells of a flow with a new VPI/VCI. However, unlike IP switching, the CSR/FANP model continues to use standard ATM Forum signaling and routing inside the ATM subnet, and it includes a technique for communicating flow-to-label mappings through intermediate ATM switches. This makes CSR/FANP much more "ATM Forum friendly" in its approach to IP switching.

10.1 Architecture

The architecture of the CSR/FANP model was developed with a desire to exploit the high bandwidth and low latency of ATM cell switching while retaining the scalability and robustness of traditional IP networking. To achieve this, an approach is taken that does not attempt to place IP traffic on traditional end-to-end ATM VCs across the entire ATM network, as is the case with MPOA. Rather the CSR/FANP model preserves the traditional Internet architecture of subnetworks interconnected by IP routing devices. But in this model, the underlying data link technology associated with each subnet is ATM, and the routers, CSRs to be exact, are capable of conventional IP packet forwarding *and* ATM cell switching. By having the CSRs relay cells between subnets it is now possible to construct a shortcut path that bypasses intermediate per-packet processing and thus delivers the high bandwidth and low latency that ATM affords.

The CSR/FANP model is illustrated in Fig. 10-1. Ingress and egress components consisting of hosts or edge devices provide a means of entering and exiting the CSR network. The CSR interconnects multiple ATM subnets, performs basic IP forwarding functions, and runs standard unicast and multicast routing protocols. Packets can be forwarded on a hop-

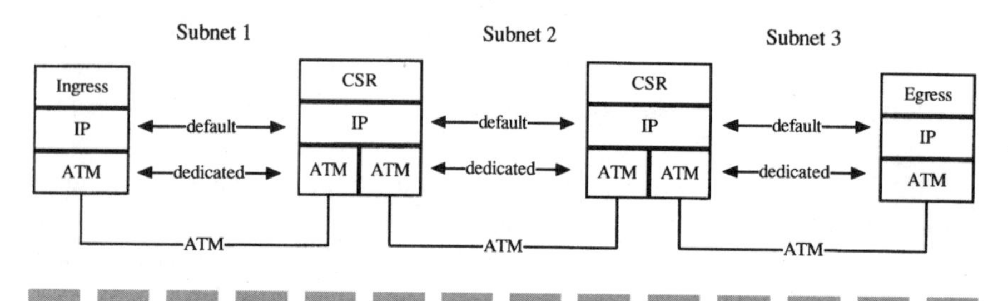

Figure 10-1 FANP model.

by-hop basis by traveling over a series of default channels that are initiated and terminated between CSR devices. Alternatively packets can be placed on an end-to-end shortcut path that consists of a series of concatenated dedicated VCs that are initiated and terminated between the same CSR devices. The concatenation process occurs in each CSR after it has associated a particular flow that is present on an incoming and outgoing dedicated VC, respectively. The process of mapping packets or more specifically a flow of packets to a particular dedicated VC is the function of FANP.

There are several unique aspects about the operation of the CSR/FANP model. First, ATM Forum signaling and routing operations are confined to the subnet only. These mechanisms are used by the CSR and any other ATM subnet members to establish both the default and any dedicated VCs that may be required. It is also quite reasonable to assume that an ATM subnet may contain multiple ATM switches running normal ATM Forum protocols. So although the entire, extended CSR network may utilize ATM data link technology to provide connectivity and optionally high bandwidth service, the operation of ATM VC management remains inside the borders of the IP subnet. Second, the flow-to-VC mapping operation of FANP only occurs inside the subnet and between adjacent or VC-connected components. The control operation does not cross subnet boundaries as is the case with NHRP. And finally, the decision to switch the cells of a flow rather than perform per-packet layer-3 processing is a local decision made by each CSR along the IP routed path.

Preserving the standard operation of the Internet model in supporting IP over ATM has several important advantages over other approaches. First, it is scalable—the existence of the Internet bears this out. Logically speaking the CSR/FANP is no different from a traditional router-based network with the exception that it happens to use ATM technology as the data links for the participating subnets. Second, it is quite feasible and relatively straightforward to map an RSVP flow to a shortcut path to provide some level of QoS. Observe that the RSVP PATH and RESV messages flow over the routed path in the normal manner. The CSRs can simply interpret the RSVP messages and map the flow to an appropriate dedicated VC with QoS within a subnet and then provide the cell-switching services (with appropriate QoS) between the incoming and outgoing dedicated VCs.

Another standard IP service that is supported with no modification is IP multicast. It is the job of IP routers to build and maintain delivery trees based on group membership information, and this does not change with the CSR/FANP model except that the CSR performs the role of multicast router. Attempting to support large-scale multicast over an ATM network without routers and just using point-to-multipoint VCs requires that the root maintain explicit information about each of the leaf members. This introduces scaling problems and puts a big burden on the source.

Another significant benefit is the fact that dynamic IP routing is used to compute not only the default but also the shortcut path through the network. Indeed the same set of subnets, links, and CSRs will be present in the default path and in the shortcut path. For a large ATM network, this approach reduces the burden on PNNI (or on network operators to configure static routing table entries à la IISP) to maintain a consistent and accurate topology over one or more levels of hierarchy. Subdividing a large ATM network into a collection of smaller ATM subnets also enables a heterogeneous mixture of ATM switches running both standards and proprietary-based routing protocols. For example, one ATM subnet may have switches capable of supporting PNNI, whereas another may have switches that can only run IISP. Note also that dynamic IP routing is just that—dynamic. If a CSR or link fails, the traffic can be quickly rerouted over an alternate path. Of course there will be some delay in executing the FANP and concatenation processes on only those transit subnets and CSRs of the new path. However, this does not require the reinitialization of the entire end-to-end connection as is the case with PNNI if a VC unexpectedly terminates.

10.2 Components

The definitions and components relevant to the CSR/FANP model as defined in RFC2098 consist of the following:

- *Flow.* A flow is a sequence of packets from a particular source to a particular destination (unicast or multicast) that is associated by a particular path and the service (default or nondefault) that these packets will receive from each router over that path. The packets of a flow typically share the same source address, same destination address, and perhaps the same TCP/UDP port number.
- *Flow identifier (Flow ID).* A set of IP and transport layer fields common to the packets of a flow. An example of a Flow ID is a source/destination IP address pair as shown in Fig. 10-2. Another example is a source/destination IP address/port pair.
- *CSR.* The CSR is a device that is capable of performing conventional IP forwarding at layer 3 and cell-switching at layer 2. It attaches to one or more ATM subnets using any number of standard ATM interfaces and performs standard ATM Forum UNI signaling over that interface. A generic CSR is illustrated in Fig. 10-3. It consists of a CSR control point and an ATM switch fabric. The CSR control point acts as the "IP router" for the device and therefore runs standard IP unicast and multicast routing protocols. It

Figure 10-2
Flow ID.

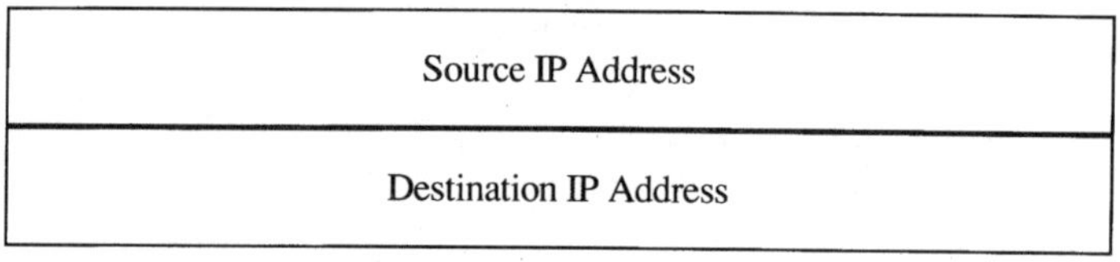

Source IP Address
Destination IP Address

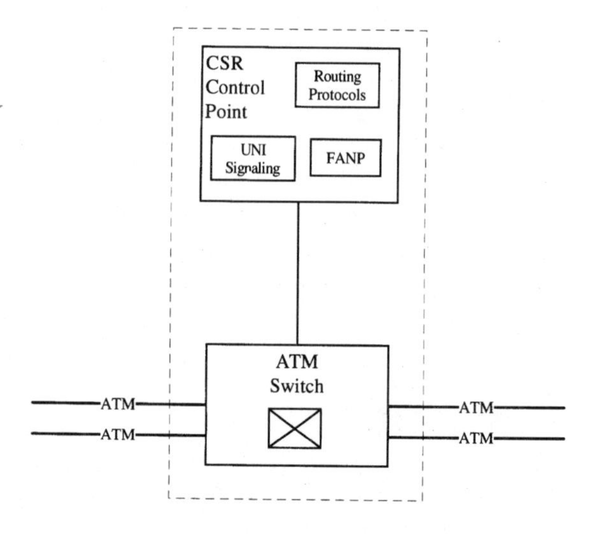

Figure 10-3
Cell switch router.

also manages connection state in the ATM switch component. In addition the CSR control point must be capable of identifying an IP flow and then exchanging FANP messages with peer CSR(s) in the same ATM subnet to map that flow to a VC. Peer CSR devices manage a default VC to exchange control information and forward packets in a default-routed manner. Peer CSR devices may also manage one or more dedicated VCs that transport specific IP flows and form the segments of an end-to-end shortcut path.

- ✓ *CSR ingress / egress.* To interoperate with external networks and to allow traffic to benefit from a CSR-based network, ingress and egress components must be positioned at the edges of a CSR network. A CSR ingress/egress can be a host, edge device, router, or a CSR itself. It serves as the source and sink of a shortcut path that passes through one or more CSR devices. A CSR ingress/egress is also capable of exchanging FANP messages to map a flow to a dedicated VC.
- ✓ *Default VC.* This is a general-purpose VC established between two CSRs within the same ATM subnet. The default VC provides a connection over which IP packets can be forwarded hop by hop at layer 3 through a network of CSRs. Incoming cells on the default VC are assembled into packets and processed at layer 3. CSRs may also use the default VC to exchange control information such as FANP messages and routing protocol updates.
- ✓ *Dedicated VC.* A dedicated VC is established between two (or more in the case of multicast) CSR devices within an ATM subnet to transport the cells of a specific IP flow. Local CSR policies determine which IP flows a FANP message exchange will map to dedicated VCs. A dedicated VC can be spliced to another dedicated VC by the CSR to form a shortcut path that bypasses layer-3 processing.
- ✓ *Bypass pipe.* This is a CSR ingress-to-egress ATM cell-switched connection that consists of a series of contiguous dedicated VCs concatenated by one or more CSRs along a routed path. *Bypass pipe* is CSR terminology for a shortcut path.[2] A bypass pipe follows the same physical path that is computed by dynamic IP routing with the exception that cell switching rather than layer-3 process-

[2]The CSR literature refers to the shortcut path as either a bypass pipe or a cut-through path.

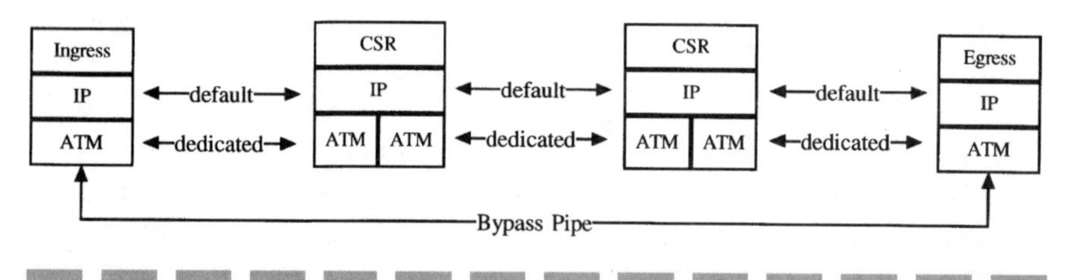

Figure 10-4 CSR connections.

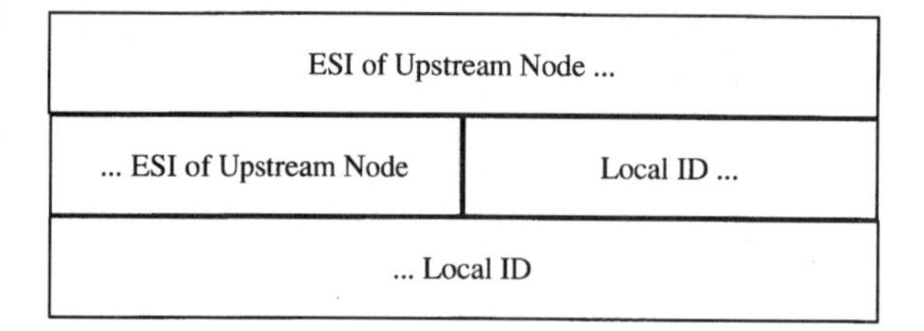

Figure 10-5 VCID format.

ing is performed at each intermediate CSR. Best-effort and QoS-based bypass pipes can be established. In addition a bypass pipe can be point to point and carry a unicast IP flow or can follow a tree topology to carry a multicast flow. The concept of the various CSR connections and the bypass pipe are shown in Fig. 10-4.

✓ *VCID.* The VCID is a unique identifier or "handle" that is used (and shared) by the CSR endpoints of a dedicated VC to associate a particular flow with that dedicated VC. The reason that the VCID is used instead of a VPI/VCI is that when communicating through one or more native ATM switches, the VPI and/or VCI label in each cell may be swapped or rewritten. It has no end-to-end significance, so attempting to map a specific IP flow to a VPI/VCI label when the latter may change would be futile.[3] The format of the VCID is illustrated in Fig. 10-5 and consists of a 6-byte ESI of the upstream node and a 6-byte unique local identifier chosen by the upstream node.

[3]Ipsilon IP switching maps a flow to a VPI/VCI label, but their architecture does not suggest the presence of intermediate ATM switches capable of swapping the labels. In the Ipsilon IP switching model, IP switches are directly connected to one another by point-to-point ATM data links.

10.3 CSR Interoperation with ATM Switches

One of the fundamental design points of the CSR/FANP model is coexistence and interoperation with ATM switches. Unlike Ipsilon IP switching, the CSR/FANP model can work with and through ATM switches. Indeed, their presence is only encouraged as a means of providing ubiquitous ATM data link coverage over a larger network geography. This extends the reach of bypass pipes and thus can improve overall network capacity and performance. It is not required that the ATM switches discard their ATM Forum software in favor of a proprietary technique as is the case with Ipsilon IP switching. In fact, ATM switching, signaling, and routing operate unmodified when CSRs are placed in the network. To accommodate ATM switching the CSR/FANP model incorporates the following special features:

- ✓ CSRs communicate over a standard ATM interface using ATM Forum UNI signaling.[4]
- ✓ Peer CSR components (e.g, ingress/egress CSR devices, CSRs) will exchange a VCID value to associate an IP flow with a dedicated VC. The procedure for communicating a shared VCID between peer CSR components is called the VCID notification process. The concept of the VCID as a means of associating a flow *N* with a dedicated VC between two CSR devices is illustrated in Fig. 10-6.

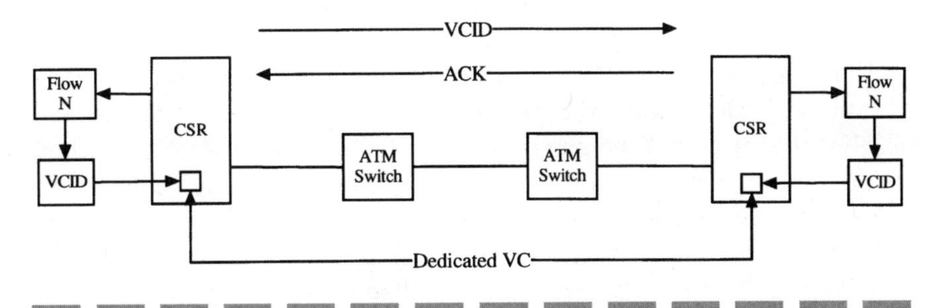

Figure 10-6 Use of VCID value in CSR networks.

[4]When operating under the CSR/FANP umbrella, the ATM switch component of the CSR does not need to support PNNI even though it is providing a cell-switching function.

TABLE 10-1

CSR Interoperation with ATM Switching.

ATM Configuration	Does VPI/VCI Value Change?	VCID Needed	VCID Notification
Point-to-point Transparent link	No	No	N/A
VP tunnel	Only the VPI value is swapped; VCI is untouched.	Suggested because VPI label is swapped; not required because VCI value is unaltered.	In-band
PVC network	Yes	Yes	In-band
SVC network	Yes	Yes	In-band or Out-of-band

CSR/FANP operations can differ depending on the type of ATM network that is currently deployed between CSR devices. The differences lie in the provisioning of dedicated VCs to transport specific IP flows and the necessity and means by which a VCID is communicated between CSR peers. Table 10-1 summarizes the four different ATM configurations that can exist between peer CSR devices and comments on the requirement for and technique used for VCID notification.

In the point-to-point transparent configuration, the VPI/VCI label is the same at both ends of a dedicated VC. An example of this kind of configuration is two CSR devices that are directly connected to each over a point-to-point link with no intermediate ATM switches in the middle. There is no need to use a separate VCID to link an IP flow to a dedicated VC. The actual VPI/VCI value itself will be sufficient because it is the same at both ends. In the next example, a VP tunnel, it is possible that the VPI value will be swapped as cells traverse an ATM switched network. Because a portion of the VPI/VCI does change, a VCID notification process is recommended although not required. It may not be required because the VCI value in each ATM cell is not altered during VP switching. So if the implementation called for a flow-to-label association in which the label is the VCI value, this could be supported through a VP tunnel.[5] In the last two examples, the VCI value is not the same at

[5]This is a standard technique supported by all of the IP switching solutions including Ipsilon, Tag Switching, ARIS, etc.

both ends because of the label-swapping process performed by intermediate ATM switches. A VCID value is required, and therefore a VCID notification process must be performed so that both ends of the dedicated VC are made aware of the specific VCID value.

The VCID notification process can be performed in one of several ways. In-band notification is the case where the VCID value is transported over the VC with which it is associating a flow. In other words, the dedicated VC is first established between two CSR devices, and then the VCID notification process is performed over that VC. Out-of-band VCID notification involves the use of a separate "channel" or alternative path in which the VCID is communicated between peer CSR components.

Out-of-band VCID notification can be performed by including the VCID (or some temporary handle) within the ATM SVC Setup message. More specifically, the VCID value is included in an information element contained in the SETUP message that is conveyed between the CSR source and the CSR destination. The VCID could be stored in a "user-specified layer-3 protocol information field" of the Broadband Lower Layer Information (BLLI) IE. However, the 7 bits allocated for the BLLI IE may be insufficient to support the required number of dedicated VCs between a pair of CSR devices. A possible solution would be to specify a temporary handle in the BLLI IE that would serve as "place holder" until an explicit exchange of actual VCID values follows, which would replace the temporary handle.

10.4 Flow Attribute Notification Protocol

The CSR/FANP model enables IP flows to be mapped to dedicated VCs to achieve higher throughput and lower latencies. In addition to the CSR device that provides both routing and cell-switching forwarding capabilities, FANP v1 defined in RFC2129 is the other important component that makes up the CSR/FANP model. FANP is a control protocol that CSR devices use to exchange the information necessary to build a dedicated VC, associate an IP flow with that dedicated VC, and then to place the cells of that IP flow onto the dedicated VC. The objective of the FANP message exchange is to establish the per-flow dedicated VCs between CSR devices within an ATM subnet so that ultimately they form the segments of an ingress-to-egress bypass pipe.

FANP is primarily a flow-driven control protocol. This means that the FANP process is triggered by the arrival of an IP flow that is similar in many respects to IFMP. What constitutes a flow or more generally what kind of data will trigger the FANP process is based on a locally configured policy within each CSR, but the policy should be consistent throughout the network. Examples of "trigger" packets could be TCP SYN packets (indicating a TCP connection setup), packets with a certain port number (e.g., HTTP = port 80), packets with a specific source and destination IP address, or a group of packets with shared header information that arrives within a specific interval. In addition it is possible for the appearance of certain control messages such as RSVP RESV messages to trigger the FANP process. Figure 10-7 illustrates the general flow of FANP messages between three intermediate CSR devices.

The arrival of a trigger packet at CSR 1 initiates an FANP message exchange with CSR 2. Independently of the FANP activity occurring between CSR 1 and CSR 2 on subnet 1, the same flow triggers an FANP message exchange between CSR 2 and CSR 3 in subnet 2. Unlike the Ipsilon approach, it is the upstream CSR that initiates FANP process. The basic FANP process consists of dedicated VC allocation, VCID negotiation, and Flow ID notification. Flow ID notification is the procedure in which a Flow ID is associated with a VCID and thus a dedicated VC. When a CSR detects that a particular flow has been placed on a dedicated VC on both the incoming and outgoing ports, the CSR control point instructs the ATM switch component to "cross-connect" the two entries in the connection table, thus forming a cell-switched path through the device. FANP also contains a mechanism that can detect if the flow over a dedicated VC has stopped for some period of time (dead-time interval). If the flow does stop, FANP removes the corresponding Flow ID and VCID entries in the CSR and releases the dedicated VC.

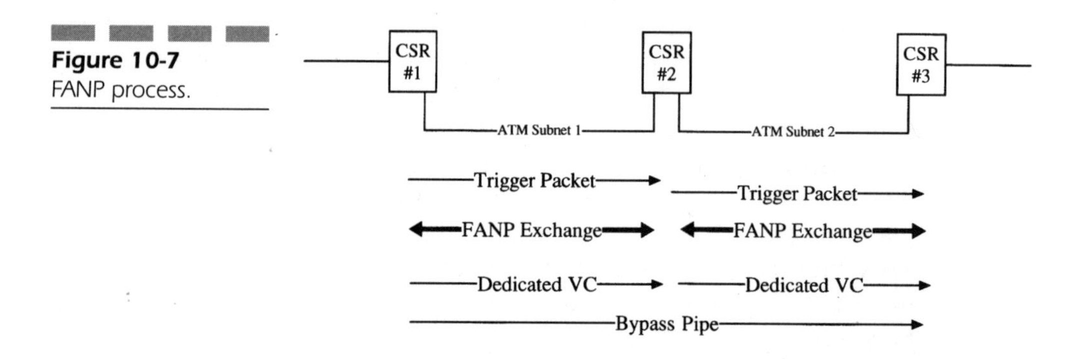

Figure 10-7
FANP process.

FANP defines five basic procedures for managing dedicated VCs between CSRs:

- Dedicated VC selection
- VCID negotiation
- Flow ID notification
- Dedicated VC refresh
- Dedicated VC removal

10.4.1 Dedicated VC Selection

Upon receipt of a trigger packet(s), the first step that a CSR must undertake is the selection of a dedicated VC. This can be accomplished in one of several ways. The first technique is called VC pool. In this approach a pool of VCs are established a priori between peer CSR devices. They can be provisioned manually (PVC) or by signaling (SVC). A CSR selects a VC from the pool of unused VCs to be the dedicated VC for a particular flow. The second technique is accomplished through the use of SVC signaling. In this approach, standard ATM Forum signaling and routing is used to establish an SVC between peer CSR devices. Note in this approach that a source CSR must resolve the IP and ATM address of the destination CSR before the SVC setup is attempted. This can be accomplished through means already discussed such as Classical IP over ATM or LANE. The third way an upstream CSR device can provision a dedicated VC is to select an unused VCI value from a preconfigured PVP that has been established between a pair of CSR devices.

One advantage of the VC pool method is that because the VCs are already built, there is no delay in moving on to the next step in the FANP process. This reduces the latency in placing a flow on the dedicated VC and thus in establishing a bypass pipe. Another advantage is that a CSR ATM subnet (CSR UNI endpoints and ATM switches) may be able to more easily absorb a flurry of separate per-flow FANP messages. This is because there is no need for UNI/PNNI signaling and routing by the CSR devices and ATM switches.[6] To ensure that there are sufficient unused VCs in the VC pool, a low-water mark for unused VCs can be

[6]This becomes less of an issue as the SVC capacity and the setup/teardown performance in the network increases as a result of faster switch processors.

configured such that if the number of unused VCs falls below a certain threshold, the CSR will restock it with additional SVCs.

The downside to the VC pool approach is that it consumes VC resources in both the CSR devices and the ATM network. This makes the dynamic on-demand SVC approach more attractive in a network where VC resources are at a premium or if it is expensive to provision VCs (e.g., in a wide area network). Another issue to consider is that the QoS of the VCs in a pool may not match the QoS requirements of a specific flow. If a specific flow requires a certain level of QoS, the easiest and most efficient technique to use would be to establish an SVC with QoS. An example is the arrival of RSVP RESV message triggering the establishment of dedicated VCs with specific QoS as illustrated in Fig. 10-8.

10.4.2 VCID Negotiation

The VCID negotiation process follows selection of the dedicated VC. This process is required if, due to the presence of intermediate ATM switches, the VPI/VCI labels are not the same at the endpoints of the dedicated VC. The VCID negotiation process begins when the upstream CSR transmits a PROPOSE message to the downstream CSR over the dedicated VC. The PROPOSE message contains a VCID and the IP addresses of the sending and receiving CSR devices. If the downstream CSR accepts the value, it transmits a PROPOSE ACK to the upstream CSR over the default VC. The upstream and downstream CSR devices now share a common VCID that is associated with the dedicated VC.

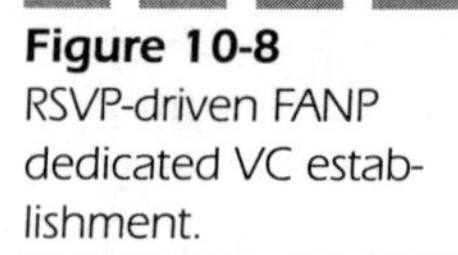

Figure 10-8 RSVP-driven FANP dedicated VC establishment.

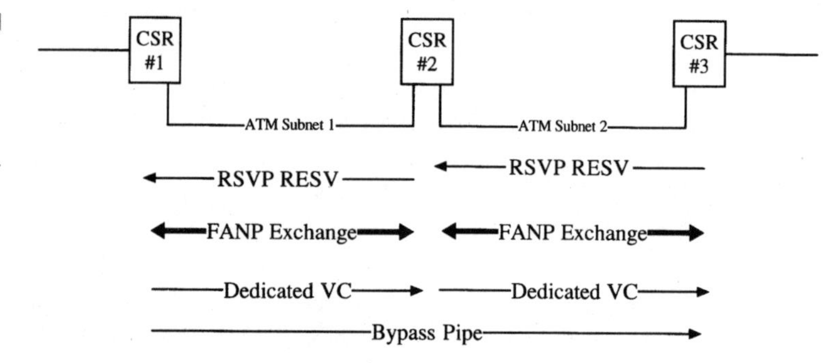

10.4.3 Flow ID Notification

The next step in the FANP process is Flow ID notification. The upstream CSR transmits an OFFER message to the downstream CSR over the default VC. The OFFER message, as shown in Fig. 10-9, contains the VCID associated with the dedicated VC and the flow ID of the IP flow that will be transported over the dedicated VC. In addition a refresh interval field is included, which determines how frequently the downstream CSR sends a READY message to the upstream CSR to indicate that the dedicated VC is still transporting cells.

10.4.4 Dedicated VC Refresh

The downstream CSR receives cells over the dedicated VC, and it periodically transmits a READY message to the upstream CSR. This informs the upstream CSR that it is okay to continue to transmit the cells of the IP flow over the dedicated VC. When the flow of cells over the dedicated VC stops, the downstream CSR will not transmit any READY messages to the upstream CSR, which is an implicit indicator that the flow is no longer active. If the upstream CSR has not received a READY message within a certain interval (deadtime), it can conclude that the flow is no longer active and that the dedicated VC can be released.

A conceptual diagram of the FANP VCID negotiation, Flow ID notification, and dedicated VC refresh flows is illustrated in Fig. 10-10.

Figure 10-9 OFFER message format.

Version	Op Code	Checksum
VCID Type	Flow ID Type	Refresh Interval
VCID		
Flow ID		

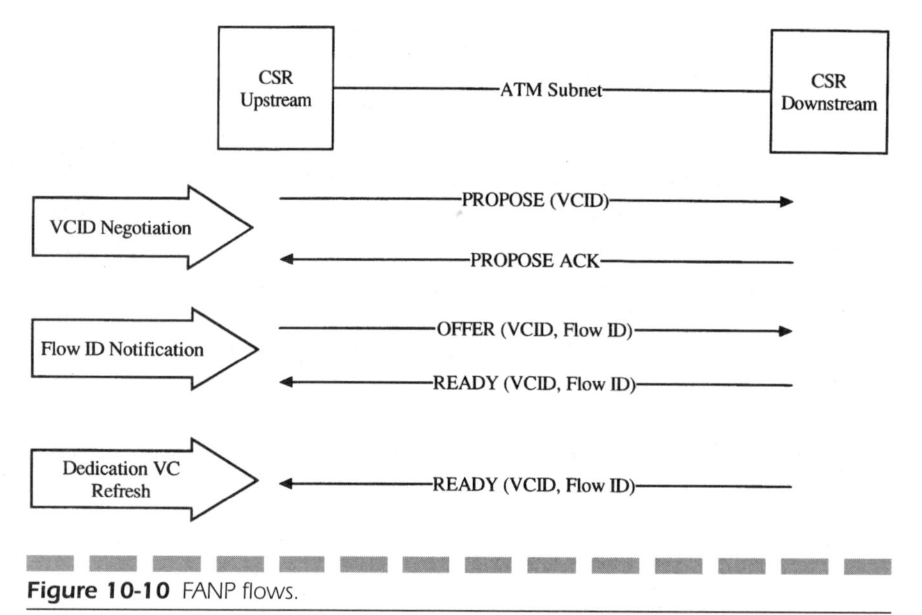

Figure 10-10 FANP flows.

10.4.5 Dedicated VC Removal

When the upstream CSR detects that the dedicated VC is no longer being used, it can take one of several actions to remove or release the dedicated VC based on the ATM configuration. If the dedicated VC is a PVC, the upstream CSR transmits a REMOVE message to the downstream CSR that contains the VCID and flow ID. This has the effect of removing the flow ID/VCID mapping associated with the VC and releasing it for use by another flow. If the dedicated VC is an SVC, ATM signaling can simply tear the connection down.

A conceptual diagram of the FANP dedicated VC removal process is shown in Fig. 10-11.

10.5 Flow Attribute Notification Protocol v2

Perhaps spurred on by other techniques described in some of the competing approaches (e.g., Tag Distribution Protocol and ARIS), the developers of CSR/FANP introduced a new version of FANP. In addition

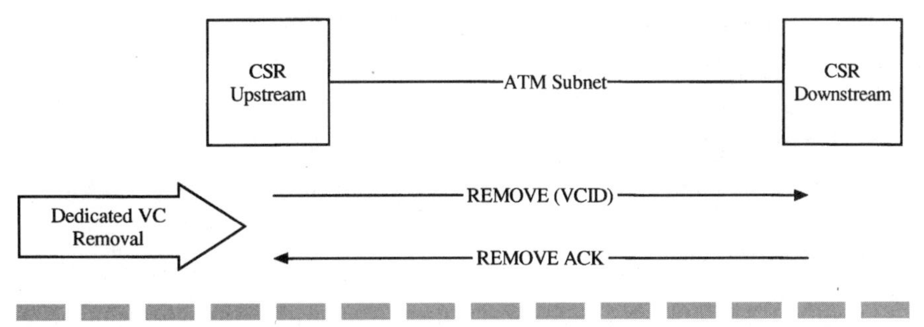

Figure 10-11 FANP dedicated VC removal.

to further specifying the behavior over specific ATM configurations discussed above, FANP v2 introduces a Neighbor Discovery (ND) protocol and outlines two operational modes for establishing a bypass pipe through a network of CSR devices. The FANP neighbor discovery protocol is used by a CSR to discover the existence of any FANP v2-capable devices on the ATM subnet, communicate specific FANP-relevant protocol attributes between CSR neighbors, and monitor and verify that consistent state is maintained between CSR neighbors. The two distinct operational modes for FANP v2 are Distributed Control (DC) and Ingress Control (IC). In DC mode FANP operations are initiated and performed in an independent and distributed manner between pairs of CSR devices along a routed path. In IC mode FANP operations are initiated from an ingress CSR device and propagated downstream over the routed path to the egress CSR. Additionally FANP v2 defines a common FANP v2 header format, several new messages, and a set of new Flow IDs that enable CSR/FANP to support flow-, topology-, and request-driven shortcut path establishment. With these enhancements the CSR/FANP model can be considered suitable for both flow-rich (campus, corporate intranet) and flow-challenged environments (large IP networks).

10.5.1 Neighbor Discovery

Prior to neighbor discovery CSR devices that shared a common ATM subnet had no way of verifying, other than through manual configuration, the existence of other CSR devices or their capabilities. For example, how is an upstream CSR device able to determine the range of VCIDs that the downstream CSR is capable of supporting? The FANP v2 ND protocol addresses this problem by enabling CSR devices that share a common network to perform the following:

- Automatically detect the existence of other FANP-capable devices on the network
- Exchange FANP protocol attributes such as support for IC or DC modes of operations
- Verify that consistent state information is maintained between the CSR neighbors

Neighbor discovery messages are exchanged between CSR devices in the same IP subnet. The ND protocol works over point-to-point and multicast-capable networks. In the case of a point-to-point link, the initiating CSR transmits an INIT message to its neighbor. The INIT message contains the local identifier (local ID) of the sender, a capability object describing the mode of FANP operation it is capable of supporting, and a VCID range object. The receiver acknowledges this information with a RECOGNIZE message that contains its own local ID, the local ID of the sender, a capability object, and a VCID range object. The sender receives the RECOGNIZE message and checks to make sure that its own local ID is contained in the message and if so transmits a KEEP_ALIVE message to the receiver that should contain the local ID of the receiver. When both CSR devices determine that each other's local IDs have been exchanged, an operational state between the two is established. It is only when both CSR devices are in an operational state that dedicated VCs can be provisioned or released. CSR devices periodically exchange KEEP_ALIVE messages with their peers to determine a change in status.

In the case of a multicast-capable network the initiating CSR transmits a HELLO message that contains the local ID of the sender. The HELLO message is addressed to the ALLNodes (224.0.0.1) if the sender is a core or interior router or, if not, to the ALLRouters (224.0.0.2) address. When a neighbor CSR receives the HELLO message, it transmits an INIT message to the sender, and the process is essentially the same as the one described above.

The general flow of ND messages over a point-to-point and multiaccess network is illustrated in Fig. 10-12.

10.5.2 Distributed Control Mode

In the DC mode of operation, FANP operations are initiated by individual CSR devices in an independent and distributed manner along a routed path. Based on the arrival of a trigger packet or some other pre-

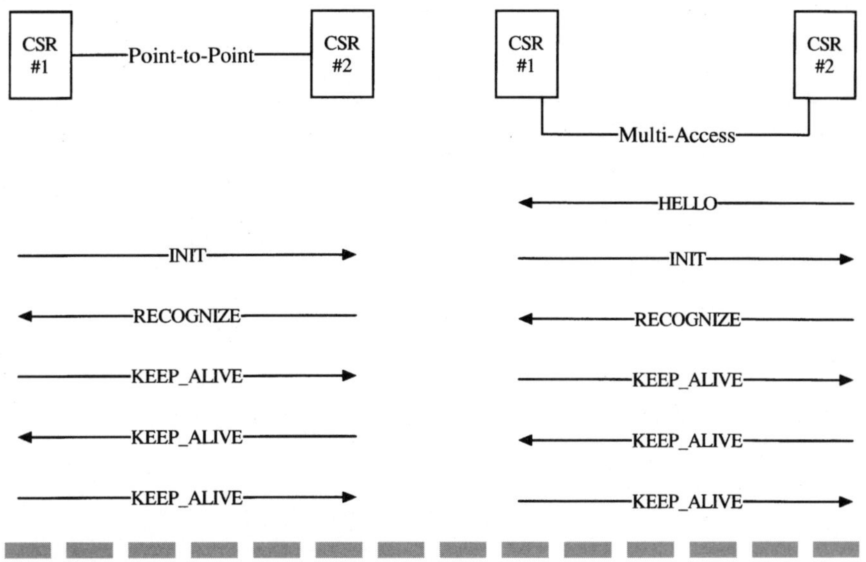

Figure 10-12 FANP v2 neighbor discovery flows.

determined triggering mechanism, an upstream CSR in an ATM subnet will exchange FANP messages with the downstream CSR to map a specific IP flow to a dedicated VC. While performing this activity, the downstream CSR is under no obligation to initiate its own FANP message exchange with its corresponding downstream neighbor. However, consistent triggering policies should be maintained throughout a CSR network or else the formation of an ingress-to-egress bypass pipe could never occur. Although there is no dependency on the FANP events occurring upstream, it is more than likely that the triggering event that precipitated the FANP process in the adjacent upstream ATM subnet will initiate a similar process on the part of the downstream CSR.

The process for establishing a bypass pipe through a network of five FANP-capable devices in DC mode is illustrated in Fig. 10-13. Initially the arrival of a specific IP flow or trigger packet (e.g., TCP SYN) kicks off the FANP independently in each of the four separate ATM subnets. The FANP process consists of the following steps:

- *Dedicated VC selection.* As discussed earlier, a dedicated VC can be established using ATM SVC signaling procedures, selected from a pool of preestablished unused dedicated VCs, or in the case of a VP an unused VCI value can be assigned. If the connection

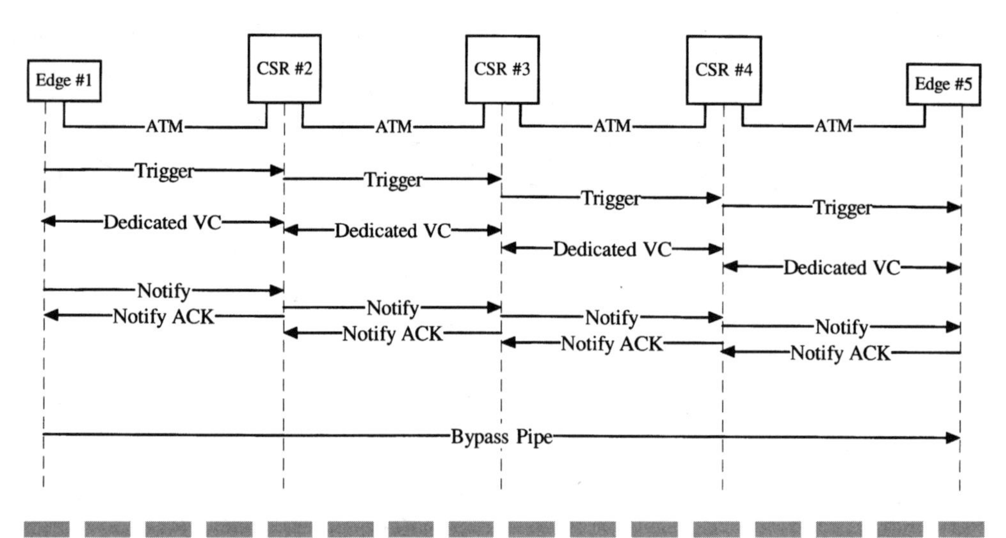

Figure 10-13 FANP v2 DC-mode operation.

between the adjacent CSR devices is a point-to-point link with no intermediate ATM switches this step is not necessary.

- *VCID notification.* This process is performed to communicate a unique VCID value that is shared between the endpoints of a dedicated VC. It can be performed in-band using a PROPOSE message exchange, or a NOTIFY message can be used to convey the VCID and a temporary VCID handle that is contained in the BLLI IE of the SVC setup request. If the connection between the adjacent CSR devices is a point-to-point link with no intermediate ATM switches, this step is not necessary.
- *Flow ID notification.* The Flow ID/VCID mapping is contained in a NOTIFY message that is transmitted from the upstream CSR device to its downstream neighbor over the dedicated VC. It is possible to group the VCID and Flow ID in the same NOTIFY message.

Once each CSR determines that an incoming and outgoing dedicated VC are associated with the same Flow ID, the two can be concatenated, and assuming that other CSR devices along the routed path accomplish this task, an ingress-to-egress bypass pipe is formed. If a change in topology occurs or it becomes necessary to remove the flow from the bypass pipe, a series of Flow ID and VCID remove procedures can be performed, followed by the release of the dedicated VC. Because the bypass pipe is established on a per-subnet basis by individual CSRs, the DC mode is quite useful in quickly reacting to changes in network topology, multicast group membership changes, and RSVP reservation state.

10.5.3 Ingress Control Model

In the IC mode of operation, a triggering event at the ingress CSR device initiates the FANP process with its downstream neighbor, which in turn prompts a similar FANP process with its downstream neighbor, and so on. This process is repeated over the routed path of CSR devices and ATM subnets until it reaches an egress CSR device. Specifically an ingress CSR generates an FANP NOTIFY message and sends it to its downstream neighbor to inform it of a flow-to-VC mapping. Rather than acknowledge it, as would be case in DC mode, the downstream neighbor records the Flow ID and dedicated VC information, regenerates the NOTIFY message, and transmits the new message to its downstream neighbor. This process continues until the NOTIFY message reaches the egress CSR. At that point, the egress CSR will send a NOTIFY ACK message that is forwarded hop by hop upstream until it reaches the ingress CSR. At each CSR along the return path, the incoming and outgoing dedicated VCs related to the same Flow ID are concatenated to form an intra-CSR shortcut path and, collectively, an ingress-to-egress bypass pipe. The process for establishing a bypass pipe through a network of five FANP-capable devices in IC-mode is illustrated in Fig. 10-14. Observe that it is an ingress-to-egress process.

An important feature of IC mode is the ability to create a loop-free bypass pipe. Obviously this can be most easily accomplished by preventing the formation of loops in the first place. A CSR device operating in FANP IC-mode uses one of two possible techniques:

- *Hop count checking.* If the hop-count value in an FANP message exceeds a specific predetermined threshold, the CSR will reject the message.
- *Flow ID checking.* If a NOTIFY message contains information about a Flow ID that is already registered, the CSR will reject the message.

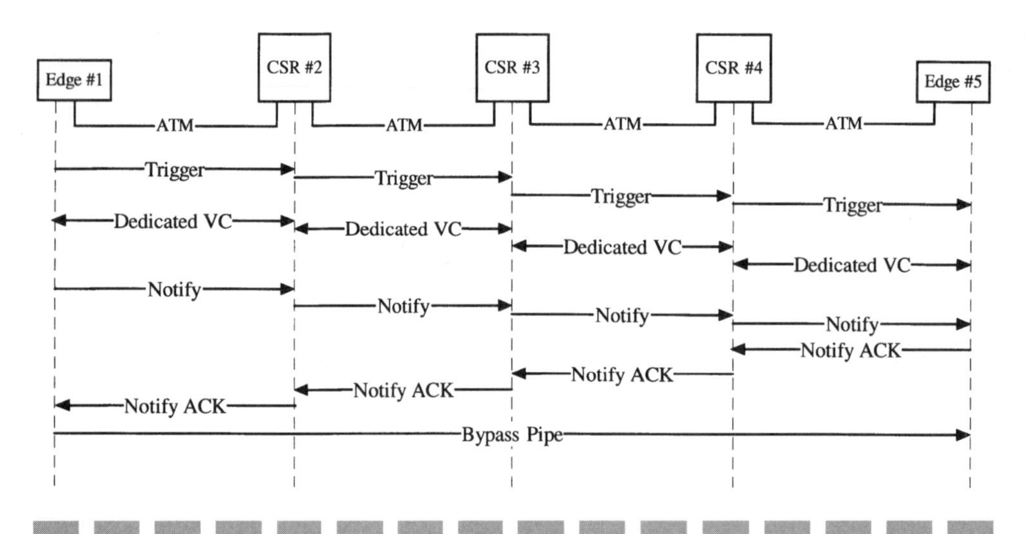

Figure 10-14 FANP v2 IC-mode operation.

10.6 CSR and Multicast

Because CSR/FANP subscribes to the Internet model, support for IP multicast is possible. Within an ATM subnet, the upstream CSR can establish either point-to-point or point-to-multipoint VCs to the one or more downstream CSR devices (and hosts) that are members of a particular multicast group and thus will become branches on a multicast delivery tree. As is the case with unicast traffic, when the CSR detects an incoming dedicated VC and one or more outgoing dedicated VCs associated with a Flow ID (which includes a multicast group address), the VC concatenation process is performed, resulting in an intra-CSR shortcut path. The trigger to add leaves to the point-to-multipoint VC can be based on the arrival of multicast data, PIM-JOIN messages, IGMP reports, or MARS JOIN messages if the CSR contains a MARS client. A multicast bypass pipe is shown in Fig. 10-15.

10.7 CSR Network Examples

CSR devices can be effectively deployed in a number of different network environments. The first one that comes to mind is a large number of border routers connected to a common ATM network. Typically a full

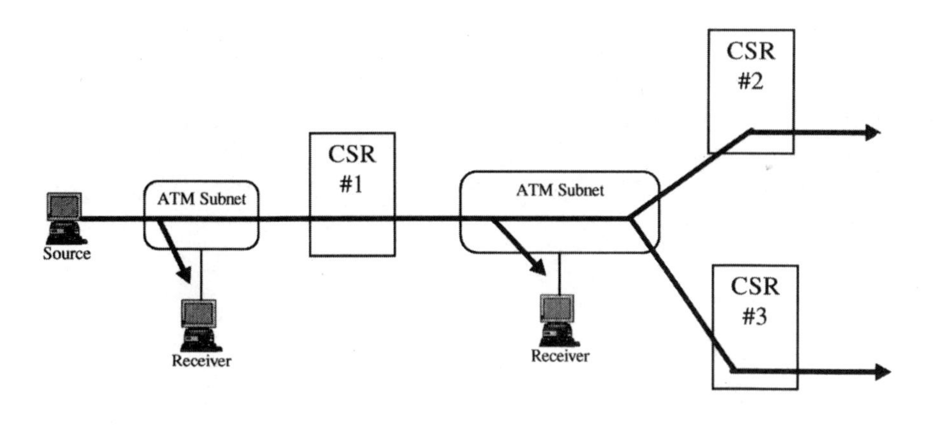

Figure 10-15 CSR multicast bypass pipe.

mesh of PVCs is required to provide optimal performance between any ingress-egress router pair. However, this requires $O(N^2)$ VCs and proportional resource consumption within the router and switches as well as operational staff involvement. Placement of CSR devices in the core of the network reduces the requirement for a VC mesh and the amount of control traffic propagated throughout the network and presented to the routers, and it still provides the ability to establish a direct VC (bypass pipe) between any ingress-egress router pair. Of course as is the case with all of the proprietary approaches, the shortcut path can only extend as far as the presence of the control mechanisms used to establish it. Therefore, the ideal scenario, as illustrated in Fig. 10-16, for this environment requires that the border routers speak FANP.

Another possible environment where CSRs could be of some benefit is an extended campus backbone or metropolitan area network (MAN) where both high IP forwarding and native ATM services are required. For example, a number of different campus workgroups consisting of switched 10/100/1000 Ethernet are attached via ATM uplinks to one of four hub locations consisting of a CSR. The CSR devices in turn form a backbone interconnected by standard ATM links operating at OC12 and possibly OC48. Observe that native ATM services are available to each of the four campus locations as well. In this scenario IP traffic is forwarded onto the backbone and can, depending on the traffic type and network configuration, follow a standard hop-by-hop routed path or be placed on a cell-switched bypass pipe.

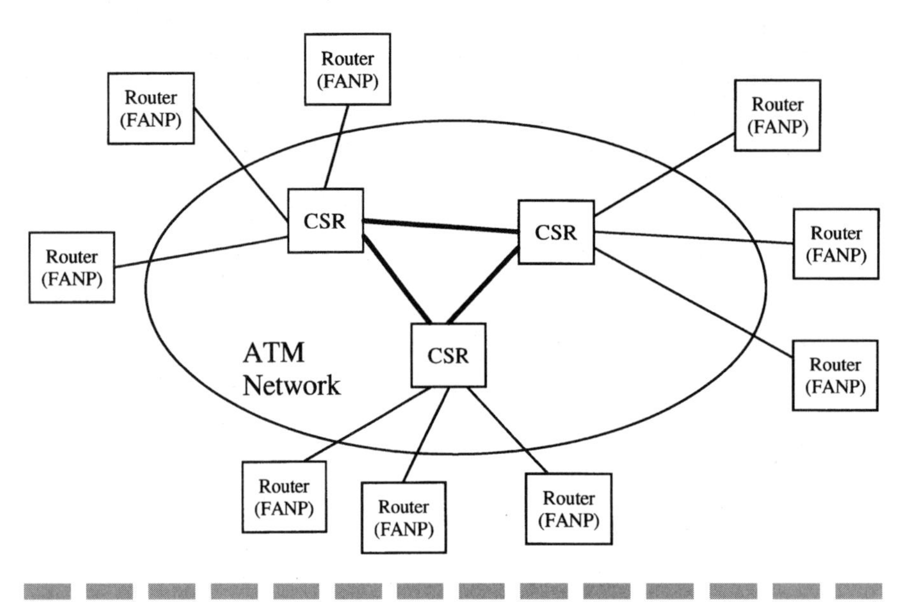

Figure 10-16 CSRs in a large ATM network.

MORE INFORMATION

Esaki et al., "Internetworking Based on Cell Switch Router—Architecture and Protocol Overview," *IEEE Proceedings,* vol. 85, no. 12, Dec. 1997.

ftp://ftp.wide.toshiba.co.jp/pub/csr/

CHAPTER 11

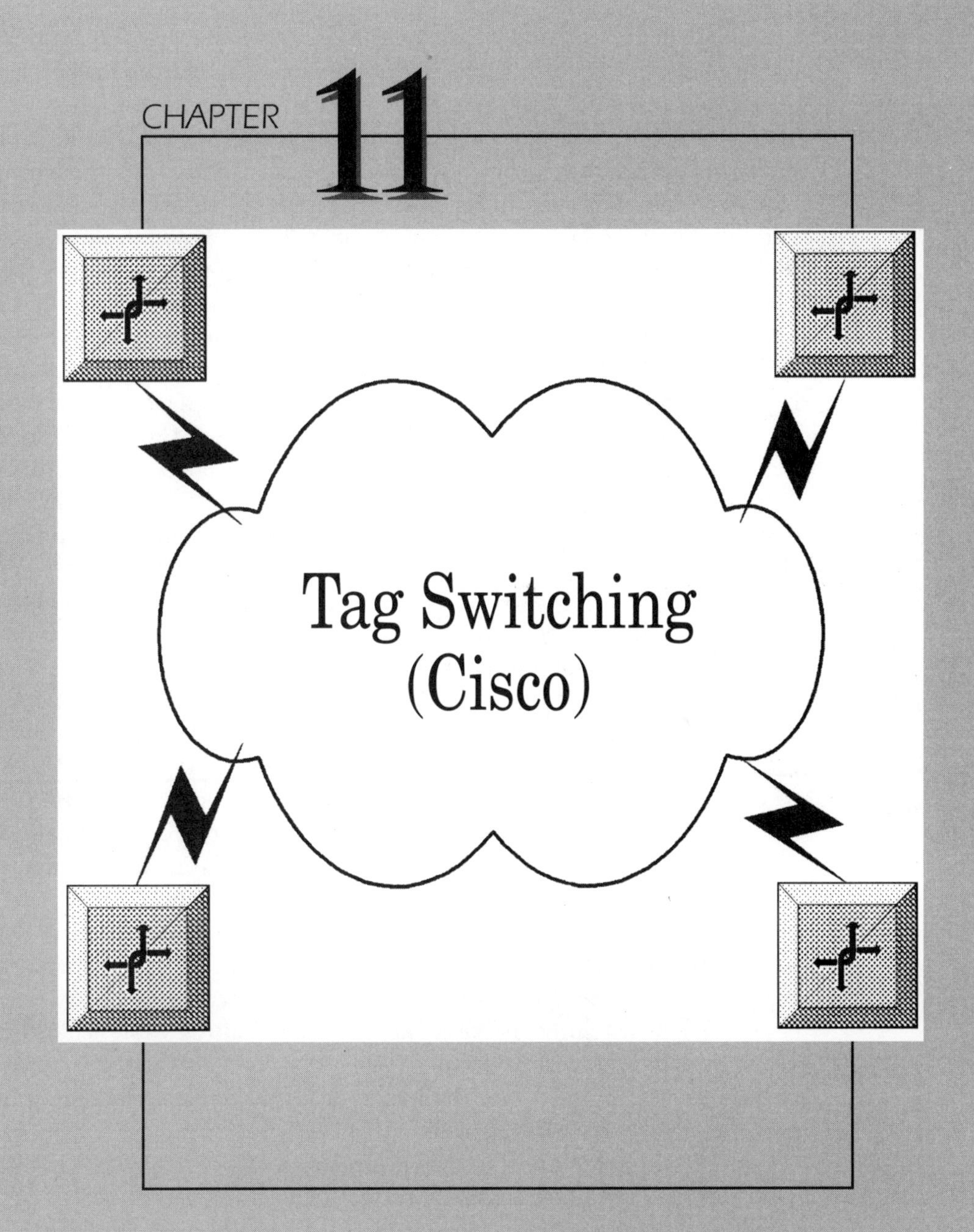

Tag Switching (Cisco)

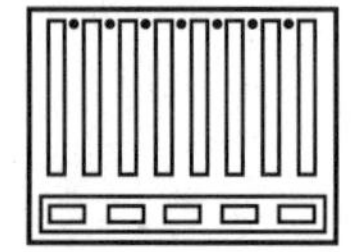

Ipsilon IP Switching and Toshiba CSR/FANP are flow-driven IP switching applications. That is, they set aside switch resources based on the arrival of individual traffic flows. Specifically, the action of allocating a label (VPI/VCI) and the exchange of signaling messages (IFMP and FANP) to map a flow to a VC is performed independently for each individual IP flow. Both retain the scalable properties of the Internet model and forward all traffic hop by hop at layer 3. And both invoke a special control protocol to redirect an IP flow over an end-to-end shortcut path. Consciously or unconsciously the architects of these solutions may have selected the flow-driven approach for one of several reasons. First it is a way to accommodate the Internet model in which the default operation is hop-by-hop routing. Another reason is that it is conceptually straightforward to classify individual IP traffic flows, peel them off from the routed path, and associate them with the connection-based paradigm of ATM. Whatever their reasons, Ipsilon IP Switching and CSR/FANP produce the desired results of providing high bandwidth and low latency for IP traffic flows. The appearance of CSR/FANP model and the subsequent industry frenzy over Ipsilon IP Switching coupled with exponential Internet expansion seemed to indicate at least a very strong interest in solutions that improve IP routing performance.

In the fall of 1996, Cisco proposed a more general and fundamentally different approach to the solution suite of integrated routing and switching. Rather than apply per-flow classification and mapping to establish dynamic shortcut paths, tag switching, as it is called, uses control protocol information to map IP traffic to a switched path. Specifically tag switching takes routing protocol information such as destination prefixes and distributes it along with associated tags (labels) to the participating devices along a routed path.[1] Packets going to a particular destination are appended with the appropriate tag and then forwarded through a network of tag switches based on the contents of the tag. In essence, tag switching replaces the standard routing table lookup performed at layer 3 with a simple label lookup and swapping operation performed at layer 2. The distribution of tags over a particular routed path constitutes a switched path.

Tag switching operates on a device called a tag switch router (TSR). A TSR supports standard unicast and multicast routing protocols such as OSPF and PIM and may be capable of forwarding IP traffic along a

---

[1]A tag is another name for a label. Throughout this chapter the terms *tag* and *label* will be used interchangeably.

default routed path. An additional control protocol, TDP, is used by TSR devices to distribute address prefix-to-label mappings called tag bindings. The distribution of individual tags associated with a destination prefix over a network of TSR devices generates a switched (shortcut) path leading to that destination. The enhanced performance comes from the fact that the packet is forwarded using a label-swap paradigm based on the tag it carries rather than the standard routing table lookup. Tag switching uses the same label-swap forwarding mechanism used in ATM and Frame Relay switching. And tag switching is agnostic to the network-layer service that precipitated the distribution of the tags. This enables any number of different network-layer functions to map to a simple and fast forwarding mechanism. Moreover the performance and scalability of this forwarding mechanism have already proven themselves in operation over large Frame Relay and ATM networks.

Tag switching differs from the previously discussed methods in that it is a topology-driven IP switching protocol. In other words the existence of a destination network prefix as reported by a routing table update creates a switched path leading to that destination. It also generalizes the label-switching forwarding approach to include not just ATM, but frame relay, PPP, and really just about any media encapsulation. The primary objectives of tag switching are to improve the forwarding performance of core routers by using a simple tag-based label-swapping forwarding function, associate different network-layer routing services (e.g., unicast, multicast, COS, etc.) with this label-swapping forwarding paradigm, and maintain media independence.

11.1 Architecture

The architecture of tag switching can be viewed from several different perspectives. At a conceptual level it was designed to improve the performance and services that a large-scale routing system is capable of supporting. Consider the fact that a core router in a large ISP network may have at any one time hundreds or thousands of individual flows passing through it. One common factor among all of those flows is that they are all destined for one of the destination address prefixes that the routing system knows about. Distributing tags to all routers to speed up the forwarding process for all traffic traversing the routing system is a simple technique for enhancing overall network performance. In addition tag generation and distribution do not depend on the arrival of specific data traffic flows. They are driven primarily by the appearance of control traffic, or in other words, routing protocol updates indicating a change in network topology.

It may also be desirable to forward traffic over a path that is not necessarily derived from a dynamic IP routing protocol based on the destination address. For example, traffic flows could be balanced over several equal or near-equal cost paths as a means of distributing traffic load over the entire network. A path through a series of selected nodes and links may provide a nondefault service such as security monitoring, traffic accounting, or some other special service.

Another view of tag switching is from the perspective of the forwarding function in the TSR itself. Typically routers must perform a standard routing table lookup based on the contents of the destination address and the routing table along with a TTL decrement, header checksum, and media translation. Efficient forwarding of traffic depends on the capabilities of the router itself as well as a routing protocol to compute the best next-hop toward the destination. Tag switching simplifies the forwarding function by replacing the standard routing table lookup with a simple lookup and label-swap operation based on the contents of a fixed-length tag contained in each packet. This basic concept is illustrated in Fig. 11-1. An incoming packet with a tag equal to 4 indexes a connection table that specifies the corresponding outgoing port and tag. The tag is swapped, and the packet is transmitted out port 6. Observe that it is the tag in the packet itself, and not the destination prefix, that determines the outbound link the packet is placed on.

An important architectural feature of tag switching is the fact that network layer control functions are segregated from the tag switching forwarding operation. The separation of the two was a deliberate design decision. It enables network providers to associate a number of current and future network services with a simple and scalable forwarding mechanism. Specific services such as destination-based routing, multicast routing, and explicit routing can be associated with a set of tags

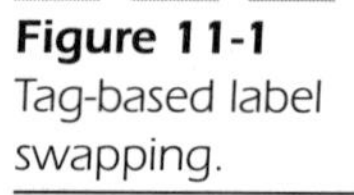

Figure 11-1 Tag-based label swapping.

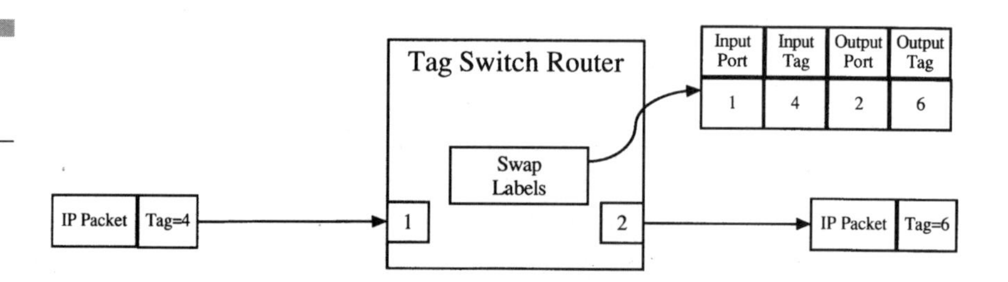

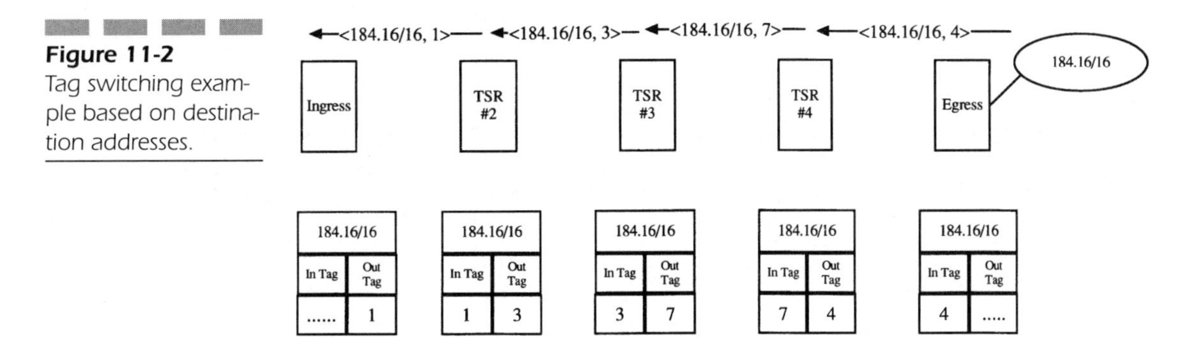

Figure 11-2 Tag switching example based on destination addresses.

that when distributed throughout the network, form per-service end-to-end switched paths. Although the services may vary, the basic forwarding mechanism remains unchanged. Thus if new network layer control functions are introduced, it is not necessary to reoptimize or upgrade the components and machinery in the forwarding path. It also provides some level of future proof protection against unforeseen but necessary changes at the network layer. For example, a sudden requirement to introduce IPv6 because of its larger address space could be achieved without any change to the existing forwarding path.[2]

Tag switching consists of two fundamental components: control and forwarding. The control component is responsible for generating and maintaining a correct set of tags among the participating tag switch devices. Distribution of tags can be accomplished via a separate control protocol such as TDP, or it can be done by an existing control (e.g., RSVP, PIM) protocol that has been modified to carry tags between TSR devices. The forwarding component uses the tag information contained inside the packet and the tag information stored in a tag switch device to perform the packet-forwarding process. Although a tag can take on different appearances depending on the physical media (e.g., VPI/VCI, DLCI, etc.), the basic forwarding mechanism remains unchanged. This means that tag information and thus tag switching can operate over any data link technology and are not bound to one specific data link implementation.

An example of the tag switching operation as applied to basic destination-based routing is illustrated in Fig. 11-2. The network consists of

[2]This is probably an oversimplification. In practice, devices placed at the edge of a tag switching network are still required to forward packets at layer 3 and thus will need to support any incremental changes introduced at the network layer.

three interior TSR devices and both an ingress and egress TSR device. Each TSR contains a routing table entry for destination network 184.16/16. The downstream TSR in each segment along the path leading to the destination network allocates a tag on the incoming port and transmits a tag binding consisting of <184.16/16, tag> to its upstream TSR device. The upstream TSR device receives the message and places the tag in the outgoing port. The result of the individually performed tag binding exchanges is a shortcut path from the ingress to the egress carrying all traffic destined for network 184.16/16.

Several important functions of tag switching can be pointed out from this example. First, the shortcut path is established based on control traffic—not data traffic. In other words, switched resources are allocated even though there may be no data flowing to that particular destination network over a tag switched path. However, when packets bound for the destination network do arrive, they are immediately forwarded over a preexisting shortcut path. There is no requirement to perform flow classification to determine if a shortcut path should be established, and no latency in setting up the switched path—it is already present. Second, the ingress and egress TSR devices are required to perform standard layer 3 processing. The ingress TSR device consults the forwarding table before appending a tag to a packet transmitted into the network, and the egress TSR must remove the tag and forward the packet to the next hop as computed by the routing protocol. Third, there is no per-packet layer-3 processing performed by the interior TSR devices. This presumably provides a forwarding performance boost and enables tags to be associated with a number of different network layer services.

11.2 Components

The tag switching model comprises a number of different functional, component, and device definitions that are explained in the following:

- *Control component.* The control component is responsible for generating and maintaining a consistent set of tags among a set of TSR devices. Generating a tag involves allocating a tag and then binding it to a particular destination. The destination can be a host address, address prefix, multicast group address, or just about any network-layer information. The distribution of the tag bindings is accomplished using TDP or by "piggybacking" the tag on an existing control protocol such as RSVP or PIM.

- *Forwarding component.* The forwarding component uses the tag contained in a packet and the contents of the tag information base (TIB) contained in each TSR device to forward packets. Specifically when a packet containing a tag is received by a TSR, the tag is used to index an entry in the TIB. An entry in the TIB consists of an incoming tag, outgoing tag, outgoing interface, and any outgoing data link layer information or encapsulation. If the tag contained in the packet matches an incoming tag in the TIB, for each corresponding outgoing entry, the tag is swapped, the data link layer information is applied, and the packet is transmitted on the outbound link. An example of the forwarding operation is illustrated in Fig. 11-3.
- *Tag switching.* The architecture, protocols, and procedures that bind network layer information to tags and forward a packet using a label (tag) swapping mechanism.
- *TSR.* A forwarding device that runs standard unicast and multicast routing protocols, is possibly capable of forwarding packets at layer 3, and has either a hardware- or software-driven label swap forwarding engine. A TSR can distribute tag bindings between adjacent or interconnected TSR devices. Examples of TSR implementations include traditional routers, ATM switches, and hybrid switch/routers.

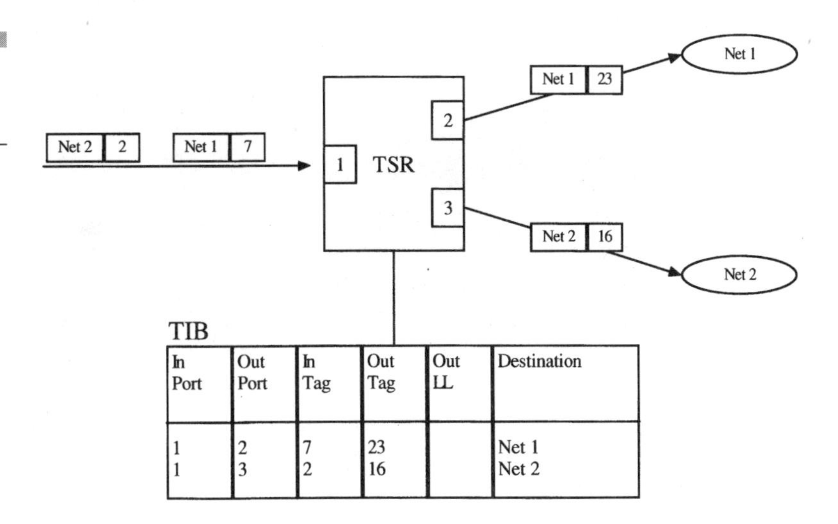

In Port	Out Port	In Tag	Out Tag	Out LL	Destination
1	2	7	23		Net 1
1	3	2	16		Net 2

Figure 11-3 Tag switch forwarding.

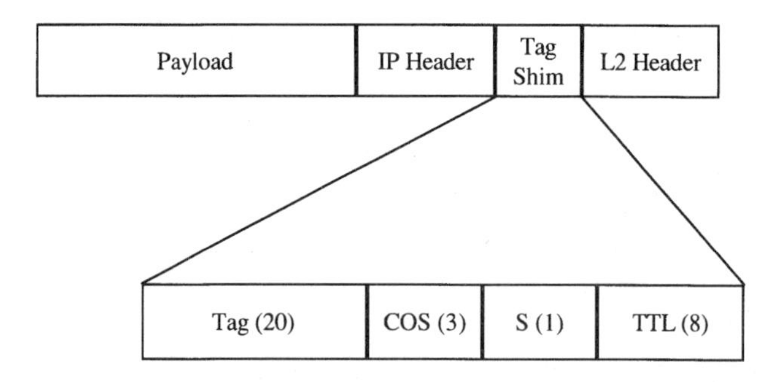

Figure 11-4
Tag shim header.

- *Tag edge routers (TER).* A TER is positioned at the ingress or egress of a TSR-based network. It is capable of allocating tags on either the incoming or outgoing port, can perform layer-3 forwarding, and runs standard unicast and multicast routing protocols. TER (or TER function) tags previously untagged packets at the ingress and removes the tag from previously tagged packets at the network egress.
- *Tag.* A tag is a short fixed-length field header that is contained in a packet. An example of a tag can be a VPI and/or VCI value in an ATM cell, the DLCI header in a frame relay PDU, or a "shim" tag that is inserted between the layer-2 and layer-3 addressing information in a packet. An example of a shim tag is shown in Fig. 11-4. It is 32 bits in length with the first 20 bits allocated for the tag, 3 bits for COS, 1 bit for a stack indicator, and 8 bits for TTL. The positioning of the tag shim header in a generic PPP or Ethernet frame is illustrated in Fig. 11-4. Tag switching requires a tag in each packet so that if the media encapsulation does not offer a "native tag," as is the case with ATM and frame relay, the tag shim header can serve as the basis for label swapping a packet through a network of TSR devices.[3]
- *TIB.* The TIB is the connection or label-swap table that is built and maintained in tag switching devices. The TIB, rather than the routing table (FIB), is the table that is indexed to forward packets through a network. An example of a TIB is shown in Fig. 11-3.

[3]Inserting an additional field into the packet increases the length of the packet. Because fragmentation is undesirable, TER devices should perform Path MTU Discovery to determine the maximum allowable MTU size over a path without requiring fragmentation.

Entries in the TIB are updated by the receipt of tags contained in TDP or other control protocols.

- *TDP.* TDP is a peer-based control protocol used by TSR devices to distribute tag bindings. TDP message exchanges only occur between peer or adjacent nodes in a tag switching network.
- *Tag stack.* One of the interesting capabilities introduced by tag switching is the notion of a tag stack. This is a technique that is roughly analogous to IP over IP encapsulation, and it enables a packet to carry more than one tag. It is supported by "pushing" a new tag (level 2) on top of an existing tagged (level 1) packet, forwarding the packet through the network based on the level-2 tags, and then "popping" the level-2 tag and resuming tag switching operations based on the level-1 tags. An example is shown in Fig. 11-5. A tagged packet arrives at TSR 1 that is configured to push a level-2 tag (A) on top of the existing level-1 tag (X). The packet is forwarded through the network until it reaches TSR 4 where the level-2 tag (C) is popped, and the packet is forwarded based on the level-1 tag (X). Tag stacking can be used when applying tag switching to hierarchical routing. For example, the level-1 tags could be associated with interdomain BGP routing, and the level-2 tags could support intradomain OSPF routing.
- *Forwarding Equivalence Class (FEC).* An FEC is a flow of packets with a set of common characteristics that can be forwarded in a similar manner through the network. For the purposes of tag switching, an FEC can be mapped to a set of labels along a routed path. For example, all packets with a shared destination prefix could be considered a single FEC and thus could be bound to a

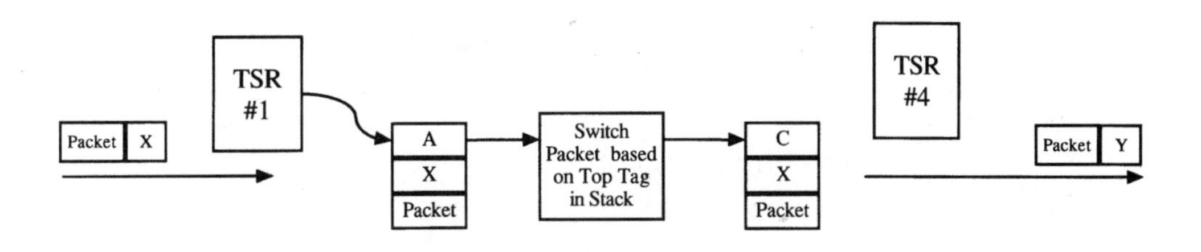

Figure 11-5 Tag stack.

single label in a tag binding. An FEC can also be as fine-grained as a host-to-host flow.

- *Tag switch path.* The ingress-to-egress switched path formed through a series of TSR devices by associating an FEC with a set of tags.
- *Piggybacking.* Tag switching allows tags to be distributed using existing control protocols such as RSVP and PIM. The term for distributing tags is such a manner is *piggybacking.*

11.3 Tag Allocation Methods

An essential component of tag switching operations is the allocation and distribution of tags and tag bindings. The distribution of the tags is handled by either TDP or is piggybacked on existing control protocols. There are three defined techniques that determine how the tags are allocated and by which TSR device: downstream, downstream on demand, and upstream. The primary difference between the upstream and downstream approaches is that with the upstream approach the tag binding is created by the TSR device that will apply the label (tag the packet). With the downstream approach, the tag binding is created by the TSR device that will be receiving the tagged packet and performing the TIB lookup. It should be noted that the tag allocation schemes described below are independent of whether TDP or piggybacking is used to distribute the tag bindings.

11.3.1 Downstream

In the downstream tag allocation scheme, tags are allocated by the downstream TSR device and conveyed to the upstream TSR neighbor. The basic steps for the downstream tag allocation scheme (based on routing table entries) consist of the following and are illustrated in Fig. 11-6.

- For each entry in the routing table, the downstream TSR device allocates a tag and updates the incoming entry in the TIB. A tag binding consisting of <address prefix, tag> is transmitted to the upstream TSR device.

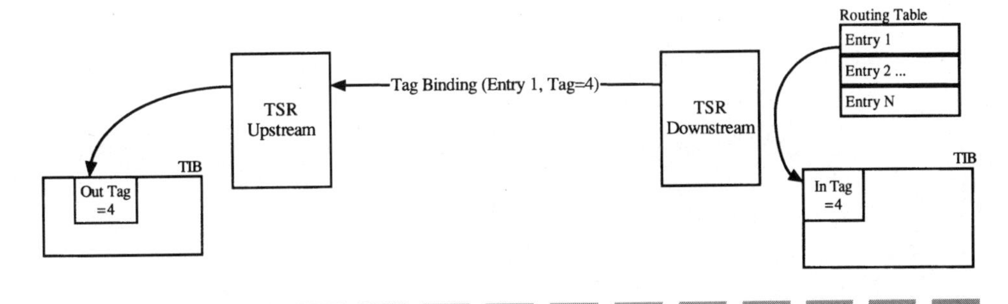

Figure 11-6 Downstream tag allocation.

- When the upstream TSR receives the tag binding and determines that it came from the next hop toward the destination, it places the tag in the outgoing tag entry of its TIB along with any outgoing link-layer information.

11.3.2 Downstream on Demand

In the downstream-on-demand tag allocation scheme, tags are allocated by a downstream TSR device and conveyed upstream as with the downstream technique. But the allocation of tags by the downstream TSR device is only performed upon a specific request from the upstream TSR device. This technique is well-suited for TSR devices that contain an ATM switch component. This is because ATM switches typically have a finite set of tags (VPI/VCI labels) that they can support. Real implementations of tag switching in an ATM environment will most likely support tag switching and native ATM services. This means that the entire VPI/VCI label space will have to be divided between the two, further reducing the amount of tags available to tag switching. Therefore it makes good sense to allocate a tag only when it is needed. In addition the downstream tag allocation schemes are consistent with the direction that routing protocol updates originate from (the actual destination network) and the flow of receiver-driven control protocols such as RSVP and PIM.

The basic steps of the downstream on-demand tag allocation scheme consist of the following and are illustrated in Fig. 11-7.

- For each entry in the routing table, the upstream TSR device generates a request for a tag binding and transmits it to the next hop toward the destination.

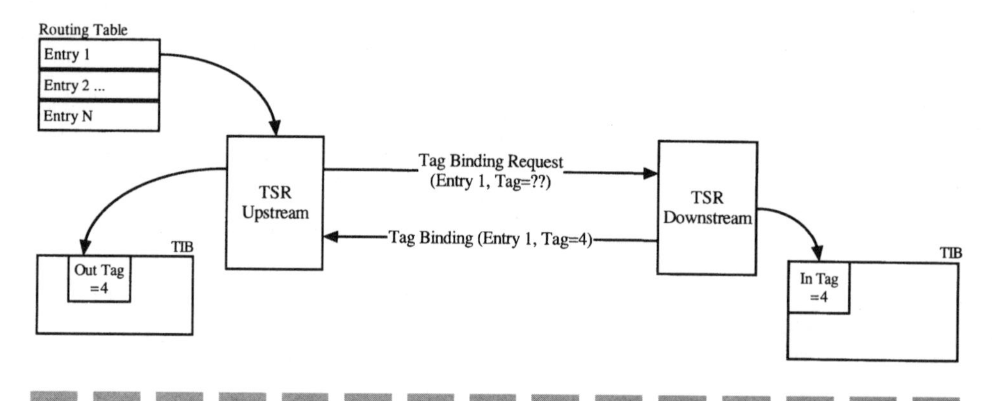

Figure 11-7 Downstream on demand tag allocation.

- After the downstream TSR device representing the next hop toward the destination receives the request, it allocates a tag and updates the incoming entry of its TIB. A tag binding consisting of <address prefix, tag> is created and transmitted to the upstream TSR device.
- When the upstream TSR receives the tag binding, it places the tag in the outgoing tag entry of its TIB along with any outgoing link-layer information.

11.3.3 Upstream

In the upstream tag allocation scheme, tags are allocated by the upstream TSR device and communicated over a point-to-point link to the downstream TSR device. The upstream approach is limited to a point-to-point configuration because the outgoing TIB entries maintained by the upstream TSR device must be unique on a given outbound port for a particular tag binding.

The basic steps for the upstream tag allocation scheme consist of the following:

- For each entry in the routing table that contains a next-hop address reachable by a point-to-point link, the upstream TSR first allocates a tag. It next updates the appropriate outbound TIB entry by placing the allocated tag in the outgoing tag entry field along with specific per-interface link-layer information. It then creates a tag binding consisting of <address prefix, tag>.

- The upstream TSR transmits the tag binding to the downstream TSR device representing the next hop toward the destination.
- The downstream TSR device receives the tag binding and places the tag in the incoming TIB entry for the destination network.

11.4 Tag Distribution Protocol

Tag Distribution Protocol (TDP) is one of several control protocols permitted by the tag switching architecture to carry tag binding information among participating TSR devices. It is, however, the only control protocol designed specifically for this purpose. TDP operates on TSR and TER devices in conjunction with normal unicast and multicast routing protocols. Its operation is independent of the event or events that require the generation and distribution of tag binding information. When called upon though, TDP can efficiently and reliably distribute tag binding information to TSR devices.

11.4.1 TDP Functions

The primary functional objectives of TDP are to support the distribution, request, and release of tag binding information among participating TSR devices. TDP operates over a TCP connection established between peer TSR devices. TCP is used as a transport for several reasons. First TDP operates on the notion of incremental updates in which only a new or modified state is propagated. This is similar in concept and operation to OSPF and BGP. To achieve this, the information must be reliably delivered to the destination in proper sequence. TCP provides this capability, so all TDP messages are transported over a TCP connection. Second, the reliable transport mechanisms provided by TCP obviate the need to engineer this capability into TDP, thus making its design and extensibility much simpler.

Two TSR devices that wish to exchange tag binding information first establish a TCP connection with each other. The TCP connection is bidirectional, so TDP messages can flow in either direction. Each TDP message contains a fixed header along with one or more variable-length Protocol Information Elements (PIE). A PIE consists of one or more TLV fields. The general structure of a TDP packet and two PIE structures is illustrated in Fig. 11-8.

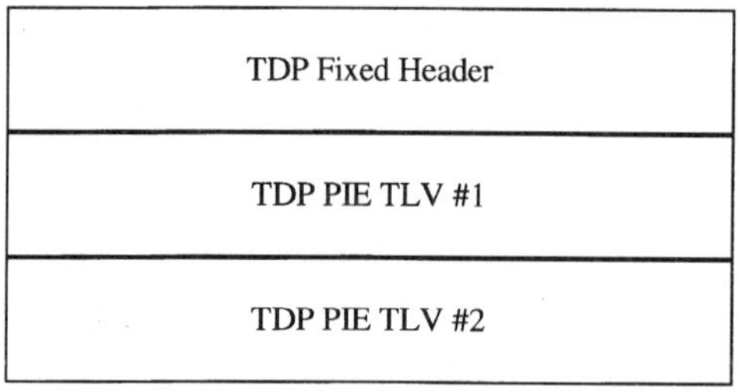

Figure 11-8
TDP packet format.

TDP Fixed Header
TDP PIE TLV #1
TDP PIE TLV #2

After a TCP connection is established, a series of initialization messages are exchanged, placing the peer TSR devices in an operational state. After that, the exchange of tag bindings using TDP may begin. Specifically tag binding information is packed into TDP PIE structures and framed in a TDP fixed header for transmission over the TCP connection. TDP messages have been defined to flow in the context of the different tag allocation schemes defined in the previous section. If the TCP connection is lost, the tag binding information is removed and the tags are deallocated.

Given that ATM is a natural label-switching technology, it makes sense where possible to accommodate tag switching operations over an ATM network. TDP, and more generally tag switching, provides the following support for ATM networks:

- TDP messages flow over VPI/VCI = 0/32.
- Default encapsulation for TDP messages is based on LLC/SNAP encapsulation defined in RFC1483.
- Tags can consist of either a VPI and/or VCI value.
- TDP enables two TSR devices to decide on the ATM encapsulation type (LLC/SNAP or VC-Mux) and range of supported VCI values.[4]

11.4.2 TDP Protocol Information Element Types

The initial version of TDP supports the following PIE types:

- *TDP_PIE_OPEN.* This is the first PIE sent by a TSR after a TCP connection is established with its peer. The TSR that receives this

[4]Tag switching assumes that neighbor TSR devices will be directly connected to each other or, if intermediate ATM switches exist, by a preconfigured ATM Virtual Path.

PIE should respond immediately with a TDP_PIE_KEEPALIVE or with a TDP_PIE_NOTIFICATION. The TLVs supported in this PIE include downstream-on-demand desirability, the VCI range supported, and the VC encapsulation type.

- *TDP_PIE_BIND.* This PIE is sent from one TSR device to another to distribute tag bindings. It may be generated based on an event (e.g., routing table update) or in response to a *TDP_REQUEST_BIND.* The tag binding information is contained in a TLV structure and can consist of a tag value, address prefix, and length. The format of a TDP_PIE_BIND is illustrated in Fig. 11-9.
    - *Request ID.* Used to correlate a response to a TDP_BIND_REQUEST
    - *AFAM.* Specifies the address family of network-layer addresses contained in tag bindings. See RFC1700 for details on address families.
    - *BLIST_TYPE.* Specifies the format and semantics of the BLIST entries in the BINDING_LIST. Values for BLIST_TYPE are shown in Table 11-1.
    - *BLIST_LENGTH.* Length of the binding list.
    - *BINDING_LIST.* Variable-length field consisting of one or more BLIST entries of the type indicated by the BLIST_TYPE. BLIST entries generally consist of a 32-bit tag value and the associated unicast address prefix or multicast group address. In the case of unicast address prefixes, a prefix length is also included. The hop count (HC) value defined in BLIST types 5 and 6 is used to indicate the number of router hops that a tagged packet would traverse were it being forwarded hop-by-

Figure 11-9 TDP_PIE_BIND format.

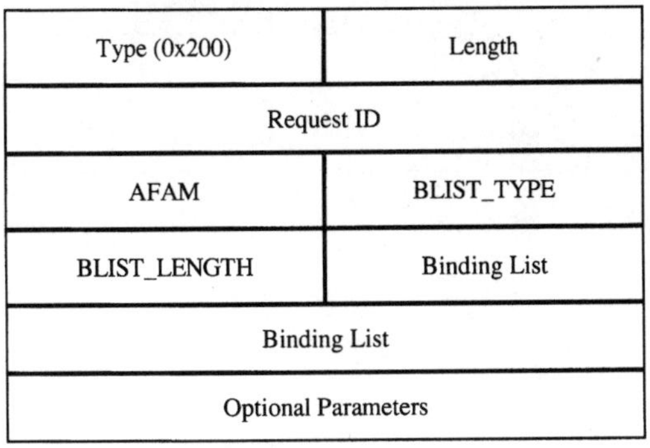

Type (0x200)	Length
Request ID	
AFAM	BLIST_TYPE
BLIST_LENGTH	Binding List
Binding List	
Optional Parameters	

TABLE 11-1

BLIST_TYPE Table

BLIST_TYPE	BLIST Entry Format
0	Null
1	32-bit upstream assigned
2	32-bit downstream assigned
3	22-bit multicast upstream assigned (*,G)
4	32-bit multicast upstream assigned (S,G)
5	32-bit upstream assigned VCI tag

hop.[5] Figure 11-10 shows the format of the various BLIST entries.

- *TDP_PIE_REQUEST_BIND.* Used to request a tag binding for a specific address prefix or all bindings that the requested TSR device contains.
- *TDP_PIE_WITHDRAW_BIND.* Issued by a TSR device that allocated a tag binding to inform other TSR devices that allocated tag binding should not be used anymore.

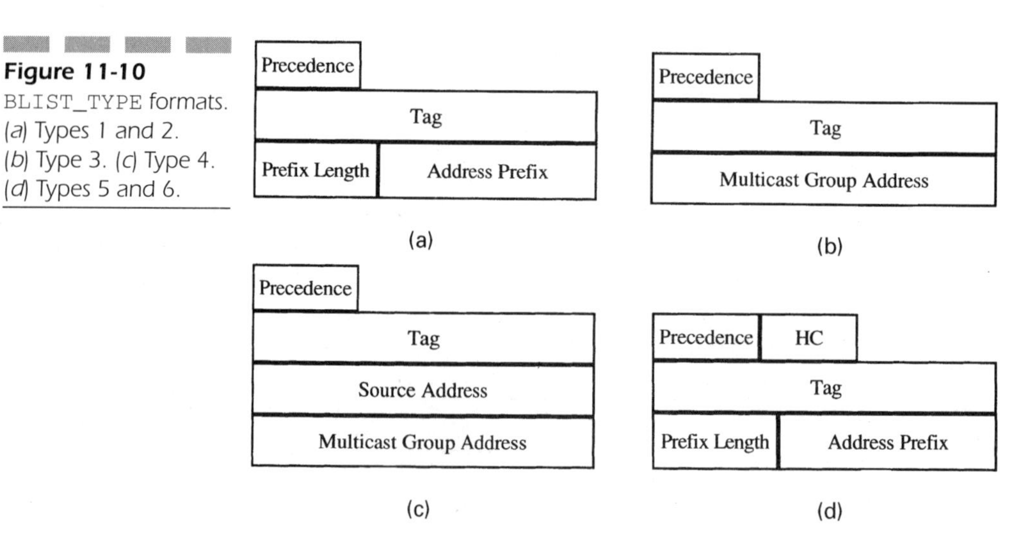

Figure 11-10 BLIST_TYPE formats. (*a*) Types 1 and 2. (*b*) Type 3. (*c*) Type 4. (*d*) Types 5 and 6.

[5]Note that in some cases such as ATM, there is no TTL field to decrement. This can be done at the network ingress or egress.

- `TDP_PIE_KEEP_ALIVE`. This is used in lieu of an actual TDP message to reset a hold timer. If the hold timer expires, the peer TSR devices will close their TCP connections.
- `TDP_PIE_NOTIFICATION`. Notifies peer TSR device of a "significant" event. Examples include errors and changes to either TSR capabilities or operational state.
- `TDP_PIE_RELEASE_BIND`. Issued by a TSR that no longer needs a tag binding that it received from an earlier upstream request/downstream assignment operation.

11.5 Tag Switching Services

The tag contained in each packet is used to index a TIB contained in each port or switch. It only has local significance as far as the network is concerned, and there is nothing about the tag that indicates what kind of IP traffic is being tag switched through the network. Therein lies one of the most powerful features of tag switching: the ability to associate any network-layer traffic or service with a set of tags and thus utilize the same label-swap forwarding paradigm.

11.5.1 Destination-Based Routing

Destination-based routing for unicast traffic is probably the most straightforward application of tag switching. TSR and TER devices run normal unicast routing protocols and exchange routing protocol updates with their peers in the network in a normal fashion. Following a change in network topology, routers will converge and build a new forwarding information base (FIB) reflecting the current topology of the network. Address prefixes are mapped to tags and distributed among the participating TSR and TER devices in the network using TDP and one of the tag allocation schemes discussed earlier. As a participant in the tag allocation process, the ingress TER device will associate a tag with each entry in the FIB. Untagged packets entering the network are processed at layer 3 by the ingress TER device and appended with the appropriate tag based on the FIB lookup. Packets are then tag switched through the network and exit at the appropriate egress TER device.

An example of destination-based tag switching was illustrated in Fig. 11-2. Observe that TDP operates independently between any two TSR devices. This enables the process of tag allocation and distribution to operate in a distributed manner. This is more a general feature of TDP than something that is specific to destination-based routing, but it is important in this context because the faster a tag switched path can be put in place, the sooner packets will flow over it to their respective destinations. Another point is that all traffic destined for a particular network will be tagged and switched through the network—not just the tail of the flow as is the case with IFMP or FANP.

11.5.2 Hierarchy

The IP routing architecture deployed on the Internet is hierarchical in nature. Routing between autonomous systems (ASs) is handled by exterior gateway protocols (EGPs) such as BGP, and routing within an autonomous system is supported by interior gateway protocols (IGPs) such as OSPF. However, there is not a complete separation between the two. Routers inside of a transit AS are required to maintain information on the interior (within the AS) and exterior (all other ASs that are transiting through) topologies. This is so a router inside the transit network can forward the packet to the appropriate border router and on to the next AS. In this case the amount of routing information that must be maintained by the interior transit routers may not be trivial and could lead to longer routing table lookups and slower convergence following a topology change.

Tag switching provides a technique that separates interior and exterior routing information so that routers inside of a transit AS are only required to maintain routing information about their respective AS. The border routers running an EGP such as BGP would continue to maintain routing information about the external topologies. Tag switching provides this clean separation between interior and exterior topologies with a tag stack. One set of tags is generated and allocated based on the routing table entries maintained by BGP border routers. Another set of tags is generated and allocated based on the routing table entries maintained by the interior transit routers running an IGP such as OSPF. Packets inside of the transit AS would carry a tag stack consisting of EGP and IGP associated tags.

The notion of applying routing hierarchy to tag switching and more specifically to tag stacks is illustrated in the conceptual diagram in Fig.

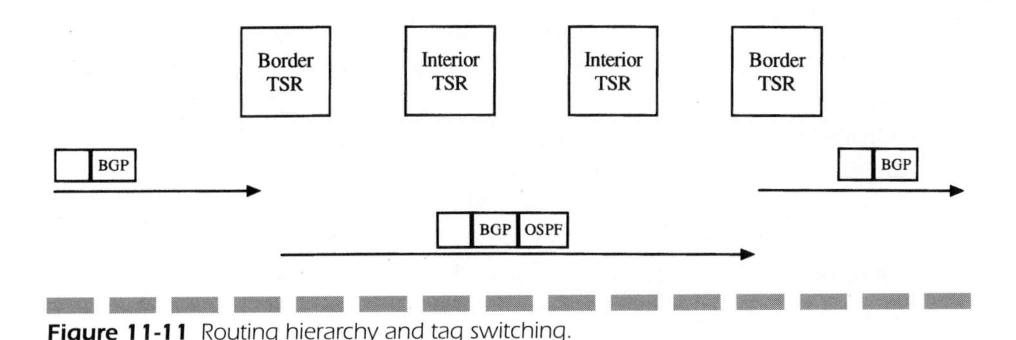

Figure 11-11 Routing hierarchy and tag switching.

11-11. Observe that prior to entry into the transit AS, the packet contains a single tag that is associated with BGP managed routes. These tags are only swapped at BGP border routers. To get across the transit AS to the next-hop BGP border routers requires the use of an IGP and more specifically an IGP derived set of tags. Therefore the ingress TSR will push an IGP-specific tag on top of the BGP tag and forward the packet through the network using standard tag forwarding mechanisms. The BGP tag is not swapped or acted upon because it is not at the top of the stack—the IGP tag is. When the packet reaches the egress TSR, the IGP tag is popped and the BGP tag is used to index the TIB and forward the packet onward.

11.5.3 ATM

For a number of reasons, tag switching and ATM switching are a natural fit. First, ATM switching employs the label-swap forwarding mechanism to forward cells through a network. Second, the VPI and/or VCI values contained in each cell can serve as the tag (or stack of tags) to provide one or two levels of tags to which network-layer information can be bound. And in keeping with the design objectives of tag switching, ATM provides a high-bandwidth transport that is independent of the network-layer traffic (FEC) associated with the tags and thus with a tag switched path.

To support tag switching, an ATM switch must first operate an instance of a routing protocol such as OSPF and exchange routing table updates with peer TSR/TER devices and routers. It must also operate a control component that is capable of allocating and distributing tags based on network-layer information. This means that either TDP or a control protocol capable of carrying tags must be supported. And if the ATM-TSR device is providing an ingress or egress function, it should be able to forward IP packets as well. In effect, the implementation of these additional router-specific functions along with the TDP control component transforms a run-of-the-mill ATM switch into a TSR device. At the same time, it is not necessary for the converted ATM switch to use or even need ATM Forum protocols.

Tag switch operations that employ ATM switching as the tag forwarding mechanism is relatively straightforward. Tag bindings consisting of address prefixes, and now VPI/VCI values are generated and distributed using TDP. To prevent the unnecessary consumption of VPI and/or VCI labels, the suggested method for tag allocation is downstream on demand. This is so a VPI/VCI label is not allocated on the incoming port of an ATM-TSR device unless it is the next hop toward a destination network for an upstream ATM-TSR device. Normal layer-3 processing along with the requisite segmentation and reassembly functions are performed at the ingress and egress ATM-TER devices, respectively.

Generally speaking, tag switching scales proportionally with the number of entries in a routing table. In other words, the number of tags that a tag switch maintains in its TIB is equal to the number of entries in the TSR routing table.[6] However, because ATM is used as the tag-forwarding mechanism, there may be situations where a TSR is required to maintain more than one tag for a particular routing table entry. This situation occurs when multiple upstream ATM-TSR devices share the same next-hop ATM-TSR for a particular destination network. As illustrated in Fig. 11-12, two ingress ATM-TER devices share a common next-hop ATM-TSR for destination network (NET X). To prevent cells from different sources (ATM-TER 1 and ATM-TER 2) from interleaving over the same downstream VC, separate VCI values must be installed in the ATM-TSR TIB as shown. This ensures that cells from the different sources will have unique VCI values when they arrive at the egress ATM-TER. It then becomes a simple matter for the egress ATM-TER to assemble the cells into packets and perform standard layer-3 processing.

Overcoming the cell interleave problem can be accomplished in one of several ways[7] First the VPI value can be used to forward cells through a network. Individual VCI values would be used to maintain the identity

---

[6]If a tag represents a group of routing table entries, even fewer tags are needed in the TIB.

[7]An additional technique called VC merging will be discussed in Chap. 12.

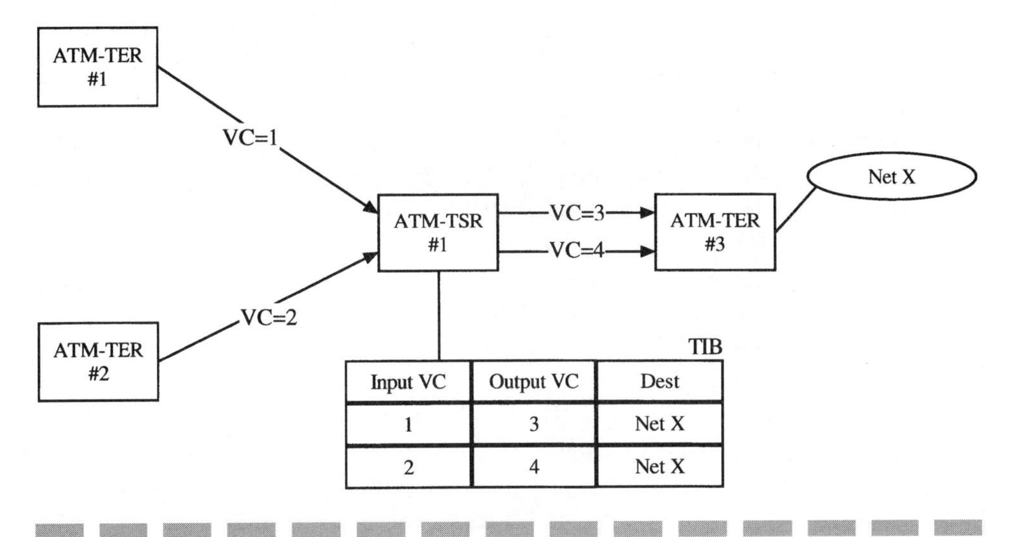

Figure 11-12 Tag switching and ATM TIB example.

of the source at the egress ATM-TSR. However, because all of the VCI labels (representing multiple different sources) may share a single VPI label upon arrival at the egress ATM-TER, some way is needed to prevent potential VCI collisions. One solution might be to configure a unique range of VCI labels that each ingress ATM-TER can use. Note also that the scalability of this approach is limited by the size of the VPI label space, which allows only 4096 unique labels.

Implementing the tag switching control and forwarding planes over an ATM switch fabric provides a number of possible advantages when compared with traditional IP over ATM approach of attaching a large number of routers to the perimeter of an ATM cloud. First, the burden of establishing relationships with multiple peer routers is reduced because just like a normal router, an ATM-TSR will simply exchange routing protocol updates with its adjacent neighbors. Second, one does not need to operate standard ATM Forum protocols. A shortcut path (tag switch path) can be established between an ingress and egress device using the control procedures defined in tag switching. Dynamic IP routing can also take into account the real network topology when selecting a path. In addition there is no need to classify individual flows; the shortcut paths to particular destinations are already in place when and if any and all data flows arrive in the network. And finally, given the separation between the control and forwarding components that tag switching provides, it is possible to associate a number of different network-layer services such as destination-based routing, hierarchy, explicit routing, and so on, so that the high-capacity, high-performance ATM transport is best leveraged.

Several other points should be noted about tag switching operations over an ATM network. First, changes in topology can result in the distribution of erroneous routing information that could then result in the formation of a looping tag switch path. Because there is no TTL field in an ATM cell header, this kind of loop could consume both link and TSR resources. Another way of viewing this situation is that the forwarding loop is formed at layer 2 but escapes detection by the loop mitigation process (e.g., TTL decrement) at layer 3. The solution to this problem is to either not form a loop in the first place or hope that the damage performed by the looping traffic is contained by mechanisms like per-VC queuing and EPD and does not in turn affect topological convergence or nonlooping data traffic. Second, unlike the overlay model supported by the IP over ATM solutions, tag switching is still dependent on dynamic IP routing to compute a path through the network. Even if the forwarding path consists of an ATM transport, the path selection is still based on the current set of metrics advertised by existing IP routing protocols such as number of hops or lowest link cost. They do not take into account available capacity, offered load, or path delay when computing a path. However, for performance or load balancing reasons, it is possible to direct traffic over an explicit path by associating a set of tags with a nondefault path through the network.

11.5.4 Multicast

In addition to unicast IP traffic, tag switching can support multicast IP traffic as well. Specifically a tag is associated with a multicast delivery tree. When a packet is received from an upstream TSR device, the tag is used to index the one or more TIB entries associated with downstream branches on the delivery tree. A swap of the tag(s) is performed, and the packet is directed further down the delivery tree. If there is no matching entry in the TIB, the packet is discarded. The benefit that tag switching brings to multicast forwarding is speed and simplicity. Traditional multicast forwarding involves two table lookups: one upstream in the direction of the sender and one downstream toward the receivers. As is the case with unicast traffic, tag switching involves a direct exact match lookup in the TIB and a simple swap of the tag values. And the RPF check is vastly simplified—if there is no matching tag value in the TIB, the packet is discarded.

In some cases a branch or branches of a multicast delivery tree may have inherent multicast capabilities such as a LAN. For example, an Ethernet segment might have an upstream and multiple downstream routers on the same delivery tree. And if tag switching is deployed in this environment, the upstream TSR device should use the same tag when forwarding the multicast packet to multiple downstream TSR devices over the multicast-capable LAN segment.

To support this environment, the TSR devices attached to a common LAN must first determine a method of tag allocation. In other words, who or what is responsible for allocating a tag that represents a group-specific delivery tree that all of the TSR devices may be part of? The solution is to allocate a nonoverlapping portion of the entire tag space to all TSR devices on the LAN. Therefore each individual TSR device is responsible for allocating tags from a specific range of tags that it is responsible for. Once the tag space has been partitioned, it is up to the individual TSR devices to allocate a tag from their range of tags, bind it to a multicast group, and then distribute the tag binding information to all other TSR devices attached to the LAN.

Distribution of tags representing multicast groups can be accomplished using either the upstream or the downstream approach. In the former, the upstream TSR device selects the tag and multicasts it to its downstream neighbors on the LAN. Although this is probably the simplest solution, there are several issues to consider. An upstream TSR may have multiple upstream multicast sources, thereby requiring more than its fair share of allocated tags. Also a change in topology may result in a new upstream TSR and therefore a requirement for tag rebinding. And finally, this approach is inconsistent with the direction that multicast group membership originated from, which is the downstream direction. The preferred approach is for the downstream TSR device to perform the tag binding and distribution function. This is consistent with the flow of multicast group membership information and with the tag allocation/distribution approach to unicast routing. It also enables tag information to be piggybacked onto PIM multicast routing messages because they flow in the upstream direction. And if there is a change in topology resulting in a new upstream TSR device, a tag rebinding is not necessary.

The partitioning of the tag space can be accomplished using an extension to PIM Hello messages. A LAN-attached TSR device multicasts a

PIM Hello message on the LAN that contains the range of tags it wishes to claim for allocation. If there is a collision with another TSR device, the one with the higher IP address wins and the loser must reallocate. PIM Hello messages can also inform the TSR devices of the existence of nontag-capable routers attached to the same LAN.

Tag distribution is performed by placing the tag binding inside of a PIM_JOIN message generated by a downstream TSR device. As illustrated in Fig. 11-13, TSR 1 multicasts a PIM_JOIN containing tag binding information about the multicast group to all other TSR devices on the LAN. The upstream TSR device (TSR 3) receives the tag binding information and places the tag in the outgoing entry of the TIB for the specific multicast group. The other TSR device (TSR 2) on the LAN may have downstream links belonging to the delivery tree, so it will place the tag in the incoming entry of the TIB for the specific multicast group. Observe in this case that it is a single downstream TSR device that is responsible for allocating the tag, binding the tag to a multicast group, and then distributing the tag using an existing control protocol (PIM).

It should be noted that this implementation of multicasting support for tag switching requires use of the PIM Sparse Mode (PIM-SM) multicast routing protocol. TDP is not used to distribute tag bindings because of its use of a TCP connection-oriented transport. If a tag switching

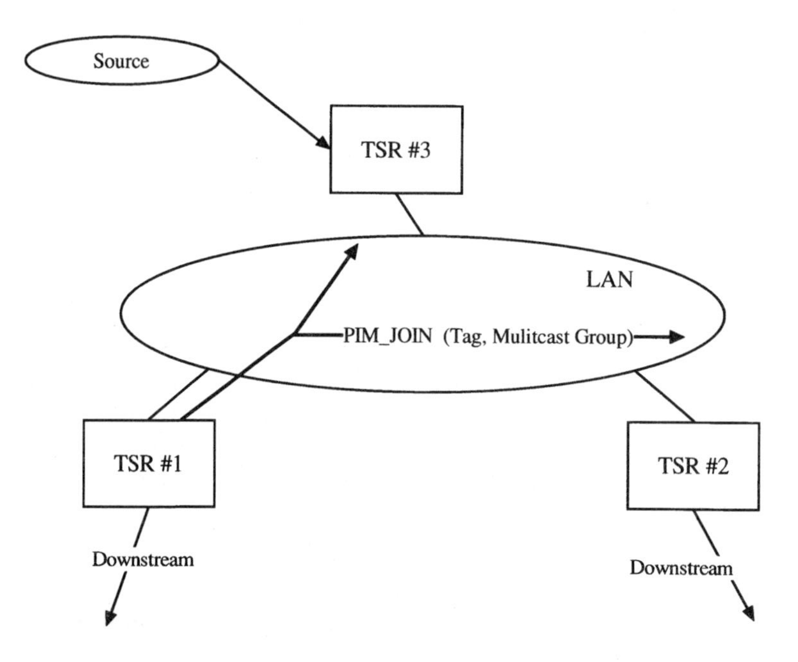

Figure 11-13 PIM_JOIN tag distribution.

domain is not running PIM-SM for various reasons, it is possible to support this capability using the simpler PIM Dense Mode (PIM-DM). However, that would require that downstream PIM-DM routers generate `PIM_JOIN` messages, which would be a departure (and definite modification) from its fundamental flood and prune approach to multicasting.

11.5.5 Class of Service

In addition to improving the performance of IP routing for unicast and multicast traffic, tag switching can support service differentiation. Several techniques can be employed to achieve this. In one case, the tag itself includes a Class-of-Service (COS) indicator that is used by the TSR device to schedule the packet for outbound transmission. In the second case, multiple tags and by extension multiple tag switch paths are established on a per-class basis. In either technique, the simple tag forwarding mechanism remains unchanged from that used to provide normal best-effort service. Tag switching support for COS is consistent with the notion of DiffServ (discussed in Chap. 3) in which multiple flows are mapped to a few service classes, and how the network treats the packets of the flow depends on either their explicit or implicit class membership.

To support COS in the tag switching architecture, a packet must first be classified to determine what kind of service it should receive. This is based on a local policy decision that is performed at various points in the network between and including the origin host and the ingress TER. After, classification, the packet may be marked in some way to indicate the level of service it should receive from the network. Again, this will occur somewhere between, but including, the originating host and the ingress TER. The packet will then enter the tag switching domain and flow over a tag switch path that presumably provides the class-based differentiated service.

The DiffServ model provides a very coarse service in which all flows are mapped to just a few classes. If a more fine-grained or even per-flow service is needed, tag switching can employ the capabilities of RSVP to build and reserve resources along a tag switch. To accomplish this, one tag per session is allocated at each TSR device and is installed along with any QoS state when RSVP messages arrive. The distribution of tag information for RSVP flows requires the introduction of a tag object that is transported in RSVP PATH and RESV messages. The fact that RSVP, a control protocol, is responsible for communicating tag information

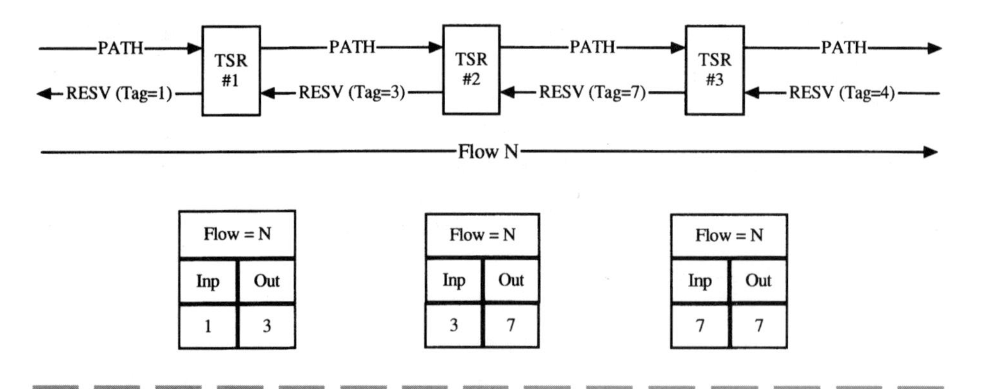

Figure 11-14 RSVP and tag switching.

between TSR devices is another example of the piggyback approach for tag distribution. This simple case is illustrated in Fig. 11-14. Note that RSVP PATH and RESV messages are processed at each hop (TSR device) in a normal fashion. In this example the RSVP RESV message for a particular Flow *N* carries the tag object upstream back toward the sender (not shown). This is an example of the downstream allocation technique for tag distribution. As the RESV message arrives at each TSR, its contents are passed down to the traffic control components of the TSR device. A check is made to ensure that the TSR device can support the bandwidth and QoS characteristics of the flow, and then assuming the check passes, the appropriate resources are set aside to support the flow in the downstream direction. At the same time, the tag object is placed in the outgoing entry of the TIB for that flow. One of the important benefits of RSVP and tag switching is the fact that the per-packet multifield processing overhead of the classifier is replaced by a simple tag lookup mechanism at each TSR. Another benefit that applies to tag switching in general is that both coarse and fine-grained services can be deployed over the same infrastructure using the same forwarding mechanism.

11.5.6 Explicit Routes

Another potentially very useful function that tag switching can support is explicit routing. An explicit route can be thought of as a preconfigured sequence of hops that a specific subset of traffic is directed through. It is, in fact, a nondefault path that overrides the destination-based paths derived from dynamic IP routing computations. Explicit routes can be useful for several reasons. A network provider may wish to better balance the load over the backbone by directing traffic over multiple paths. A special service (e.g., RSVP) may require its own dedicated path that is distinct from the normal default routed path. A customer-defined set of policies (e.g., security, billing, VPN connectivity) may require that certain traffic pass through specific links and nodes. These along with many other reasons not discussed make the notion of explicit routing and more generally the ability to actively and efficiently engineer traffic around the network quite compelling.

To support explicit routing, tag bindings that do not necessarily correspond to normal default routing must be installed in the TSR devices along the specified path. A classifier must also be configured and placed at the explicit path ingress to identify the subset of traffic that will be traveling over the explicit path. This can be accomplished in one of several ways. First, one could manually configure the tag bindings in each TSR device. Second, TDP messages can be "steered" through the series of TSR devices that form the explicit route. Although technically feasible, this approach may involve some local configuration given the current scope of TDP processing, which is confined to adjacent TSR devices[8] A third approach is to modify RSVP to basically follow a sequence of hops that form the ingress-to-egress explicit route.

Specifically RSVP is augmented with an explicit route object. The explicit route object provides a sequence of TSR devices that an RSVP PATH message must pass through on the way to the destination. One of the purposes of the PATH message is to lay down path state along the route that is traversed. When the RESV message is forwarded in the reverse direction back toward the sender, the tag object defined in the previous section is installed in the TSR devices, thus forming an explicit tag switched path.

The use of RSVP to establish explicit paths over an tag switching network is illustrated in Fig. 11-15. In this example ingress TER device A initiates an RSVP PATH message with an explicit route (ER) object. The ER object directs the PATH message through a sequence of TSR devices. The RESV messages flowing in the reverse direction over the same path

---

[8]Because an explicit path runs ingress to egress, it may be simpler to initiate its construction at the edge of the network. TDP does not operate from the edge of network; its operation is distributed and autonomous between pairs of TSR devices. Therefore it is not well-suited for this particular solution.

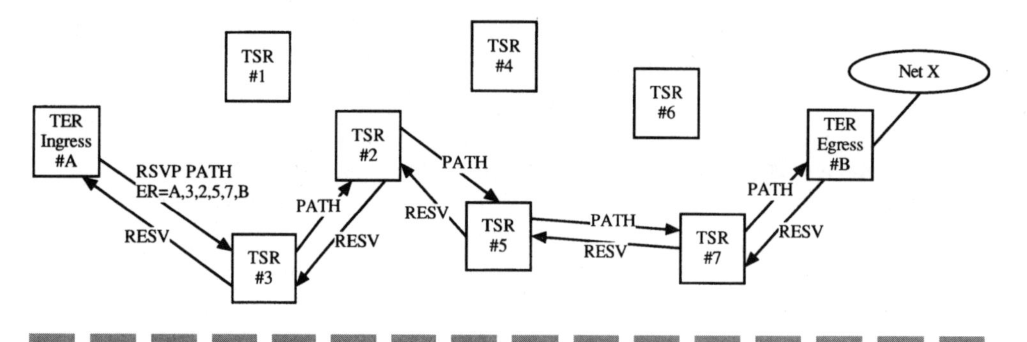

Figure 11-15 Explicit path setup using RSVP.

will install the tags, thereby creating an explicit route for traffic destined for Net *X*. Observe that the source and destination in this case is the ingress and egress TER devices, respectively—not the source and destination hosts as is the usual case with RSVP. Indeed although it is possible to lay down QoS state as well, the primary objective for RSVP in this scenario is to convey tag bindings along an explicit path. In addition the classifier that identifies which traffic will be using the explicit path is installed at the ingress TER device. This could be derived from a combination of several existing RSVP objects (e.g., sender template) manually configured or installed through some other mechanism. It should also be noted that there are no clearly defined mechanisms for dynamically selecting which sequence of hops will form an explicit path. This information must still be manually configured or computed based on network policies. RSVP only provides a mechanism to establish the switched path—not actually define what links and nodes it consists of.

11.6 Tag Switching Network Examples

Given its routing and topology orientation, tag switching is best suited for a large IP-based network such as corporate intranets or ISP networks. It can be deployed in software on existing routers or as an augmented control component for establishing VC-like switched paths in a multiservice ATM network. An example of a basic tag switching network is illustrated in Fig. 11-16. The backbone consists of a number of core

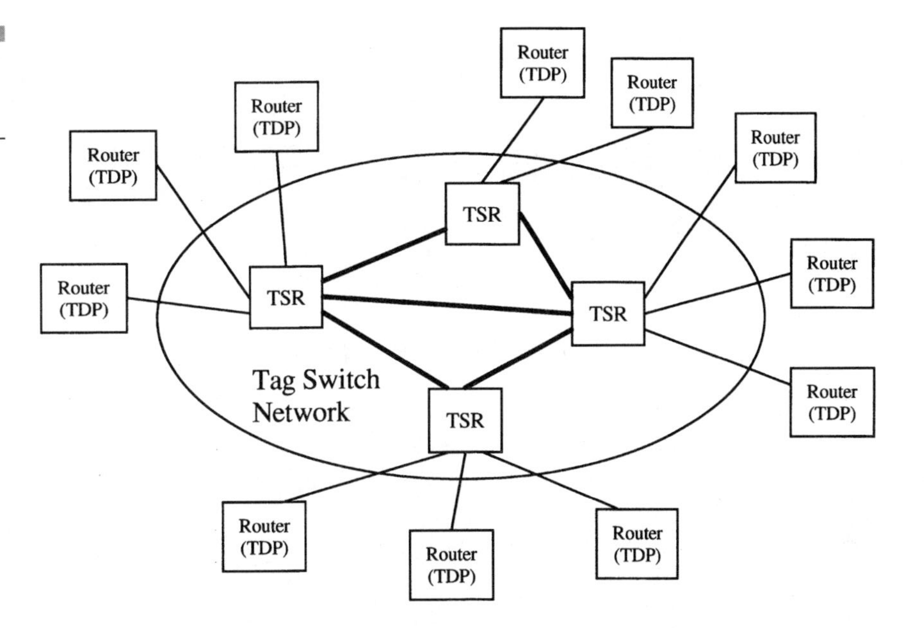

Figure 11-16 Tag switching network example.

TSR devices surrounded by multiple TER devices. This is not unlike the concept of a large switched cloud (frame relay or ATM) serving as the core for a larger number of edge routers. The difference with tag switching is that all of the components shown participate in computing and distributing routing table information. In fact, tag switching is an example of the peer model, so only one address and routing protocol is required. The extra component, TDP, is required to map routes to tags and then distribute the tags among the TSR devices. It also provides the benefits in terms of performance and explicit path control that switched virtual connections provide without requiring frame relay or ATM.

MORE INFORMATION

Doolan et al., *Switching in IP Networks,* Morgan Kaufmann, 1998.

http://www.cisco.com.

Rekhter et al., "Tag Switching Architecture Overview," *IEEE Proceedings,* vol. 85, no. 12, Dec. 1997.

CHAPTER 12

ARIS (IBM)

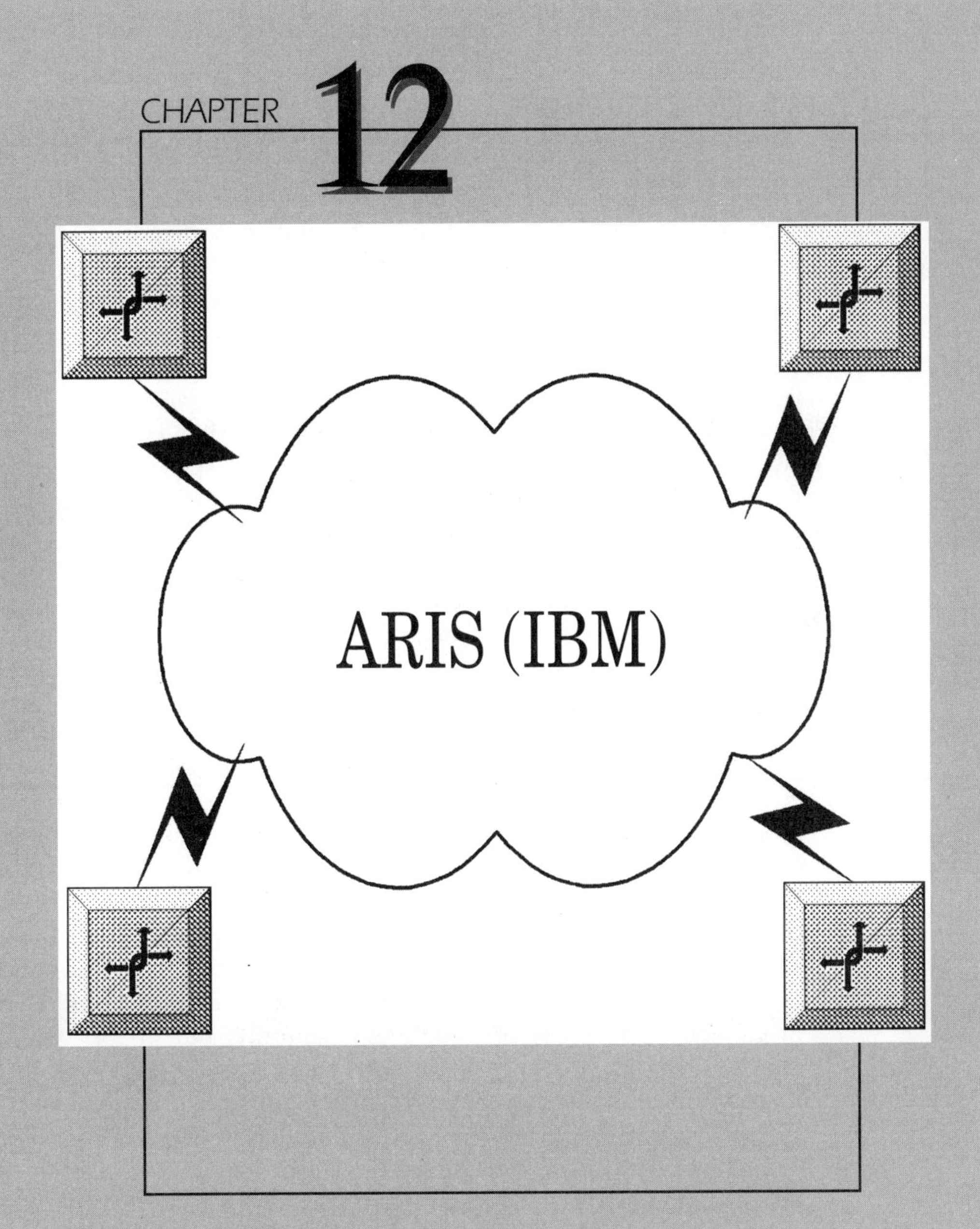

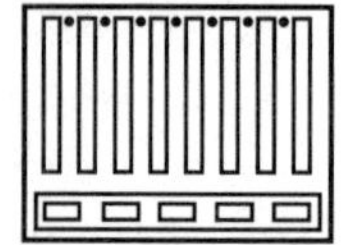

The introduction of tag switching by Cisco signaled a departure from the close association of IP flows with ATM virtual connections as defined in the Ipsilon IP switching and Toshiba CSR/FANP models. Rather than map individual IP flows characterized by host-to-host or application-to-application communications onto what is essentially an ATM virtual connection, tag switching significantly broadened the scope of IP traffic that is placed on a switched path. Indeed the switched paths, constructed from the generation and distribution of tag bindings derived from entries in the routing table, placed all traffic leading to a particular destination on a switched path. This topology-driven approach to IP switching enhanced the scalability of the routing system and allowed all traffic to benefit from the faster performance provided by the label-swapping forwarding paradigm. In addition, other network services such as nondefault explicit routing that are not specifically related to destination-based routing can be mapped to tags, thus allowing specific traffic flows to traverse a different path with no performance penalty.

Following the introduction of tag switching in the fall of 1996, IBM shortly afterward announced a somewhat similar topology-driven IP switching solution with the catchy acronym of ARIS. ARIS stands for Aggregate Route-Based IP Switching, and its primary function is to build and maintain switched paths for aggregate flows of IP traffic based on their intended destinations. It accomplishes this task by taking information maintained by routing protocols and associating it with a switched path that is established between the network ingress and egress. The switched path can take the form of a multipoint-to-point (MPT-PT) tree that is rooted at the egress router and branches out to all ingress routers. The topology-driven approach employed by ARIS is in fact quite similar to the function provided by tag switching. However, there are some important (and not so important) differences between the two that in the short term generated some interesting (and heated) discussions within the IP switching community but in the long term improved the functionality of both and contributed to an overall better understanding of topology-driven IP switching.

Specifically ARIS is a control protocol that runs on a network of integrated switch/router (ISR) devices. An ISR is a layer-2 switch (e.g., ATM) augmented with IP routing support. ARIS uses the information contained in routing protocols (e.g., OSPF, BGP) to map IP packets onto layer-2 switched paths that traverse a network of ISR devices. ARIS was the first IP switching scheme to introduce the notion of switched path merging so that MPT-PT switched paths (trees) could be established between many different sources and a common destination point, thereby minimizing the consumption of switched resources. A destination point is represented by an egress (exit) ISR that has associated with it one or more destination prefixes. Traffic flows that share a common egress ISR flow over the branches of the MPT-PT tree and exit at the egress ISR. Thus an ARIS network is capable of switching *all* traffic to any number of egress ISRs using only O(*N*) switched paths (MPT-PT trees). This unique scaling property enables an ARIS network to grow with many more routers and frees up switched resources to support additional services such as QoS-based connections.

Central to the operation and understanding of ARIS is the notion of an Egress Identifier (egress ID).[1] The egress ID is a configurable value that is used to identify which IP traffic will flow over a specific MPT-PT tree to an egress ISR. An egress ID is configured on an egress ISR and can be dynamically generated using existing routing protocol information or can be manually configured. The values of an egress ID range from a single IP host address all the way up to a group CIDR prefix or even a BGP `NEXT_HOP`. The appearance of an egress ID on an egress ISR is used to initiate the establishment of a switched path from the egress ISR. ARIS builds switched paths from the edge (egress or ingress ISR) of the network, not locally as is the case with tag switching and IFMP. In addition the ARIS protocol provides a loop-prevention mechanism so that only loop-free switched paths are built. This is especially important for operation over ATM networks because the cell does not contain a TTL field to act as a loop control mechanism.[2]

12.1 Architecture

The ARIS architecture is built on some simple and sometimes surprisingly subtle observations about the behavior of IP routing systems. First, unsurprisingly, all traffic flowing through a transit routing

---

[1]An egress ID is essentially an FEC in tag switching terms. However, with the latter there were no specifics given on how and where an FEC is defined or used. The egress ID clarifies this issue by stating that it is configured on the egress ISR and is used to establish exactly one switched path for all traffic destined for the value specified in the egress ID.

[2]Recall that the native ATM routing protocol, PNNI Phase I, prevents the formation of loops with a source routing mechanism.

domain is intended for some destination. It enters the domain through an ingress router and exits through an egress router. Even if the traffic originated from many different sources and thus entered through a similar number of ingress routers and is targeted for multiple destinations behind an egress router, it must flow through that single egress router. Therefore the common element in the forwarding path is not necessarily one or more destination prefixes contained in the routing table of the transit routers but rather is the egress router itself. Moreover if one was to devise an IP switching scheme to accelerate the flow of packets through a transit domain, a convenient and logical point to configure and initiate the process would be at the egress router.

Another point about the flow of packets through a transit domain is that they follow a sort of reverse multicast tree that is rooted at the egress router and branches out to all ingress routers. Consider the topology in Fig. 12-1. Networks A, B, and C can be reached through egress router 4. All traffic destined for those networks will enter the transit domain at different ingress points, and the routing protocols will select a path that takes the packets over the fewest hops to the egress router. Invariably some packets from separate upstream links will be placed on a single downstream link by a router. This N-to-1 upstream-to-downstream multiplexing capability is a natural function of routers and allows router-based networks to scale accordingly. Again if one was to devise an IP switching scheme for faster performance through a transit domain and allow it to scale like normal IP routing, providing this "merge" function at layer 2 would be an attractive and desirable feature.

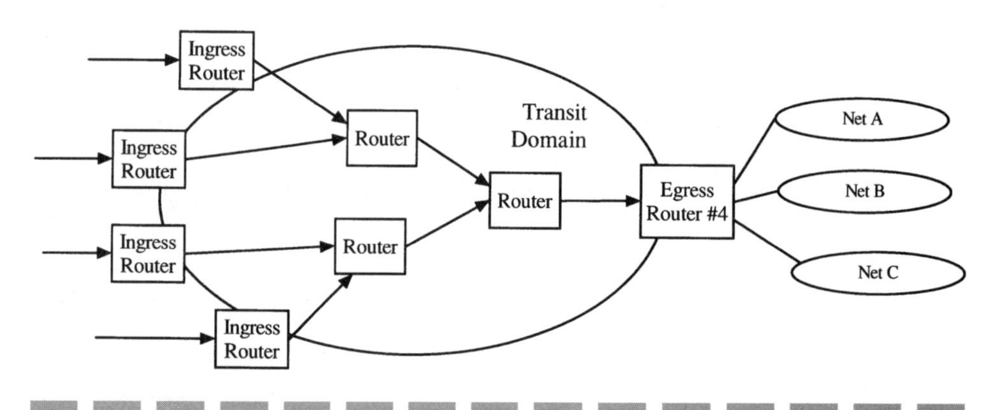

Figure 12-1 Transit domain egress router.

From a practical perspective, large networks do and will contain switching technology, frame relay or ATM, along with, of course, IP routers. And given the price and performance advantages that switching offers, there will always be a desire to exploit it. It cannot be argued that the quickest path between an ingress and egress router is a switched path—not through an intermediate router. In large networks though, as the number of feeder, access, and edge routers grows, the number of switched resources (labels, buffers, memory, VC endpoints, etc.) required to maintain any-to-any switching will grow exponentially. Every new router that is attached to the switched cloud with N other routers must establish (N-1) switched connections with its router peers. In addition, there will be cases where there is a requirement for multiple per-service switched paths between any two routers. This will further deplete the finite amount of switched resources. If an IP switching solution is to scale in size without exhausting the capacity of the switched network, it should attempt to minimize the use of switched resources needed to switch all or most of the traffic. The conservation of switched resources will leave more available for special nondefault services such as QoS-based connections.

And finally, it is noted that ATM is playing an increasingly larger role at the core of large ISP and carrier networks. The reasons are clear: manageable bandwidth on deliverable technology at a decent price. The introduction of an IP switching solution that dispenses with traditional ATM signaling and routing and instead is based on the contents of a routing table should take special care to minimize or prevent the impact caused by transient IP routing conditions. IP routing is very dynamic and quick to react to change—but not every router in the network receives information about a change in topology simultaneously. This results in the formation of temporary routing loops. The damage caused by looping traffic is controlled by the TTL decrement process in each router. Therefore if a topology-based IP switching solution is introduced into an ATM switch fabric, the formation of loops must be avoided.

Keeping in mind the aforementioned points, the developers of the ARIS solution opted to design an IP switching scheme that employs an egress or edge-based orientation for configuration and switched path establishment. They also engineered into the protocol the ability to support switched path merging so that switched resources are conservatively utilized. ARIS can also be considered an IP switching solution that is biased toward and sensitive to operations over an ATM network. The egress/edge orientation of ARIS is based on the observation that all traffic exiting a transit domain must, of course, pass through an egress router (ISR). Therefore the notion of an egress ID is introduced and configured on the egress ISR. It serves as a "link" between the routing protocol information and the switched path. The idea is that it is simpler and much more straightforward to configure a single egress ID on an egress ISR that represents one or more destinations behind the egress ISR (or even the egress ISR itself). Moreover the egress ID is used to establish a switched path for all traffic leading to those destinations through the egress ISR.

The concept of an egress ID and the role that it plays in the ARIS architecture is conveyed in the flow diagram in Fig. 12-2. In this example all IP traffic for destinations 1 through N is associated with an egress ID that can be extracted from routing protocols or configured manually on the egress ISR. ARIS uses the egress ID to establish and identify a switched path through a network of ISR devices that will terminate at the egress ISR. This switched path will carry all traffic targeted for destinations 1 through N. Thus an egress ID can be thought of

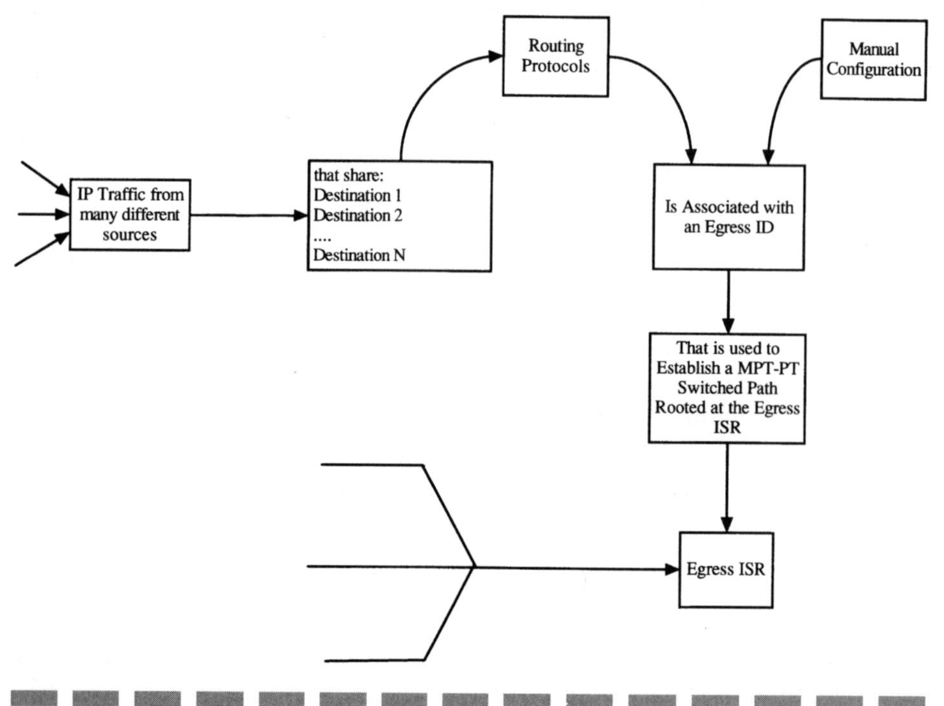

Figure 12-2 Concept of ARIS egress ID.

as an identifier for traffic sharing one or more destinations (e.g., routing table entries) and a layer-2 switched path. The end result once ARIS creates a switched path is that aggregate traffic flows that share a common set of destinations traverse a single layer-2 switched path. One switched path is created for each egress ID, and of course an ISR can support multiple egress IDs. In a practical sense, the egress ID is a configurable value that a network provider can use to select the granularity of traffic that will flow over the switched path.

ARIS is a lightweight control protocol that operates on an ISR. An ISR device is basically an ATM switch that runs IP routing protocols and can possibly forward packets at layer 3. An ISR is similar to an IP switch, CSR, or TSR device with the exception that it is also capable of supporting native ATM switching (via ATMF UNI/PNNI signaling and routing) in a ships-in-the-night configuration. Indeed ARIS is designed to operate on a multiservice infrastructure where all best-effort IP traffic is switched and special services are provisioned through any number of different mechanisms including native ATM and/or RSVP over ATM.

The egress orientation and more specifically the presence of an egress ID on the egress ISR makes it natural to originate the establishment of a switched path from the egress ISR.[3] In other words, when a switched path needs to be established based on a change in network topology or the configuration of a new egress ID, the ARIS path setup messages initially flow from the egress ISR. They travel then hop by hop away from the egress ISR, finally reaching all possible ingress ISR devices that forward traffic to the destinations associated with the egress ID. It is only when the path setup messages reach the ingress ISR devices that a complete ingress-to-egress switched path is established. The ARIS path setup messages contain an egress ID and a label that is installed in the switch component of the ISR. The edge-based approach makes it possible to build switched paths with certain ingress-to-egress characteristics. For example, loop-free paths or explicit paths can be easily provisioned by an edge-based setup mechanism. In addition this approach is consistent with other scalable IP control protocols such as PIM-SM and RSVP. However, the disadvantage is that there is some delay in building the path and there is a dependency on the "other" ISR in passing the path setup message.

---

[3]The term *edge* is used to account for the fact that ARIS path setup messages for multicast originate from the ingress ISR—not the egress ISR. Throughout the chapter, though, the terms *edge* and *egress* will be used interchangeably.

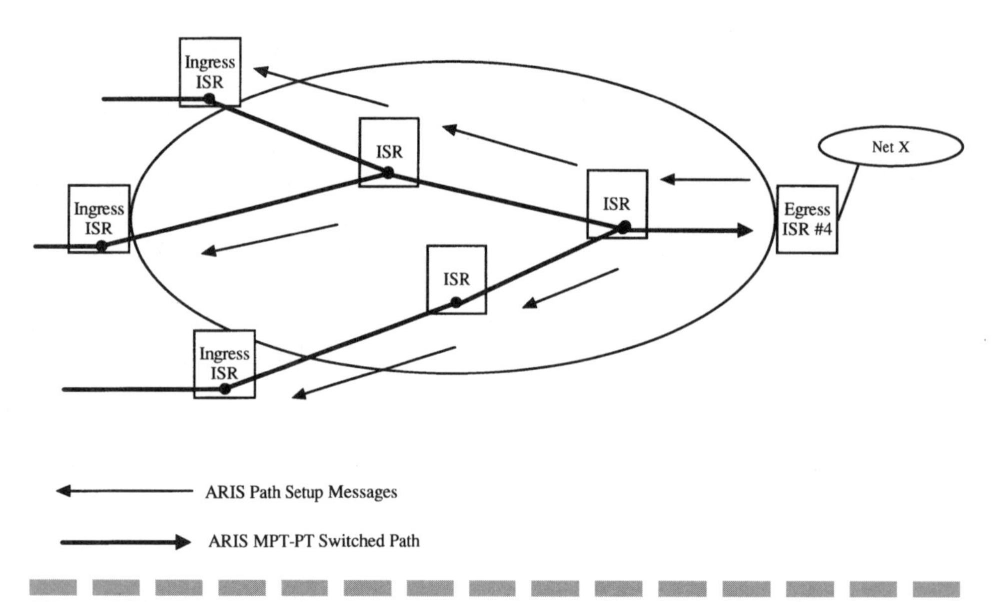

Figure 12-3 ARIS architecture.

ARIS addresses the requirement for switch resource conservation by supporting switched path merging. With switched path merging, the cells from multiple upstream links are merged onto a single downstream link. By supporting switched path merging, an MPT-PT tree can be built that is rooted at an egress ISR and branches out to all ingress ISR devices. Therefore it is possible to provide an any-to-any switch paradigm and use O(N) switched paths where N is the number of ISR devices in the network.[4] It should be noted that switched path merging does not necessarily carry QoS-based traffic. The merging function itself can introduce some additional jitter and delay. Rather switched path merging enables the construction of MPT-PT trees that will consume fewer switched resources while transporting all best-effort traffic. If one needs to support a QoS-based connection, ARIS permits the use of non-ARIS connection setup methods such as RSVP or native ATM signaling and routing.

The basic architecture and functions provided by ARIS are illustrated in Fig. 12-3. A domain of ISR devices running ARIS is shown. A specific ISR (ISR 4) has been configured with an egress ID of "Net *X*" because it

[4]In this case, N MPT-PT switched paths are used.

is the egress ISR for all traffic destined for Net *X*. ARIS path setup messages flow in reverse path multicast (RPM) style outward from the egress ISR and toward each ingress ISR that is in the forwarding path for packets destined for Net *X*. Those ISR devices along the routed path to the egress ISR will process the path setup message by allocating labels (associated with the egress ID) in the switch component. At the ingress ISR(s), the standard FIB is augmented with the egress ID so that when packets enter the network, a standard routing table lookup reveals the existence of an egress ID along with the associated switched path to the egress ISR. Packets targeted for Net *X* are then label swapped over the switched path all the way to the egress ISR with no intermediate layer-3 processing required. The result of all this is a routing system with better overall performance because *all* traffic is switched.

12.2 Components

The ARIS solution consists of the following components:

- *ISR.* Layer-2 switch (e.g., ATM) that runs an IP routing protocol, the ARIS control protocol, and is capable of forwarding packets at layer 3 and switching packets at layer 2. An ISR can concurrently support native ATM signaling and routing (Q.2931/PNNI) in a ships-in-the-night configuration and possibly serve as an MPOA Route Server. If an ISR is supporting PNNI, the VPI/VCI label range must be split between PNNI and ARIS. An example of an ISR is illustrated in Fig. 12-4.
- *ARIS.* Control protocol that enables layer-2 switched paths to be built based on IP routing protocol information. ARIS runs in addition to IP routing protocols on an ISR. Another way of explaining what ARIS does is to say that it maps many IP destinations to an egress ID, which in turn is associated with a set of labels that form a switched path through a network of ISR devices.
- *Egress ISR.* An ISR that provides an exit point for IP traffic traveling from an ARIS domain to a directly attached IP subnet (e.g., Ethernet LAN with IP hosts), a non-ARIS domain, or another ARIS domain. Note that an ISR can be configured to operate concurrently as an ingress, egress, and transit node. ARIS path setup messages originate from an egress ISR.

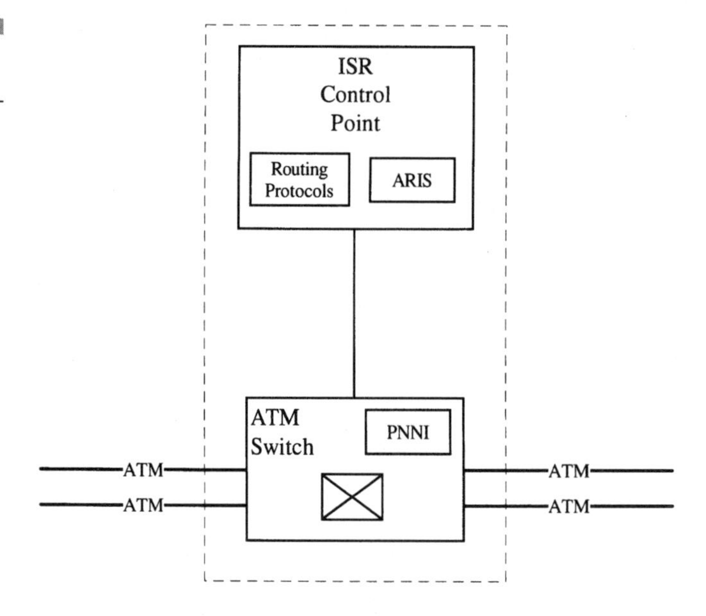

Figure 12-4
ARIS ISR.

- *Egress identifier.* Configurable value that identifies and associates a switched path to one or more resources (destinations) "behind," associated with or reachable through an egress ISR.
- *VC merging.* Layer-2 switch function that enables multiple upstream VCs to merge into a single downstream VC without interleaving cells from different frames. A VC merge-capable switch must collect all the cells of a frame on the inbound VC and then transmit the frame's cells in a contiguous fashion with a new VCI value on the outbound link. The cells transmitted on the new VC must be grouped contiguously by frame so that the destination device can distinguish cells from different sources. VC merging may require special hardware or a per-VC buffering capability in the switch. Figure 12-5 illustrates an example of an ISR (ISR 3) performing a VC-merging function. Note that because the cells with VCI = 7 are grouped by frame, the egress ISR can assemble them into packets and forward them on to the destination. The advantage that VC merging offers is a lower consumption of VC resources (e.g., labels) while still providing a layer-2 switched forwarding function. The savings in VC resources comes from the fact that N upstream VCs only need a single downstream label versus N downstream labels as the case would be without VC merging.

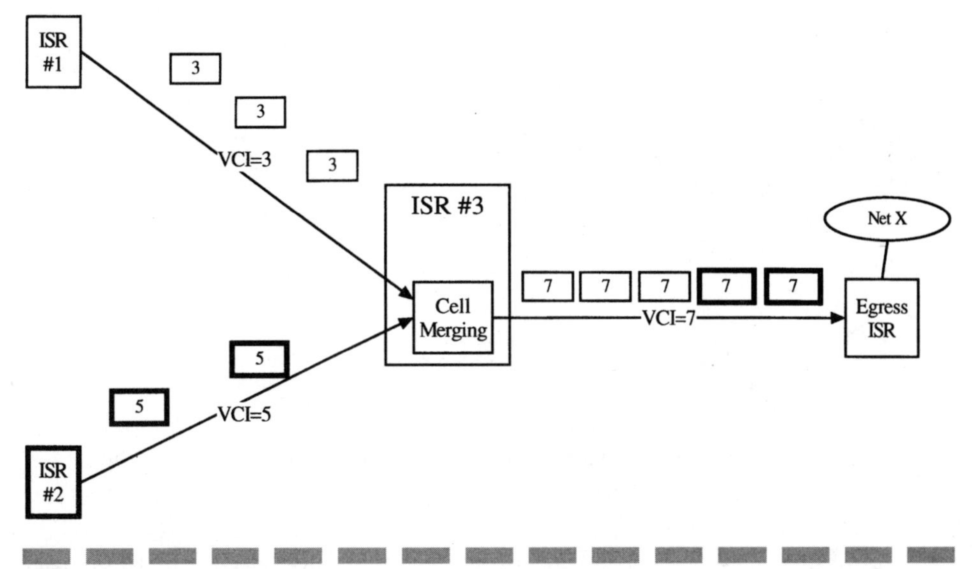

Figure 12-5 VC merging.

- *VP merging.* Standard ATM switch function that enables multiple upstream VPs to merge into a single downstream VP. The inbound and outbound VCI values remain unique and unchanged, so no special VC merging function is required on the switch. The unique VCI value is used to retain the identity of the source (e.g., ingress ISR). However, because the VCI values may share a common VPI, a unique range of VCI values must be communicated to the ingress ISR devices so as to prevent VCI collisions.
- *Multipoint-to-point (MPT-PT) switched path.* Another name for an ARIS-built switched path that is rooted at the egress ISR and branches out to all ingress ISR devices. Traffic flows in the reverse direction from the leaves of the tree back toward the root, which is the egress ISR. An MPT-PT tree is associated with an egress ID. MPT-PT trees make use of VP or VC merging to limit consumption of switched resources.
- *VC information base (VCIB).* The VCIB is a table of egress IDs and associated layer-2 label information (e.g., VPI/VCI) that is maintained in the switch component of an ISR. An example of a VCIB for ISR 3 in Fig. 12-5 is depicted in Fig. 12-6.

Figure 12-6
VCIB example.

Egress ID	Inp Port	Inp VC	Out Port	Out VCI
Net X	1	3	3	7
Net X	2	5	3	7

12.3 Egress Identifiers

To provide maximum flexibility and control over the selection of traffic that will traverse a switched path, the ARIS specification supports a number of different egress identifiers. An egress ID can be extracted from routing protocol information or can be manually configured. The value of the egress ID is communicated from the egress ISR to all ISR devices along the routed path by the ARIS protocol during switched path establishment. It is installed in the VCIB (along with layer-2 switching information) of the ISR devices along the routed path. In addition the FIB is extended to include the egress ID for each next-hop ISR.

The following egress identifiers are defined in the ARIS specification:

- *IPv4 address prefix.* This identifier is used to establish a switched path to a specific host or CIDR prefix. This type of egress ID is recommended in environments where no aggregation is provided by the routing protocol (e.g., RIP) or in networks where the number of IP destination prefixes is small. It is also useful if a separate switched path to a specific subnet or well-known host needs to be established.
- *BGP Next_Hop.* This contains the value in the BGP Next_Hop attribute. The `NEXT_HOP` attribute is a well-known mandatory attribute that is used to identify the IP address of the BGP router that announced the route. Conceptually it represents the next-hop address toward what could potentially be a large number of destinations. This egress identifier could be the IP address of a BGP border router connected to the AS or the IP address of an external peer.[5] Either way one can achieve a significant level of aggregation

[5]In the latter case it was suggested by an ARIS developer that this would be analogous to throwing one's leaves into the neighbor's yard.

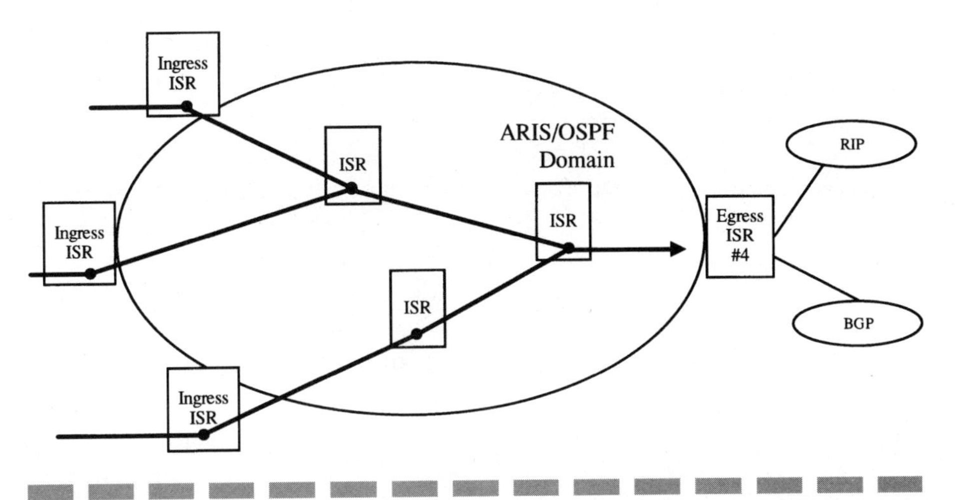

Figure 12-7 ARIS of the OSPF Router ID as an egress ID. Egress ID = OSPF router of egress ISR.

because all traffic flowing to the BGP Next_Hop router would flow on a single switched path.

- *OSPF Router ID.* This identifier contains the OSPF Router ID of the ISR that initiated a link-state advertisement. ARIS interacts with the information contained in the OSPF link-state advertisements (LSAs) to establish switched paths to egress ISRs that connect to resources outside of the OSPF domain.[6] In the first case, remote networks may be injected into an OSPF routing domain through a specific ISR either by static configuration or redistribution from other routing protocols (e.g., RIP, BGP). These remote networks may be represented in the routing table as many destination prefixes, which cannot be aggregated. Rather than using many egress IDs (and an equal number of switched paths) to cover the remote networks, ARIS makes use of the OSPF Router ID (of the egress ISR) as the egress ID to build an MPT-PT tree to the egress ISR. In other words, ARIS can abstract all of the remote networks injected into the OSPF routing domain into a single value—the OSPF Router ID of the advertising ISR. This capability is illustrated in Fig. 12-7. This results in a much simpler form of aggregation and consumes fewer switched resources while providing a

[6]To clarify, a domain of ISR devices running ARIS and OSPF.

switched path for all traffic destined to remote networks through the egress ISR.

- *OSPF Area Border Router ID.* The one exception to using the OSPF Router ID is when the ISRs receive an AS External Link Advertisement with a nonzero forwarding address. The forwarding address refers to the IP address of a router that provides reachability to networks outside of the OSPF routing domain. OSPF uses the forwarding address to allow traffic to bypass the router that originated the advertisement. Therefore better use of network resources is achieved when the forwarding address is used as an egress ID.
- *Multicast (source, group).* This identifies the unique source address/multicast group address pair of a multicast delivery tree that is built using flood and prune protocols such as DVMRP or PIM-DM. In this case the establishment of a switched path is initiated by the ingress ISR, or in other words, the one closest to the source. It consists of a point-to-multipoint delivery tree that is rooted at the ingress ISR and branches out to all egress ISR devices. One switched path is built for each (source, group). An example is illustrated in Fig. 12-8.
- *Multicast (*, group).* This egress ID is used to build switched paths based on shared tree multicast routing protocols such as PIM-SM. In this case the root of the switched path is the rendezvous point (RP) ISR.

Figure 12-8
ARIS multicast support.

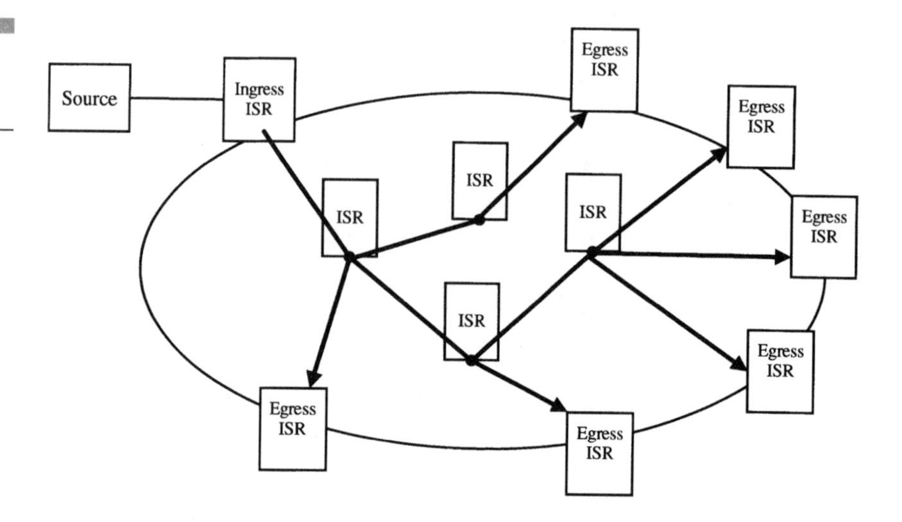

- *Flow.* This identifies a specific source and destination IP address pair and could be used to build a switched path for a specific flow.
- *CIDR group.* This egress ID contains a list of individual CIDR prefixes that happen to share a common egress ISR. This particular egress ID is useful if one needs to aggregate multiple dissimilar CIDR prefixes that cannot be aggregated using existing routing protocols.

12.4 ARIS Protocol

ARIS is a control protocol that operates in a network of ISR devices. As already discussed, it maps IP routed traffic onto switched paths. Specifically ARIS lays down switched path state over the routed path for traffic targeted for a destination or group of destinations associated with an egress ISR. The ARIS path setup message (ESTABLISH) originates at the egress ISR and is forwarded in RPM-style through a network of ISR devices until it reaches the ingress ISR(s). The ESTABLISH message contains several objects including the egress ID and a list of ISR devices that the ESTABLISH message has passed through. The latter prevents the formation of switched path loops, which can be problematic at layer 2. Unlike TDP and IFMP, ARIS path setup messages must flow all the way from the egress to ingress before a switched path is built. In addition ARIS messages are explicitly acknowledged to ensure reliable delivery.[7]

12.4.1 Message Types

ARIS protocol messages are structured based on the illustration in Fig. 12-9. The fields in the common header consist of a version, message type, length, header checksum, sender router ID, sender sequence number, and a sender and receiver session number. The common header authenticates an ARIS message and indicates the message type. Follow-

[7]ARIS uses its own acknowledgment scheme and TDP uses TCP. This was curiously a major point of contention between the two. TCP does offload the reliability function from TDP but may also require some additional configuration, whereas a protocol-specific acknowledgment scheme may be a bit more flexible.

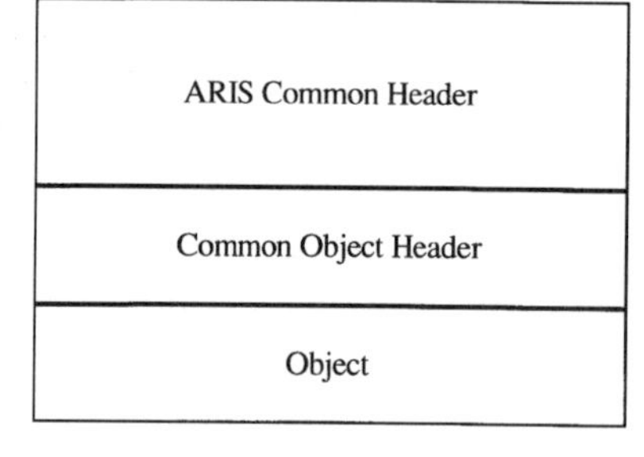

Figure 12-9 ARIS message structure.

ing the common header is the common object header that defines the object type, object subtype and the length. After that comes the object. An ARIS message typically contains multiple common object headers and objects.

The five message types defined in the ARIS specification are

- *INIT.* The INIT message is an exchange between ARIS neighbors and is used to establish an adjacency. Once a successful INIT message is received, the ARIS neighbors will transition to an active state. It is only in the active state that switched paths can be established.
- *KEEPALIVE.* The KEEPALIVE message is sent by an ISR to inform its neighboring ISR devices that it is still active. It must be sent within the bounds of the timer object that was communicated in the INIT message or the adjacency will time out.
- *ESTABLISH.* The ESTABLISH message is used to create a switched path. It is initiated by the egress ISR or sent to an ISR in response to a TRIGGER message. It flows in RPM style until all ISR devices with the egress ISR in the forwarding path (for the specified egress ID) have received the message and installed switched path state. An ISR that receives an ESTABLISH message first checks to make sure that it came from the next hop ISR toward the destination. It also verifies that it has not seen the message before by looking for its own ISR ID in the router path object. If either check fails, a negative acknowledgment is transmitted to the sender. If both checks pass, the receiver will respond with an acknowledgment to the sender. It will then update the VCIB, increment the hop count by 1, append its own ISR ID to the router path object, allocate a new label on the upstream ports leading to neighboring ISR devices, and transmit ESTABLISH messages to each of

its neighboring upstream ISR devices. The important objects contained in the ESTABLISH message consist of the following:

- *Egress ID.* This object contains the egress identifier.
- *Router path.* This is the list of ISR devices that the ESTABLISH message has passed through and is included if loop prevention is required.
- *Label.* This contains a label that is allocated on the inbound port of the ISR and communicated upstream to the neighbor ISR in an ESTABLISH message.

- *TRIGGER.* The TRIGGER message is sent by an upstream ISR to a downstream ISR to request an ESTABLISH message. This is usually the case when a change in topology generates a new forwarding path (thus a new switched path) to an egress ISR and a new one needs to be computed. When this occurs, the obsolete portion of the path is unspliced and a TRIGGER message is sent to the new downstream ISR (new next hop toward the egress ISR).
- *TEARDOWN.* This message deletes established switched paths. For example, if the routing protocols or manual configuration withdraw an egress ID from service, the ISR devices with remnants of the withdrawn egress ID should exchange TEARDOWN messages.
- *ACKNOWLEDGE.* This message is used to provide a positive or negative acknowledgment of ARIS messages.

12.4.2 Objects

The objects contained in ARIS protocol messages as defined in the ARIS specification are

- *Label.* This is the label assigned to a particular egress ID by the downstream ISR. The format of the label object is illustrated in Fig. 12-10.
- *E.* The E bit is an end-to-end acknowledge bit that is turned on if the label is allocated by the upstream ISR devices beginning with the ingress ISR(s).
- *V.* The V bit is turned on only if the VPI field is significant.
- *VPI.* 12-bit VPI field.

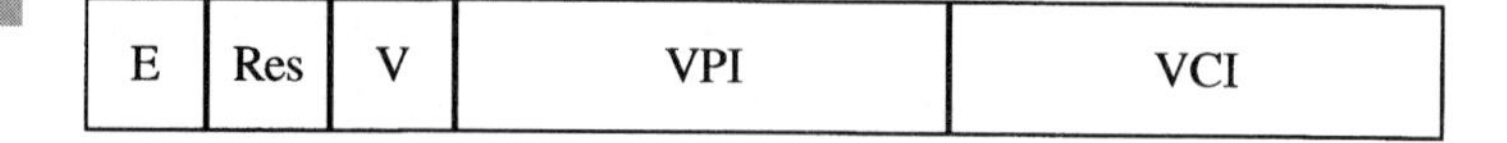

Figure 12-10
Label object.

- *VCI.* 16-bit VCI field.
- *Egress ID.* This object contains one of the egress identifiers described in Sec. 12.3. Figure 12-11 illustrates the format of several of the more commonly used values. The contents of the fields are self-explanatory.
- *Multipath.* This object is an identifier that defines one of several possible equal-cost paths for the same egress ID.
- *Router path.* This object contains a list of ISR identifiers that presents the path that the ESTABLISH messages have traversed for a given egress ID.
- *Explicit path.* The explicit path object contains a list of ISR devices that the ESTABLISH message must traverse to establish a nondefault switched path.

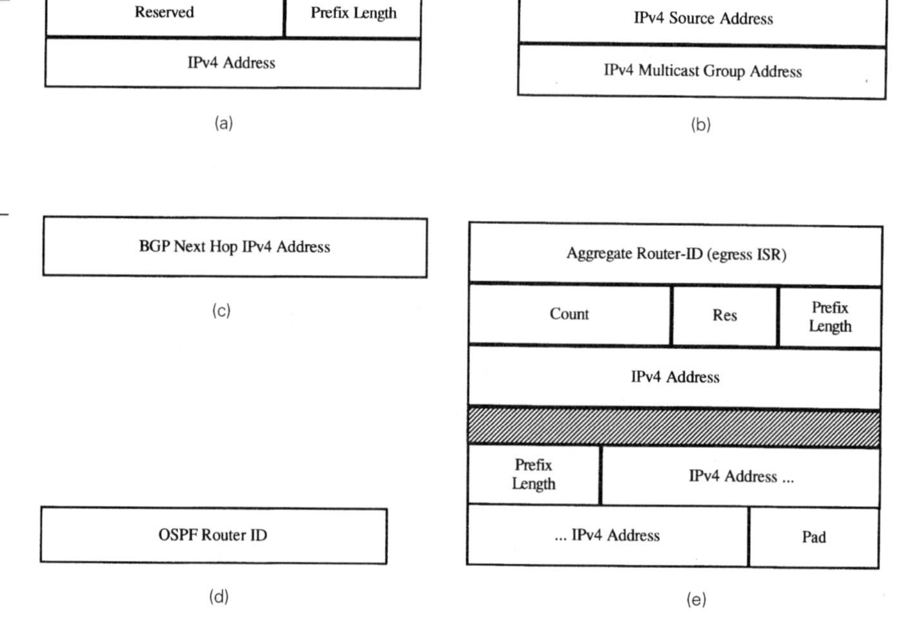

Figure 12-11
Egress ID formats. (*a*) IPv4 address. (*b*) Multicast (source, group). (*c*) BGP Next_Hop. (*d*) OSPF router ID. (*e*) CIDR.

- *Tunnel.* This object contains a link-layer label and the IPv4 address prefixes associated with that label.
- *Timer.* The timer object can take on different meanings depending on which message type it is contained in. When present in the INIT message, it is a neighbor timeout value. When present in the ESTABLISH message it is the refresh-establish interval.
- *ACK.* This object contains the sequence number of the message that is being acknowledged.
- *Init.* The init object contains a range of VPI and VCI values that an ISR is capable of supporting.

12.4.3 Switched Path Setup

The setup of a switched path using the ARIS protocol is illustrated in the example shown in Fig. 12-12. A network of ISR devices is contained in a transit domain that is connected to a non-ARIS domain. Net 1 is one of several destination networks contained in the non-ARIS domain that the ISR devices may need to forward traffic to.

- First the ISR devices exchange INIT messages with their neighbors to form adjacencies and enter the active state. This is required prior to processing any path setup messages.
- An egress ID of Net 1 is configured on the egress ISR.[8] This can happen dynamically via a "download" from the routing table or it can happen manually via network configuration. Net 1 is a network that is outside of the ARIS domain and is reachable through the egress ISR 4.
- The egress ISR transmits an ESTABLISH message upstream to ISR 3. The Establish message contains the egress ID, a new label (VCI = 9), and a list (router path object) of ISR devices that the ESTABLISH message has passed through. In this case the list will only contain the ID (4) of the egress ISR.
- ISR 3 receives the ESTABLISH message and (1) checks to make sure it came from the correct next-hop address toward the egress

[8]In this example the egress ID type is an IPv4 address prefix.

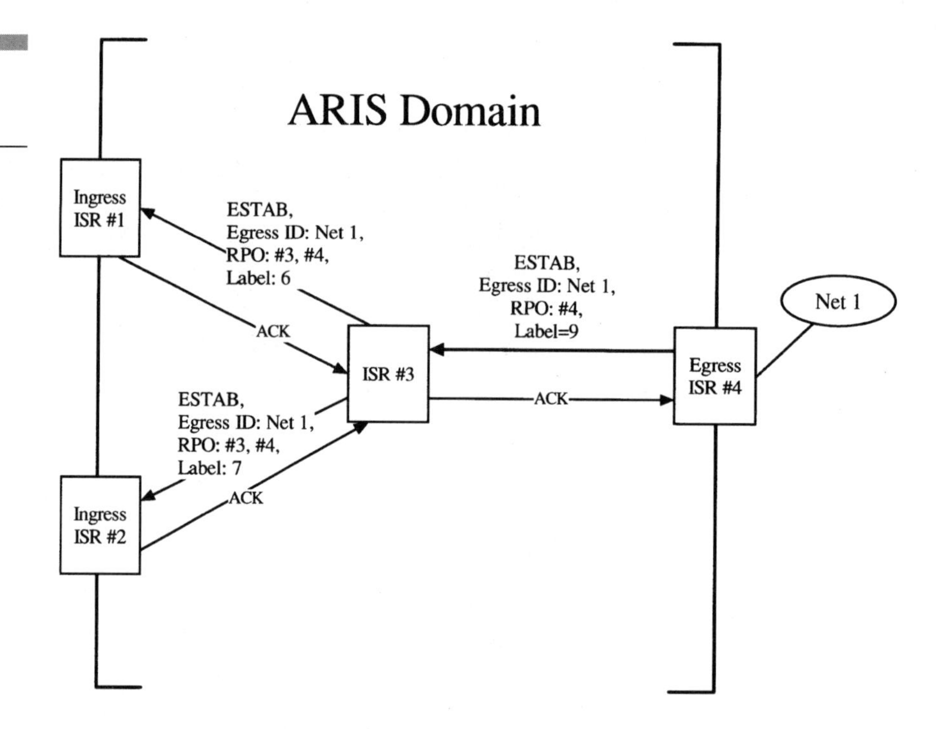

Figure 12-12 ARIS path setup example.

by looking at its FIB table and (2) checks that the path is loop-free by looking at the ISR list. ISR 3 then updates its FIB with the new egress ID and creates an entry in its VCIB consisting of the egress ID (Net 1) and the new label (VCI = 9). ISR 3 then sends an ACKNOWLEDGE message back to the egress ISR.

- Next ISR 3 allocates a new label (VCI = 6) for its upstream neighbor ISR 1 and a new label (VCI = 7) for its upstream neighbor ISR 2. It also appends its own ISR ID (3) to the router path object. It then sends a separate ESTABLISH message to ISRs 1 and 2. When ISR 3 receives an ACKNOWLEDGMENT message from each of its upstream neighbors, it updates the VCIB by splicing the upstream labels on the inbound ports to the downstream labels on the outbound port associated with the egress ID of Net 1. The process is then repeated recursively until all ISR devices in the routed path to Net 1 have updated their FIB and VCIB tables accordingly. The end result is an MPT-PT switched path from all ingress ISR devices to the egress ISR.

12.5 Loop Prevention

Routing loops occur when there is an inconsistency between the real network topology and the contents of the routing table. This results in IP packets "looping" through a router and network more than once. This situation is controlled by the use of the TTL field in the IP header. This value is decremented each time a packet passes through a router. When the TTL value in a packet reaches zero, the router drops the packet. No such mechanism exists in ATM. Because ATM hardware typically operates at much higher speeds than IP routers, a layer-2 loop could consume a significant amount of bandwidth and switch resources. This could have a harmful impact on other nonlooping traffic and may affect the ability of the network infrastructure to support new or existing traffic flows.

Given its focus on operation over ATM networks, ARIS provides a solution to the looping problem by not attempting to build a switched path loop in the first place. Because of its egress orientation and the fact that paths are built from the edge of the network, it is a simple matter for the ARIS path setup process to keep track of the path back to the egress ISR. Specifically this is accomplished by the presence of the router path object that is contained in the ESTABLISH message. It is updated with the ID of each ISR that processed an ESTABLISH message. An ISR will not process an ESTABLISH message that contains its own ISR ID in the router path object.

Preventing loops from the egress is fairly straightforward. But changes in the topology can happen anywhere in the IP network. How is the formation of loops prevented if there is a change in the interior of the network? ARIS addresses this problem by transmitting a TRIGGER to the new next-hop ISR toward the egress ISR. The next-hop ISR then issues an ESTABLISH message that essentially replaces the existing upstream switched path from the ingress ISR(s) to the ISR that issued the TRIGGER message. This process comes with some cost though because it is possible that a perfectly good upstream switched path is removed and then rebuilt upon the arrival of the new ESTABLISH. In other words it may not have been necessary to tear down the original upstream switched path in the first place.

A solution to this problem is achieved by the use of a diffusion computation to compute a new loop free path following a change in topology.[9]

---

[9]J.J. Garcia-Luna-Aceves, "Loop-Free Routing Using Diffusing Computations," *IEEE/ACM Transactions on Networking,* vol. 1, no. 1, Feb. 1993.

The idea of the diffusion computation is to first determine and then prune any upstream links and nodes that would form a loop if a downstream ISR switched to a new switched path. Until the diffusion process is completed, the old switched path is used. Specifically, when an ISR (A) receives an ESTABLISH message from a new next hop, it checks to see if it is in the new path. If it is, there is a loop. If it is not, it initiates a diffusion process. In this case a query message is issued to the upstream ISR neighbors to determine if a loop would form if the upstream neighbors and ISR A are spliced on to the new switched path. If a loop does form, those upstream neighbors are pruned (removed from further consideration during this process) from the new switched path. If a loop does not form, the upstream neighbors respond with okay messages and are added to an upstream tree. When ISR A has received okay messages from all of its neighbors, it discards the old path and splices the new upstream tree onto the new switched path. The advantage of this solution is that the upstream switched paths are not unnecessarily affected by a change in the downstream topology.

12.6 ARIS Services

The notion of the egress ID and the edge-based orientation (ingress or egress) provide an ARIS network with a single point of control and flexibility in supporting network-layer services. Varying levels of traffic granularity can be mapped over different MPT-PT switched paths by adjusting the egress ID on a single egress ISR. In addition, ARIS supports both unicast and multicast routing without requiring any modifications to existing routing protocols. And the resource conservation mechanisms supported by VC merging and MPT-PT switched paths free up more switched resources for specialized nondefault connections that can be built through ARIS or non-ARIS means. Indeed it is envisioned that ARIS is one of just several control protocols operating over a multiservice ATM infrastructure. As in tag switching, the layer-2 label-swap forwarding function is independent of the type of network layer services that can be applied.

12.6.1 Egress Aggregation

The core of a large IP-based network is expected to support hundreds and thousands of simultaneous IP flows. Aggregating IP flows that share a common egress onto a single switched path consumes fewer network resources than a system that maps a single IP flow or even a single routing table entry to a single switched path. It requires fewer control messages to set up a switched path and requires less switched path state in the network. The use of a configuration egress ID as a means to associate a group of destinations with an egress ISR allows an ARIS network to switch all egress bound traffic flows while utilizing a minimum amount of switched resources.

12.6.2 Multicast

ARIS can be used to set up switched paths for IP multicast traffic. The establishment of a switched point-to-multipoint delivery tree is initiated at the root or ingress node. The switched path tree carries all multicast traffic from the ingress ISR to all egress ISRs, using hardware-based switching in the transit ISR devices.[10] This offers a substantial improvement over software-based multicast routing performed by routers.

ARIS is independent of any underlying multicast routing protocol. ARIS supports the establishment of data-driven, source-rooted delivery trees for each (source, group address) pair as required in DVMRP and PIM Dense Mode (PIM-DM). And it supports the establishment of receiver-driven, shared delivery trees for each (*, group address) as required in PIM Sparse Mode (PIM-SM).

12.6.3 Explicit Routing

The ability to engineer and provision nondefault routes through a network to provide a special service (e.g., QoS, performance, security, etc.) or just for simple load balancing is an important and desirable function. The use of the egress ID along with its edge orientation enables ARIS to provide this function quite naturally. ARIS supports this function through the use of an explicit path object. The contents of the explicit path object consist of a vector of ISR devices that the ARIS ESTABLISH messages must pass through for a given egress ID. The result is a loop-free nondefault switched path between the ingress and egress ISR devices. The contents of the explicit route object can be manually

[10]Standard layer-3 multicast forwarding is still required at the ingress ISR and all egress ISRs.

defined, computed from the routing protocol (e.g., OSPF), or installed through some other means. ARIS supports explicit routing for point-to-point, point-to-multipoint, and multipoint-to-point switched paths originating from the egress or ingress node. An example of a explicit path for traffic destined for Net B is illustrated in Fig. 12-13.

12.6.4 ATM

ARIS is designed to work over any layer-2 switching technology. Frame relay, ATM, and IEEE 802 LAN switching are examples. However, it is ATM that deserves special attention for several reasons. It is QoS-capable and scalable in speeds up to and beyond OC12, and ATM switches are quite affordable from a number of different vendors. It is also true that ATM is being deployed in many large IP networks to provide additional network capacity and performance.

The accommodations that ARIS has made with respect to ATM consist of the following:

- *Loop prevention.* ARIS has a built-in mechanism that prevents loops from forming at ATM layer 2. Again this is an important

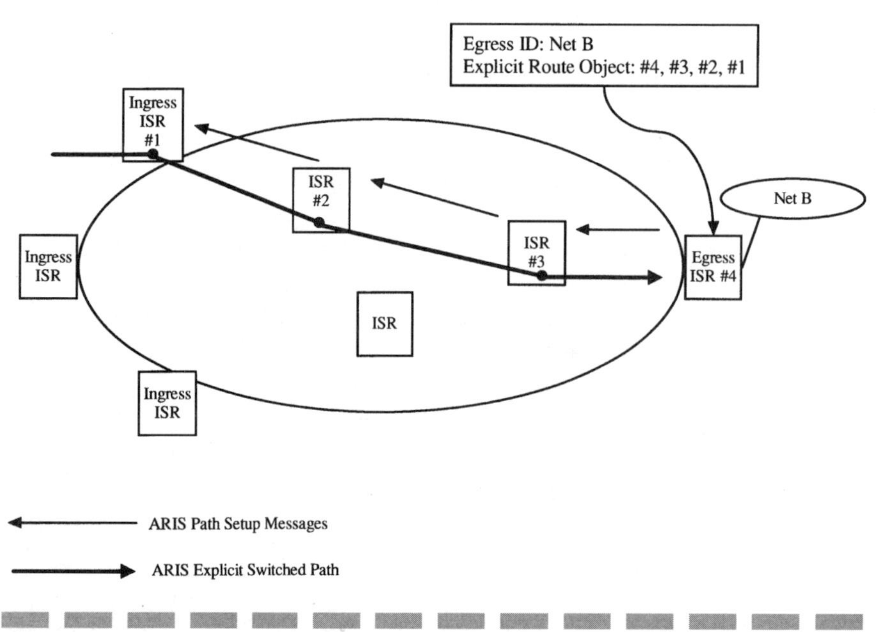

Figure 12-13 ARIS explicit routing.

capability because ATM does not have a loop prevention mechanism.

- *Ships-in-the-night coexistence.* An ISR with an ATM switch component can support ATM protocols such as ATMF signaling and routing as well as PVCs and PVPs.
- *Interoperability.* ARIS ISR devices can communicate with each other through a network of native ATM switches as illustrated in Fig. 12-14 using one of two possible techniques.[11] In the first approach a VP is established between adjacent ISR devices through a network of ATM switches. All data and control traffic could flow over this VP, and labels associated with egress identifiers would be selected from unused VCI values contained within the VP. In the second approach a PVC or SVC is established between adjacent ISR devices. This VC, the default VC, carries routed traffic and control traffic including ESTABLISH messages. When an ARIS ESTABLISH message flows upstream, the downstream ISR builds an SVC for the associated egress ID through a network of ATM switches. The ESTABLISH message and the ATM SVC setup message both carry information (called a handle) that enables the receiving (of the ESTABLISH message) ISR device to correlate an egress ID with a newly established ATM SVC.

Figure 12-14 ARIS ATM support.

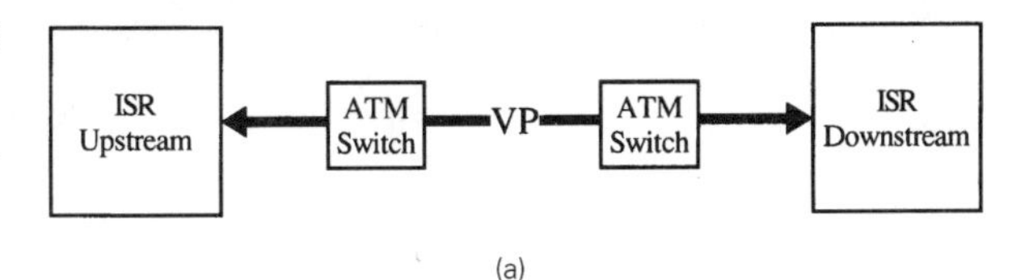

(a)

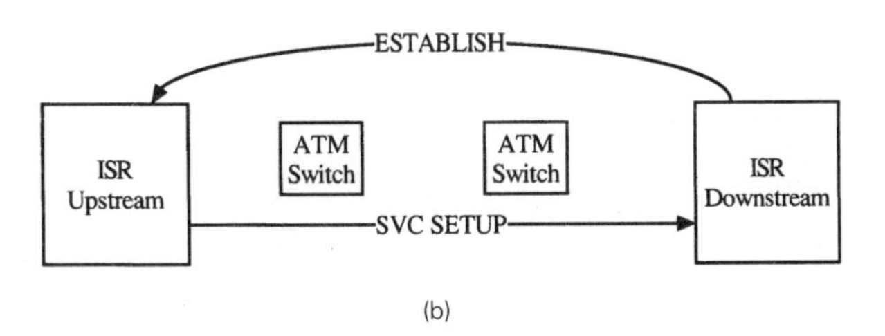

(b)

[11]These techniques are similar to the mechanisms defined in the CSR/FANP model discussed in Chap. 10.

12.6.5 Layer-2 Tunneling

ARIS supports the establishment of switched path tunnels. A switched path tunnel switches data at layer 2 to an egress ISR or tunnel endpoint. Data traveling through a tunnel will carry a stack of labels of which only the top label in the stack is processed (label-swapped). At the tunnel endpoint, the top label (e.g., VPI) is "pushed aside," and the data is switched based on the next label (e.g., VCI) in the stack. Therefore no layer-3 processing occurs at the egress ISR. Again note that when the data is inside the tunnel, switching only occurs based on the top label in the stack—the rest of the labels in the stack, if present, are untouched. Possible applications for ARIS-built layer-2 tunnels include virtual private networks, building switched paths based on routing hierarchy (interdomain routing uses one label, intradomain routing uses another), DVMRP tunnels, and aggregating/deaggregating flows at layer 2.

Figure 12-15 shows an ARIS network supporting a layer-2 tunnel established between an ingress and egress ISR. Data is forwarded through the tunnel based on the VPI value (in the cell header). When the data reaches the egress ISR (or tunnel endpoint), the VPI value is popped, and the next label in the stack, the VCI value, is used to forward the data at layer 2. ARIS can tell the ingress ISR(s), using a tunnel object contained in the ESTABLISH message, what labels (e.g., VCI values) to include in a stack of labels. These values can be used to switch traffic to specific destinations upon exiting the tunnel. In this example, VCI = 1 is used to switch data toward Net 1, and VCI = 2 is used to switch data toward Net 2. The gain in performance comes from avoiding the need to deaggregate and process the packets at layer 3.

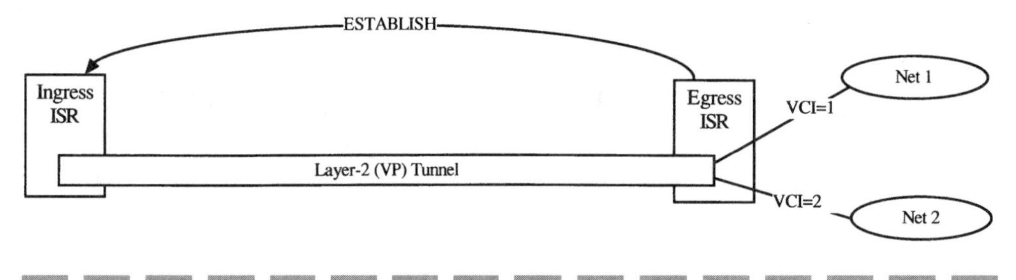

Figure 12-15 ARIS support for layer-2 tunneling.

12.6.6 Class of Service

An ARIS-based network can support COS/QoS-based connections using several different techniques. First, RSVP can be used to install QoS and switched path state along a path of ISR devices. This approach is quite similar to the technique used in tag switching, described in the last chapter. A label object is included in the RSVP RESV messages that flow upstream toward the sender along the route established by the RSVP PATH message. The RESV message contains a label and the QoS parameters that are then installed in the switch component of the ISR. Once a label has been allocated on both the inbound and outbound ports for a particular flow, the ISR can splice the two together to form a switched path. Once this occurs in each ISR along an ingress-to-egress path, IP flows can exploit the high bandwidth and QoS capabilities of the underlying layer 2 (e.g., ATM) infrastructure.

Another technique, unrelated to ARIS but nevertheless the result of its conservative use of switch resources, is to use native ATM signaling and routing. This requires that the ISR support ATM Forum signaling and routing (PNNI Phase I) in the switch component. A network provider could therefore use ARIS to switch all best-effort IP traffic and simultaneously offer and deliver ATM support for those applications and end users requiring ATM QoS.

Yet another approach would be for ARIS to support multiple MPT-PT switched paths to a particular destination. For example, the default for any traffic destined for a particular destination is switched path 1. Using a separate egress ID, a second switch path, switched path 2, is established to the same destination for traffic originating from a source. In this case ARIS enables one to set up per-source switched paths to the same destination, thus providing some level of service differentiation.

12.7 ARIS Network Examples

ARIS is a topology-driven IP switching solution that is best suited for large IP-based networks. By design ARIS minimizes the amount of network resources required to switch all IP traffic between any two nodes. This means that more network (switch and router) resources can be allocated to providing nondefault special services. These services might

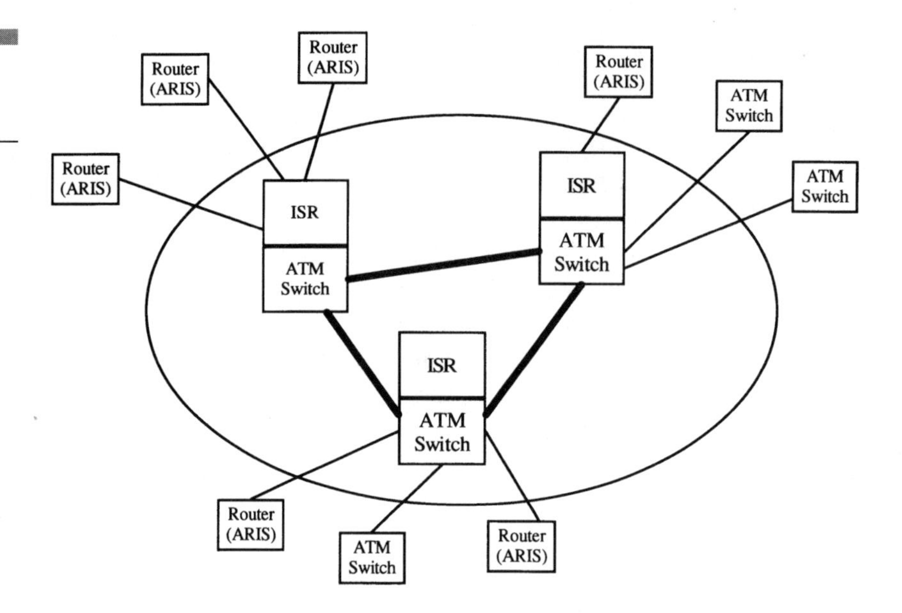

Figure 12-16 ARIS network example.

include virtual private networks (VPNs), QoS flows, explicit routing, tunnels, native ATM connections, etc. Overall, ARIS enables very large networks to switch large aggregated amounts of IP traffic and thus provide improved overall network performance.

ARIS fits best in a large multiservice ISP type that operates an ATM infrastructure. An example of such a network is illustrated in Fig. 12-16. Observe that the network of ISR devices supports IP routing, IP switching services via ARIS, and native ATM switching. The routers situated at the periphery of the network that can forward packets at layer 3 to the core ISR devices, or if the ARIS protocol is present, they can establish direct ingress-to-egress switched paths.

MORE INFORMATION

Feldman et al., "Aggregate Route-Based IP Switching (ARIS)," IBM Technical Report, TR 29.2353.

Feldman et al., "ARIS Protocol Specification," IBM Technical Report, TR 29.2368.

Metz, C., "Aggregate Route-Based IP Switching (ARIS)," IBM White Paper, Aug. 1997.

CHAPTER 13

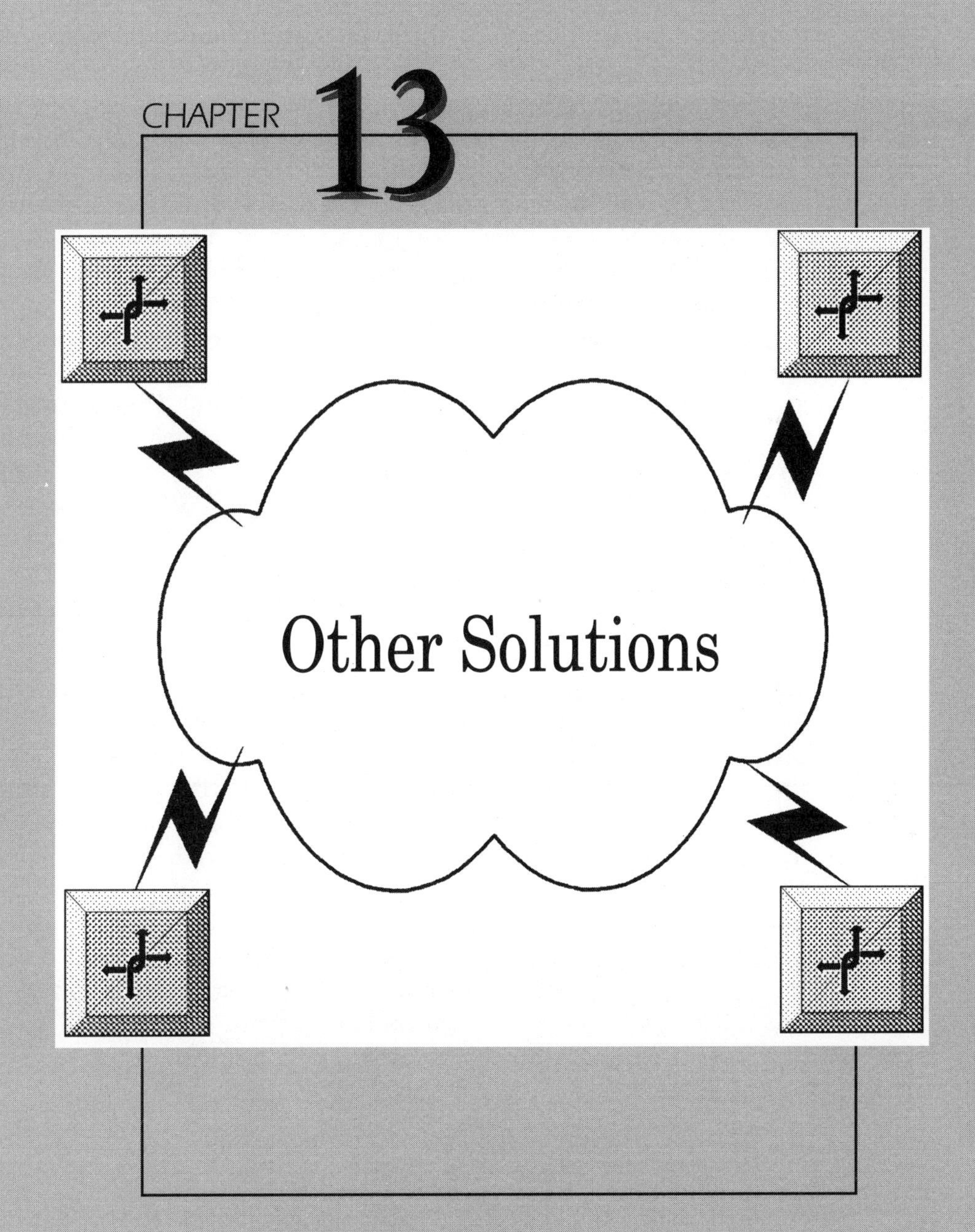

Other Solutions

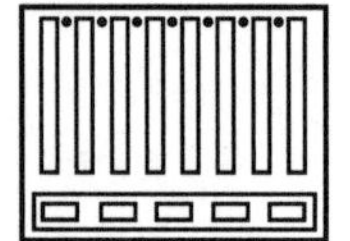

So far this book has reviewed the IP protocol suite, LAN and WAN switching technologies, the Classical IP over ATM solutions, the ATM Forum's LANE and MPOA solutions, and the four well-known IP switching solutions from Ipsilon, Toshiba, Cisco, and IBM. However, these are not the only solutions that attempt in some way to leverage and exploit hardware-based switching as a way to speed up the IP forwarding process. And certainly given the investment in research and development and the focus on ATM specifically, these are not the only solutions that redirect or map IP traffic through ATM switches. A number of other solutions have been introduced by vendors and researchers that integrate routing and switching with the express written purpose of providing a better service for IP networks and applications. This chapter will provide a brief overview of some of the other solutions that have been proposed to support the function of IP switching.

13.1 FastIP (3Com)

3Com has developed and introduced a solution for IP switching over layer-2 switched networks called FastIP (FIP).[1] FastIP employs NHRP between end stations attached to the same physical layer-2 switched network but otherwise logically affiliated with different subnets. In fact FastIP can be viewed as an extension to NHRP in that it enables Ethernet-based LAN-attached end stations on different subnets to set up direct layer-2 switched paths with each other rather than having to forward packets through a router. Instead of resolving a destination IP address with an NBMA (e.g., ATM) address, NHRP is used by FastIP to resolve a destination IP address with a MAC address. Once the requesting end station resolves the destination IP address with the destination MAC address, layer-2 switched communications between the two end stations based on the MAC address can take place. However, the router is still present in the environment so that it can still forward packets between subnets if required. It may also serve as a shortcut policy administrator. In other words, the router can be configured to select certain traffic flows that are eligible for switched path service.

Figure 13-1 illustrates the basic operation of FastIP. A client and server are attached to the same layer-2 (e.g., Ethernet) switched network but to different subnets. First both the client and the server must be con-

[1]*http://www.3com.com.*

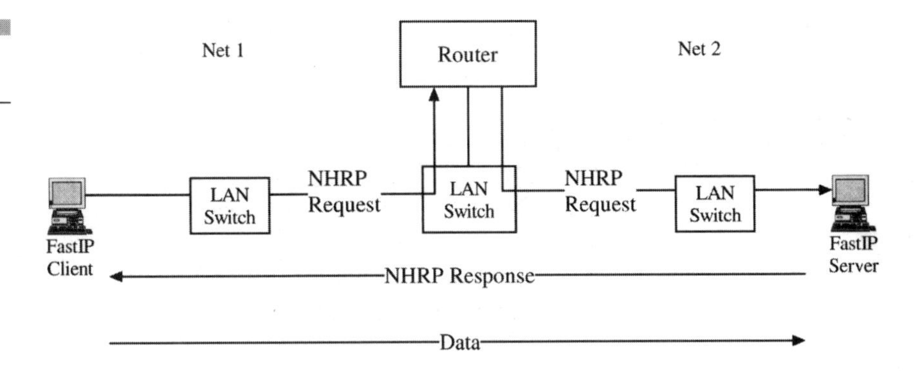

Figure 13-1
FastIP operation.

figured with a FastIP shim driver. Initial packets flow to the default gateway router. After a certain number of packets within a time interval have been sent, the client issues an NHRP request and directs it to the router. The NHRP request contains the MAC address of the source and is forwarded using conventional IP routing to the destination or, in this case, the server. However, the router that must forward the packet will first determine if it is okay for the source (client) to communicate with the destination (server) using a shortcut path. The decision to allow the NHRP request to proceed is based on a configured policy in the router. If the policy indicates that it is okay for the client to communicate with the server via a shortcut path, the NHRP request is forwarded directly to the destination or server. If not, the router may drop the NHRP request, and packets will continue to be forwarded by the router.

The server running the FastIP shim code recognizes the NHRP request and issues an NHRP response directly to the originating client using the source MAC address contained in the NHRP request. The layer-2 switches along the return data path that the NHRP response is traversing switch the packet based on the client's MAC address.

After the client receives the NHRP response, it redirects any packets destined for the nonlocal destination (server) using the server's MAC address rather than addressing the packets to the default router. Like other IP switching solutions, the objective of FastIP is to bypass intermediate router processing by enabling end stations on different subnets to communicate directly with each other over a layer-2 switched path.

Another feature of FastIP is the capability to support switched inter-VLAN communications. This is accomplished by including the IEEE 802.1Q VLAN tag of the source in the NHRP request. The destination can then address the NHRP response to the MAC address and VLAN ID of the source. All subsequent communications between the source and destination use standard 802.1Q VLAN tagging. This enables a VLAN that is essentially a closed broadcast domain to "open up" and support direct switched connections with non-VLAN members, policy permitting of course.

FastIP can be classified as a flow-driven IP switching solution that subscribes to the "route once—switch many" paradigm, which is similar to MPOA. FastIP has been extended to the Ethernet LAN media, so it is not just an ATM solution. To participate in a FastIP network, the servers and all clients must be upgraded with the FastIP shim code. And there is the issue of possible flooding of the NHRP response throughout the switched infrastructure. That aside, FastIP can deliver better performance for switched IP traffic by redirecting packet flows away from router-based processing.

13.2 Multiprotocol Switched Services (IBM)

In addition to the contribution of ARIS, IBM has developed and introduced several additional techniques that fall under the category of IP switching,[2] Multiprotocol Switched Services (MSS) is a suite of services that enhance the performance and scalability of IP internetworking over switched networks, in particular ATM networks. They are based on functional extensions to the standards-based LANE and NHRP protocols. Specifically, information maintained in the server components (LES/BUS, NHRP) is processed in an intelligent manner so as to optimize the data forwarding path and reduce network overhead. Put another way, the server components assist clients in establishing switched paths with each other while reducing extraneous broadcast traffic.

The first MSS function is called Route Switching (RS). Route Switching is a technique that, like FastIP, enables LAN-attached end stations on different subnets to establish a layer-2 switched connection with each other so that intermediate routing is avoided. Route Switching extends the capabilities of NHRP to support IP-to-MAC address resolution between Ethernet or Token Ring end stations connected to an ATM backbone. The intent of Route Switching therefore is to support zero-hop routing across an ATM backbone.

---

[2]*http://www.networking.ibm.com.*

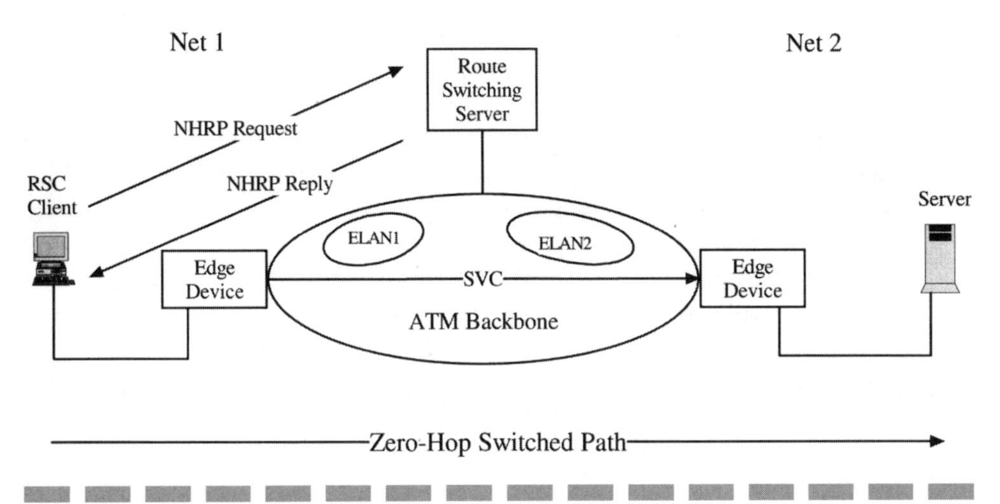

Figure 13-2 Route switching network.

A typical Route Switching network is illustrated in Fig. 13-2. A client and server are attached to different Ethernet segments, all of which are connected to an ATM backbone consisting of one or more ELANs. The client is outfitted with the Route Switching Client (RSC) software so that it may request a shortcut path to a destination. The Route Switching Server (RSS) functions as a default IP router between subnets, a LES/BUS for the ELAN-attached hosts and edge devices, and a next-hop server (NHRP server). In this example, the client first discovers the MAC address of the RSS by issuing an NHRP request to loopback IP address 127.0.0.1. The RSS responds with an NHRP reply, and the RSC saves the source MAC address contained in the reply. Initially data packets from the client and destined for the server are addressed to the MAC address of the default IP router, in this case the RSS. When the RSC detects packets addressed to the MAC address of the default IP router, it issues an NHRP request that will attempt to resolve the destination IP address with a MAC address. The RSS responds to the NHRP request with the MAC address of the server that was learned from a previous registration or through the ARP process. In addition, the NHRP response may contain additional information (e.g., egress ATM address, Token Ring route descriptors, etc.) needed by the client and/or source edge device to complete the layer-2 switched communications path from the client to the server. For example, the ingress edge device will need the ATM address of the egress edge device so that it may establish an SVC across the ATM backbone. Once this information is delivered to the source and ingress components, a shortcut path is established across the ATM backbone. The client now addresses all packets targeted for the server with the MAC address information learned from the NHRP process. The ingress edge device along the shortcut path places the packets on the appropriate SVC to the egress edge device. No routing hops are encountered along the shortcut path, hence the term *zero-hop routing.*

Route Switching is supported for Ethernet and Token Ring attached end stations; however, the source and destination must be attached to the same LAN type. Route Switching does not impose any functional prerequisites such as MPOA or NHRP client function on the edge devices. In addition the notion of asymmetrical shortcuts is supported. This enables a shortcut path to be established in only one direction. For example, the server shown in Fig. 13-2 may be configured with the RSC code and thus become capable of establishing a shortcut path to any client—even if the client does not have the RSC code installed and does not participate in the RS process. If the client does have the RSC code, all traffic in the client-to-server direction is routed through the default IP router. This particular function benefits high-volume server-to-client traffic flows with a layer-2 switched path service, and client-to-server traffic is routed and unchanged. The only change needed is the installation of the RSC code on the servers. No change is required to the clients.

Another IP switching function supported by MSS is called SuperELAN. A SuperELAN is a collection of two or more ELANs that are tied together by a SuperELAN bridging function. Each ELAN supports the standard LANE v1.0 services such as LES/BUS/LECS. A SuperELAN supports data direct VCCs between LE clients that reside in different ELANs. This eliminates intermediate bridge processing that is normally required for inter-ELAN communications. Another term for this process is *short-cut bridging* (SCB).

SCB, as illustrated in Fig. 13-3 enables LE clients to bypass the standard bridge processing when crossing ELAN boundaries. The SuperELAN bridging function can be coresident with the LES/BUS entities of the respective ELANs or run on a separate device as shown. It is configured as a traditional bridge whose bridge ports operate as proxy LEC components and provide special SCB processing. The SCB processing consists of forwarding certain LANE control frames, including `LE_ARP` requests/responses between ELANs. Initial unicast frames are forwarded between ELANs through the SCB ports as well. The net result is that a source, LEC A.1, on one ELAN can set up a data direct VCC with a destination, LEC B.2, on another ELAN.

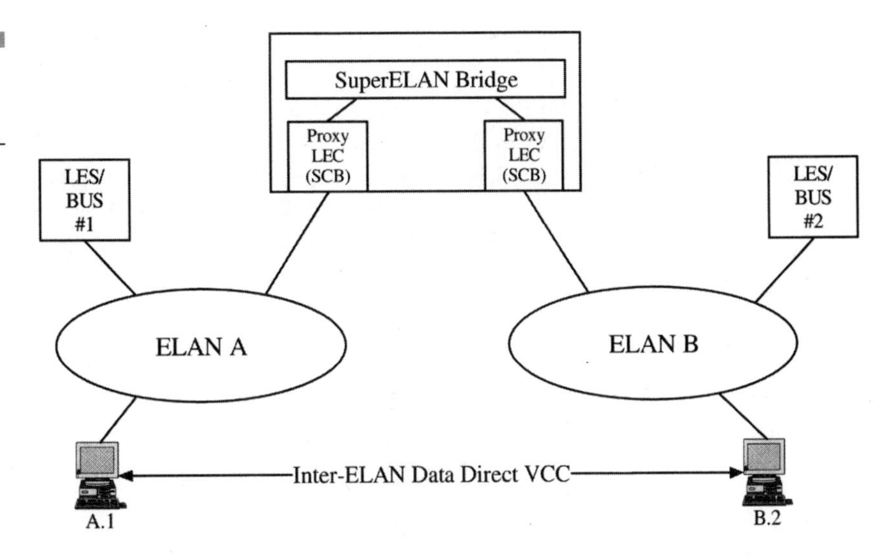

Figure 13-3
SuperELAN short-cut bridging.

To control broadcast propagation and allow IP-based ELANs to scale, MSS supports an additional function called the broadcast manager (BCM). For LANE V1.0 IP subnets, broadcast frames are sent to the BUS over the point-to-point Multicast Send VCC. The BUS then forwards the broadcast frames to all LE clients in the ELAN over the Multicast Forward point-to-multipoint VCC. The volume of IP subnet broadcasts (ARP requests) can be substantially reduced by redirecting a specific ARP request over an existing Multicast Send VCC. The BUS is able to do this because it inspects all ARP requests and replies and caches IP/MAC/LEC associations. In effect, the broadcast frame is converted into a unicast frame. This enables a large point-to-multipoint broadcast tree to be "pruned" to a single branch, thus reducing network resource overhead and enabling the creation of large IP-based ELANs. Note that the *broadcast reduction takes place within the IP subnet.*

Broadcast propagation in a SuperELAN is controlled by two additional MSS functions: Bridging Broadcast Manager (BBCM) and Dynamic Protocol Filtering (DPF). BBCM acts similarly to the broadcast management function described above except that it is performed by the SuperELAN bridge function. It reduces unnecessary IP ARP broadcast traffic and processing on ELAN segments. DPF enables the creation of protocol-specific VLANs by learning what protocols are supported on what bridge ports. The scope of broadcast packets for a particular protocol or network is confined to only those bridge ports where the protocol is known to exist. This further reduces the amount of broadcast traffic one might encounter in a large bridged network.

In summary, MSS supports several functions that enhance the scalability and performance of IP-based LAN and ATM switched environments. Route Switching enables zero-hop routing for either Ethernet or Token Ring end stations across an ATM network. The SuperELAN function enables large flat IP subnets to be deployed over an existing ATM backbone while providing the same broadcast control functions that layer-3 routing provides. Indeed it might even be better given that the BCM function will reduce broadcast traffic within a subnet.

13.3 IP Navigator (Ascend)

Ascend's IP Navigator (IPnav) is a solution that provides IP switching services over a network of Ascend WAN Frame Relay and ATM switches.[3] Traditionally routers attached to a "cloud" of switches would have to establish virtual connections with each other, resulting in $O(N^2)$ of VCs where N is the number of routers. IPnav solves this scalability issue and provides an IP switching service over a cloud of WAN switches in a number of ways. First, the WAN switch operates an instance of an IP routing protocol such as OSPF. This enables the WAN switch to discover the IP topology of the network and to exchange routing table updates with external routers. Second, the WAN switch builds MPT-PT trees that are rooted at the egress WAN switch and branch out to all ingress WAN switches. This minimizes the number of VC resources consumed and provides a switched path across the network. And finally, IPnav can exploit a VC signaling and routing protocol, supported by Ascend's Virtual Network Navigator (VNN), as a means of selecting the best path that meets the characteristics of the data flow.[4]

A network of Ascend WAN switches running IPnav is illustrated in Fig. 13-4. For a particular egress WAN switch, an MPT-PT tree rooted at this switch and branching out to all ingress WAN switches is built. It is based on the notion of an egress node identifier, which is similar to the function provided by ARIS. For an ATM-based network, the MPT-PT

[3]*http://www.ascend.com.* IP Navigator was developed and introduced first on the Cascade WAN switch technology. Ascend purchased Cascade in 1997.

[4]VNN is an extension to OSPF that runs on Ascend WAN switches. Network resources such as bandwidth and delay along with addressing information are advertised. This enables VNN switches to select paths based on available resources and traffic characteristics.

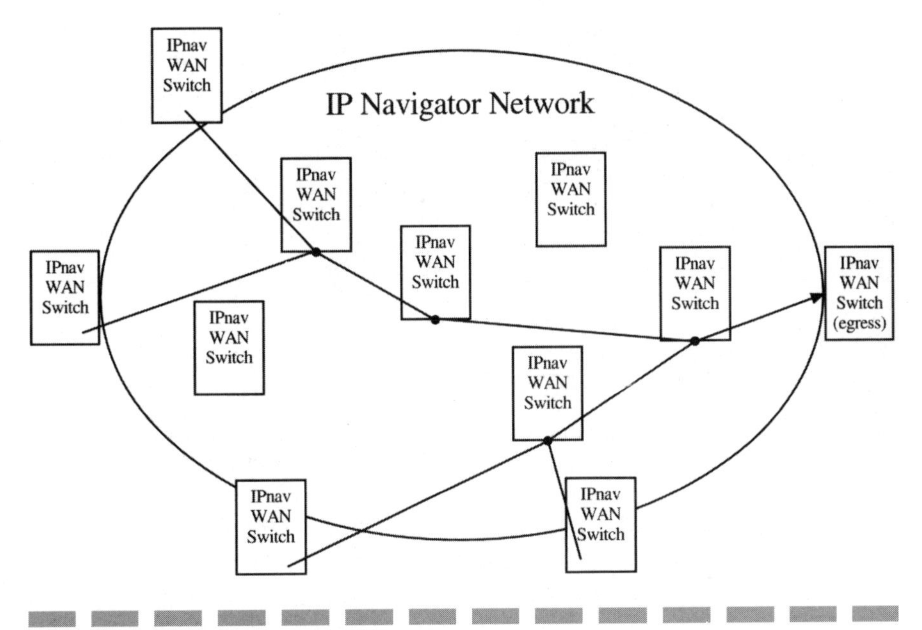

Figure 13-4 IP Navigator network.

switched path actually consists of virtual paths, and unique VCI values selected at the ingress WAN switch are used to maintain the identity of the source when the cells arrive at the egress WAN switch. When a packet arrives at an ingress WAN switch, a standard routing table lookup is performed. However, instead of forwarding the packet to the next-hop router as is the case with normal IP routing, the ingress WAN switch places the packet on a preestablished VC to the egress WAN switch. No intermediate layer-3 processing is performed within the IPnav network.

IPnav can be classified as a topology-driven IP switching solution, and it is designed for large IP networks operating Ascend WAN switches. Even though it does operate with external routers and supports the full suite of unicast and multicast routing protocols, its operation is confined to a domain of Ascend WAN switches. It should also be pointed out that the WAN switches can simultaneously support standard Frame Relay and ATM switching services in addition to the IP switching capabilities offered by IPnav. This makes IPnav a very good solution for network providers with Ascend WAN switches who wish to upgrade their network to support IP switching.

13.4 IPSOFACTO (NEC)

The very neat sounding acronym, IPSOFACTO, is derived from IP Switching over Fast ATM Cell Transport and describes a technique that enables the cells of IP flows to be cell switched through an ATM switch component. IPSOFACTO is a flow-driven solution for IP switching in that the packets of an individual IP flow are switched through a network of IPSOFACTO-capable ATM switches. Unlike IFMP and FANP, it does not define an explicit control protocol for mapping flows to VCs. Rather the IPSOFACTO entity in the ATM switch manages the allocation and assignment of VCI values for specific IP flows. In addition the IPSOFACTO entity manages the entries in the ATM switch connection table so that an inbound and outbound VC for a particular IP flow can be spliced together, forming an internal shortcut path. Assuming this is performed among a contiguous set of IPSOFACTO switches along the routed path, an ingress-to-egress switched path is formed.

To support IPSOFACTO, an ATM switch must support standard IP routing and forwarding and run an instance of the IPSOFACTO component in the switch controller. ATM Forum signaling and routing is not used in this environment. An IPSOFACTO IP switch and the operation for mapping an IP flow to an ATM cell-switched connection is illustrated in Fig. 13-5. Initially all unused VCI values on an inbound port are

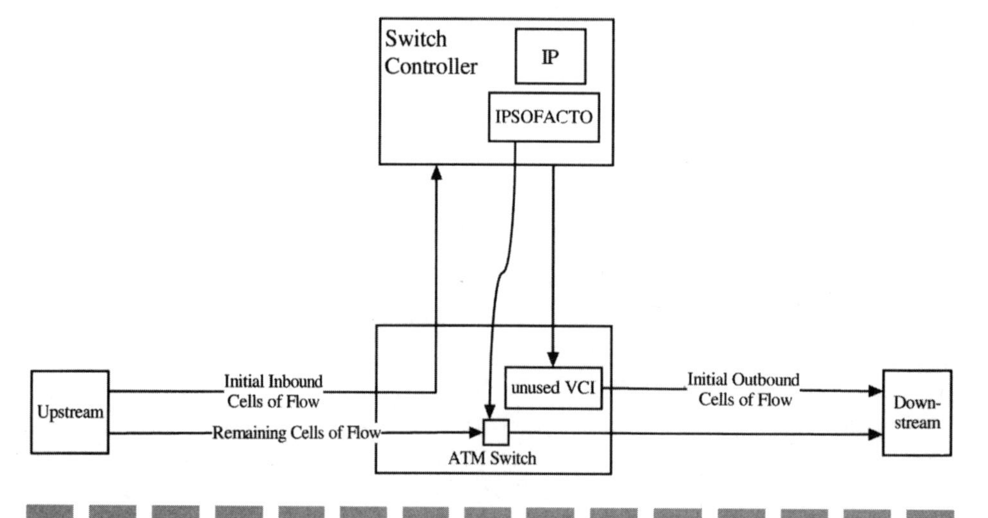

Figure 13-5 IPSOFACTO.

mapped to the IPSOFACTO entity. An upstream sender allocates an unused VCI on the outbound link and labels the cells of that flow with that new VCI. When the downstream IPSOFACTO IP switch receives the cells on the inbound port, it directs them up to the switch controller. The IP routing component forwards the packet to the appropriate downstream node but not until the IPSOFACTO component has labeled the cells of the flow with an unused VCI value. At this point the cells of the respective flow have been relabeled on the inbound and outbound ports. Therefore the switch controller can simply add an entry to the switch connection table consisting of <inbound port, inbound VCI, outbound port, outbound VCI>.

IPSOFACTO supports both unicast and multicast shortcut paths. The allocation of an unused VCI on the outbound link can be based on specific packet types (e.g., TCP SYN), control traffic (e.g., PIM Join), or the arrival rate of a particular flow.

13.5 Virtual Network Switching (Nortel)

Virtual Network Switching (VNS) is a WAN-based multiprotocol switching technology developed and introduced by Nortel.[5] It operates on Nortel WAN switches and like all IP switching solutions, confines layer-3 processing to the ingress and egress nodes. Layer-2 switching based on a destination address is used to transport packets through a VNS network. In addition, VNS supports COS-based routing, traffic engineering, virtual private networks, and multicasting. A VNS network can interoperate with a number of external logical networks and interfaces such as LANs, PPP links, LANE/MPOA networks, and RFC1483-encapsulated ATM links. An overview of VNS is contained in RFC2340.

To the surrounding IP-aware ingress and egress nodes, a VNS network, consisting of multiple VNS-capable switches interconnected by point-to-point links, appears as a single broadcast-capable LAN in which all nodes are directly connected to each other. The logical formation of the switches into this configuration is called a logical network (LN). When a packet enters the VNS network, the ingress node determines the egress node that the packet should be forwarded to. The ingress

[5] *http://www.nortel.com.*

Common Header	VNS Header	IP Header	Payload

Figure 13-6
VNS header.

node then appends a special common VNS header to the packet, as illustrated in Fig. 13-6. VNS routing, not IP routing, is used to select a path through the network to the egress node. The common header indicates that the packet is a VNS-encapsulated packet. The VNS header contains a TTL field, a Logical Network Number (LNN) that the packet belongs to, discard priority, COS, a link selection key, a source node ID, and an egress node ID header. An IP multicast packet has an additional 4-byte field for the multicast group address. It is a combination of the LNN and egress node ID values that form the label that is used to forward the packet through a VNS network. The labels are distributed to the participating nodes of an LN using OSPF-based extensions.

An attractive feature of the VNS solution is the ability to configure the network into multiple LNs and thus separate traffic onto multiple VPNs. An ingress node may contain multiple instances of a routing table, each reflecting the topology of a particular customer's VPN.

Because VNS makes use of both the IP and switch topologies to select a path through the network, it can be considered a topology-driven IP switching solution.

13.6 Responder-Initiated Shortcut Path Protocol

Responder-Initiated Shortcut Path Protocol (RISP) was suggested by researchers from Fujitsu in a 1998 Engineering Interop whitepaper.[6] Further details on RISP can found in a submission to the Internetworking over NBMA mailing list.[7] RISP is a peer-to-peer protocol that enables hosts or routers to establish shortcut paths through a layer-2 switched network. It is designed to operate over an IP/ATM overlay network in which both IP and ATM topologies and routing protocols are maintained. Unlike NHRP, which is considered an intersubnet (LIS) address resolution protocol based on the client/server model, RISP is not

[6]Ogawa et al., "Shortcut Routing for Networks with Heterogeneous Routing Mechanisms," *Interop Engineer Conference,* 1998.

[7]*http://netlab.ucs.indiana.edu/hypermail/ion/9708/0027.html.*

a client/server-based protocol and does not require specialized NHS-like servers in the network to set up shortcut paths. RISP is in fact a peer-to-peer protocol in which the communicating parties initiate and establish shortcut paths. The function of first querying an NHRP server for an IP-ATM address mapping before setting up a shortcut path is not required. It is even possible to operate an RISP-based network with no address resolution services at all.

Like all IP switching solutions, the objective of RISP is to set up shortcut paths between devices on different subnets so that intermediate routing hops can be bypassed. RISP accomplishes this by first having a host or router, called a requester, that wishes to establish a switched path (VC) to a destination (remote host or router) issue a Callback Request message. The Callback Request message is forwarded over the routed path to the destination and contains the IP and ATM addresses of the requester. If the actual destination is not on the ATM network, the egress router will intercept and process the Callback Request message. Either the destination or the egress router (*the responder*) will then set up a direct SVC back to the requester. Data from the requester to the responder will then flow over the shortcut path. This basic process is illustrated in Fig. 13-7.

The advantage of RISP over NHRP in establishing intersubnet shortcut paths is twofold. First, intermediate routers do not have to operate NHRP and maintain a table of IP-ATM address bindings. Second, the latency in establishing a shortcut path is reduced because the sender does not have wait for a response from a remote or transit NHRP server. The destination will respond to the Callback Request by establishing a shortcut path.

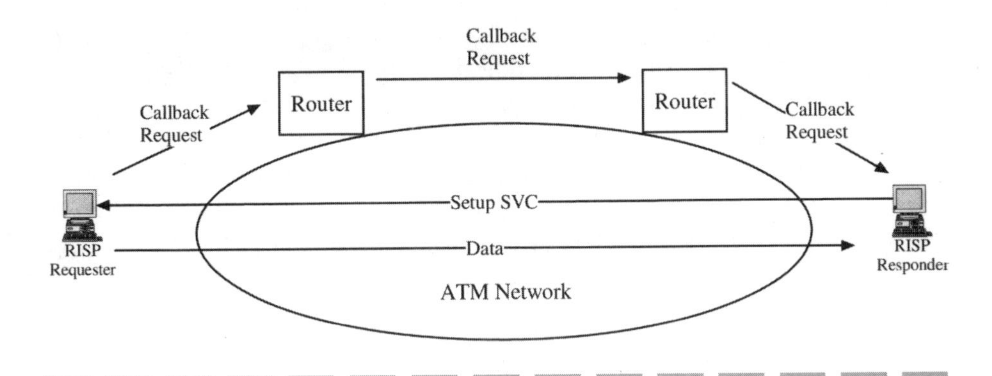

Figure 13-7 RISP.

The RISP protocol does not require that the transit routers support RISP; however, they must be capable of forwarding RISP messages. In other words, conventional ATM-attached routers can participate in the establishment of the shortcut path without operating a specialized control protocol (NHRP). The only entities required to support the RISP protocol are the requesters and the responders. RISP hosts and routers can be configured on a Classical IP logical IP subnet (LIS) that requires just an ATMARP server. The ATMARP server is used to resolve IP/ATM mappings for SVC communications within the LIS. Therefore intersubnet shortcut paths can be established without the need for NHRP to run on each transit router. It is also possible to configure an RISP host to attach to a router via a separate point-to-point link and have the router maintain (advertise) host routes for each host. In this scenario no address resolution protocol such as ATMARP or NHRP is needed, and the only control protocols are the IP routing protocols (e.g., OSPF) and RISP. It should also be noted that RISP only supports unicast connections.

13.7 DirectIP (Nbase)

DirectIP from Nbase Communications is a technique for building switched connections between LAN-attached devices on different subnets through a network of Nbase LAN switches.[8] In the DirectIP solution, the switch listens for and stores information about the IP and MAC addresses of IP devices attached to the switch. The switch then uses this information to "fool" the IP source on one subnet into thinking it is communicating with a destination on the same subnet—when in actuality the destination is another subnet. Despite the apparent chicanery, no change is required to the end stations, and intersubnet flows can benefit from wire-speed layer-2 switching. DirectIP incorporates DHCP to simplify switch and host configuration requirements and supports the notion of VLANs.

A switched connection in a DirectIP-based network is accomplished in the following manner and is illustrated in the two diagrams in Fig. 13-8. First, the sender issues an ARP request for the destination. The intermediate switch examines the destination IP address and determines if it

[8] *http://www.nbase.com/*.

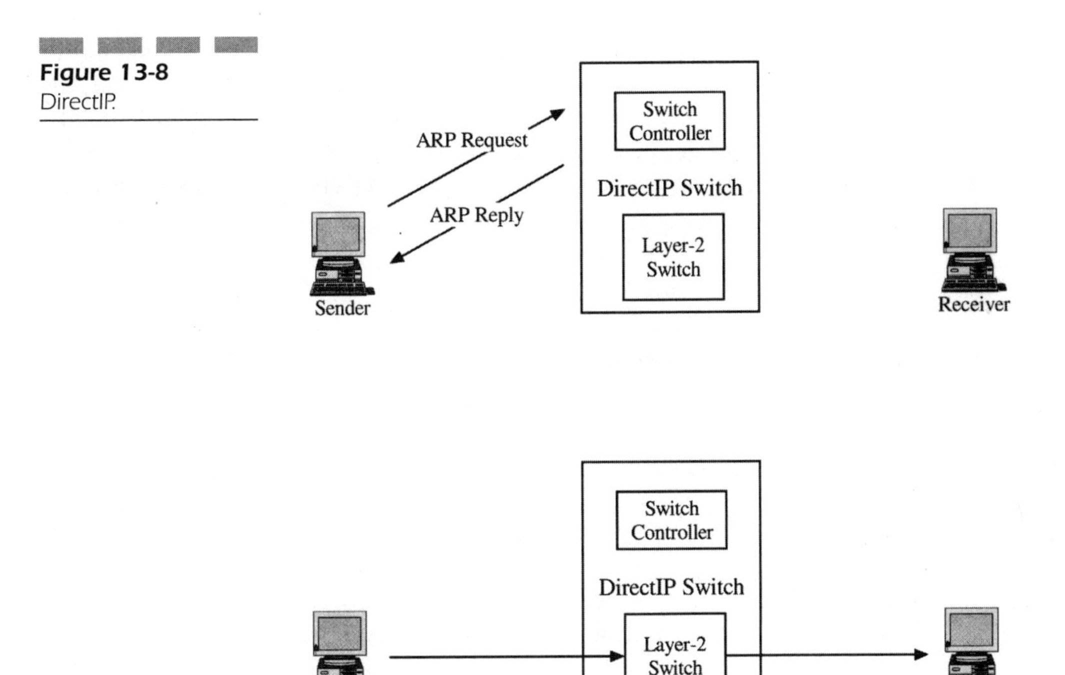

Figure 13-8
DirectIP.

can proceed with a switched connection setup based on entries in a policy database. If not, the switch will not disclose the MAC address of the destination, and subsequent packets from the source will be forwarded at layer 3. If it is okay, the switch will return the MAC address of the destination in an ARP response to the source. The source caches the IP/MAC address binding for the destination in its ARP cache. The source may now send packets to the destination using the destination's MAC address. The switch forwards the packets at layer 2—no processing at layer 3 is necessary.

DirectIP can be classified as an IP learning solution. It learns the addressing information of devices attached to the switch segments and then uses this information to build layer-2 forwarding tables between devices on different subnets. In effect an IP learning switch is acting as a proxy MAC destination to spoof the source into believing it is switching packets with another host on the same subnet. Due to their dependency on ARP broadcasts for the learning process and the requirement to store both IP and MAC addresses, IP learning switches are best suited for a campus LAN environment. Depending on the implementation, they may or may not run dynamic IP routing. If the latter is the case, some manual configuration of IP information on the switch may be required. Generally speaking, though, one can view IP learning switches as simply fast IP forwarders for legacy LAN environments. They confine broadcasts to the configured subnet and can forward packets between subnets at wire-speed, but they may not have the complete set of functions such as dynamic IP routing or multicasting associated with traditional routers. Indeed one intended scenario of DirectIP is to front-end a campus router and essentially off-load the IP forwarding function from the router.

13.8 AREQUIPA

One of the perceived limitations of IP switching, in particular when it operates over ATM, is the lack of QoS support. Recall though that the purpose of IP switching is to accelerate the IP forwarding process. It is not implicitly or explicitly mandated that IP switching do anything more than just provide a faster IP forwarding service. One technique to support QoS in the IP switching environment is by using RSVP. As outlined in the chapters on tag switching and ARIS, one could simply add an additional label or tag object to the RSVP RESV messages so that QoS and the switched path state could be established concurrently. However, this solution is confined to tag and ARIS domains and does not provide an end-to-end service.

The use of RSVP does not address the lack of QoS support in the Classical IP solution set.[9] Solutions such as Classical IP over ATM (RFC 1577/2225) and NHRP (RFC2332) establish shortcut ATM SVCs with best-effort UBR service. To address this limitation, Swiss researchers developed a solution that enables peer IP applications to establish virtual connections to each other with specific QoS[10] The solution is defined

[9]Until the Integrated Services model came about as described in Chap. 3, IP had no idea what QoS was. So there was really no basis to engineer QoS into the Classical IP solution set. However, there are mechanisms in the IP over ATM signaling specifications defined in RFC1755 (UNI 3.1) and RFC 2331 (UNI 4.0) that enable SVCs to be set up with QoS.

[10]*http://icawww.epfl.ch/*.

in RFC2170 and is called AREQUIPA, which stands for Application Requested IP over ATM.

The motivation behind the development of AREQUIPA is simple: develop a solution that can support QoS for IP over an ATM. AREQUIPA can be viewed as an extension to the Classical IP solution suite. As illustrated in Fig. 13-9, it enables two hosts and more specifically two applications to build an ATM SVC between each other with QoS. It does so by enabling applications at the socket layer to either request a socket-based connection with QoS or to expect that one will be established. This involves the definition of several new API functions (outlined in RFC2170) that enable applications to perform these functions. Once the API calls have been made and the SVC (with QoS) between the two hosts built, application data will be directed over that specific SVC. Other best-effort traffic will flow over normal best-effort connections.

As suggested by its developers, AREQUIPA can be useful as a technique to exploit ATM QoS as a means of delivering real-time Web content. In this scenario, Web documents are extended with meta-information to describe the ATM service and corresponding QoS required for transmission over an ATM network. A Web browser that wishes to access these documents includes its ATM address and possibly a socket

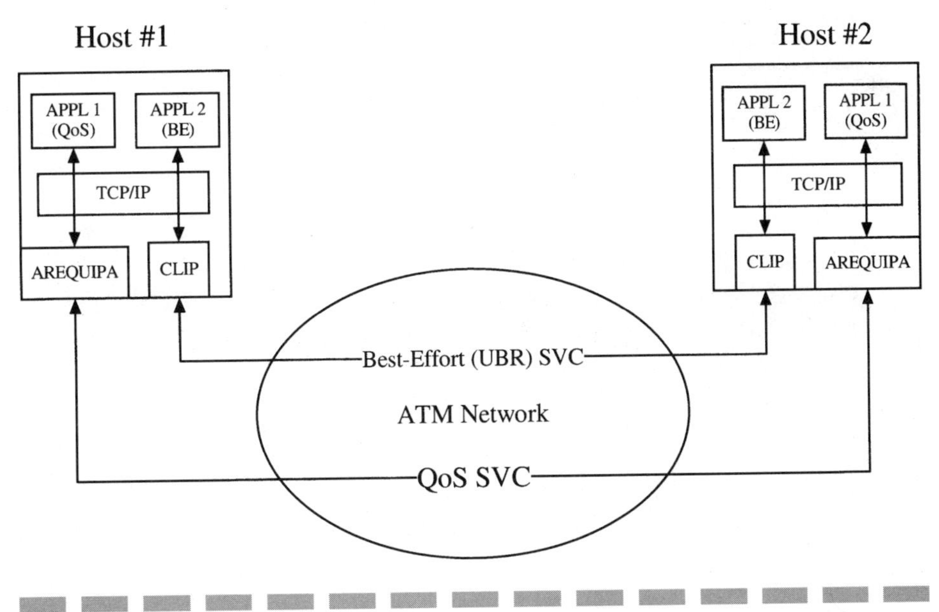

Figure 13-9 AREQUIPA.

number in the HTTP request. The server may then transmit the document using AREQUIPA. Of course this solution requires a homogeneous ATM cloud in which all clients and servers are attached. Although this scenario may not be realistic, the technique of associating QoS (special service) with a particular document (content) and then transmitting the document with the special service (QoS) over the network upon request from the client is probably something that will happen in the not too distant future.

Part 3

New Related Works

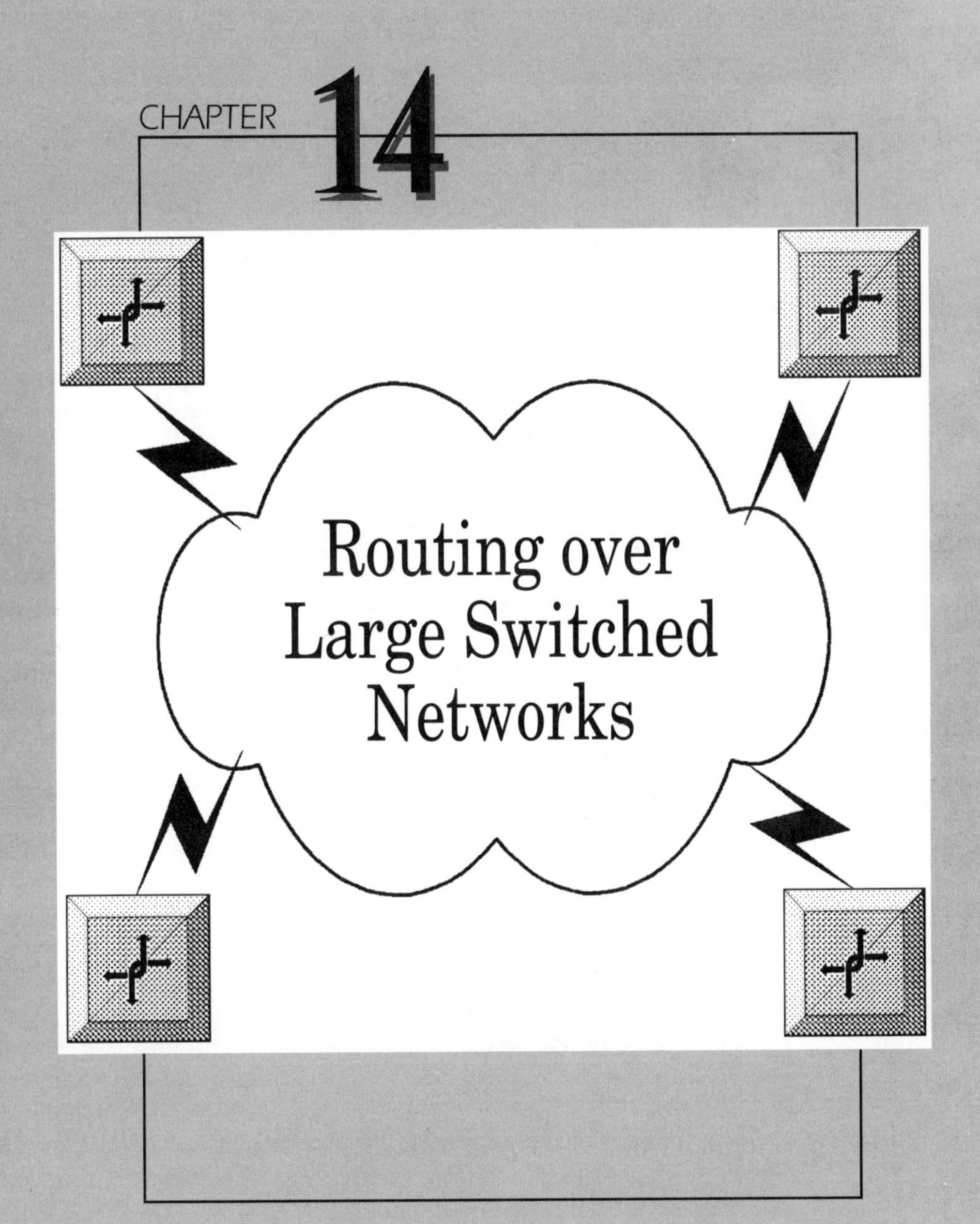
CHAPTER 14
Routing over Large Switched Networks

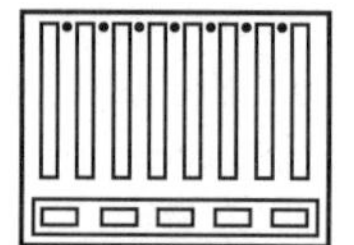

In the last section, a suite of different IP switching solutions was discussed. They all marry the control functions supported by IP routing and the high-performance switching capacity offered by ATM. They all require the integration of the IP control plane and the ATM user plane on a single device, a hybrid switch/router. They also all require the introduction of a specialized control protocol that maps IP layer constructs, whether they are source/destination address pairs of a data flow or simply destination prefixes, to a switched path. They all subscribe to the peer model in which a single IP address space is maintained and a single IP routing protocol is supported. On paper, in prototype, and even in some controlled and preliminary implementations, they seem to perform as claimed. However, due to their relative youth and proprietary nature, they are not solutions that have been or are likely to be deployed on any large scale in existing networks for some time.

Up until now and most likely for the foreseeable future, switching IP packets through a network will be accomplished using the traditional overlay model. Routers running IP routing protocols (e.g., OSPF) are attached to a Frame Relay or ATM switched cloud. Communication between routers is accomplished through virtual connections. While achieving switched performance across the network, issues such as VC capacity and routing protocol overhead must be considered as the size of the network and volume of traffic increases, as it most assuredly will. Therefore, network providers and in particular those that operate large Frame Relay and/or ATM infrastructures face a big challenge: how best to support IP internetworking over a large switched cloud that maximizes performance, leverages the switched infrastructure, minimizes control overhead and processing, and avoids unnecessary manual or operator intervention.

Peer model IP switching is one possible solution, but it is too early in the deployment cycle and there are too many variables in possible implementations to consider at this time.[1] In addition there are many network providers who are perfectly content with operating traditional switched clouds, and they may not want to move to IP switching for operational or cost reasons. Another important reason is that they currently offer a collection of networking services that depend on native switching technologies. Frame Relay and ATM PVCs are two examples, with possibly SVC support to follow. These traditional switched network

[1]It is hoped that the convergence of many of the vendor-specific solutions into a single standard, as described in the next chapter, will remedy this.

providers are simply looking for solutions that make better use of their switched infrastructures for supporting IP. To make the best use of the network infrastructure to support these and future services, the optimization that these network providers are seeking will probably call for the use of a more dynamic (on-demand) allocation of resources.

There are several possible solutions—some of which have already been discussed. Operating a partial or full mesh of PVCs between routers is one common solution. Although not dynamic it does serve a purpose—the establishment of a "lease line"-like network for interconnecting routers. Another solution, let's call it the subnet model, is to divide the large switched network into separate logical IP subnets and route between them. This solution is simple to envision and operate, but it introduces one or more routing hops in the intersubnet data forwarding path. A third approach is a relaxation of the subnet model. NHRP is used to build intersubnet (LIS) shortcut paths between hosts or routers on different subnets, thereby enabling packets to bypass intermediate router processing. However, this solution introduces a new control protocol and requires that all routers support NHRP and maintain IP/ATM address mappings.

Two newer solutions depart from the notion of an additional control protocol needed to map IP traffic to layer-2 switched paths. In both cases, extensions to existing link state routing protocols are introduced as a means of propagating IP/ATM address mappings around the network. The first solution involves an extension to OSPF called the Address Resolution Advertisement (ARA). An ARA is essentially an IP/ATM address mapping that is advertised by an ATM-attached router using normal OSPF flooding mechanisms. This enables all other ATM-attached routers in the OSPF network to store this information in their topology databases and factor it in during path calculations. When a source router needs to communicate with a destination router across the switched network, it knows a priori the IP and ATM addresses of the destination, and it immediately can initiate an SVC setup request to that destination. The only change to the network infrastructure is to update OSPF on the routers.

The second solution involves one of several possible extensions to another link-state routing protocol, PNNI. Recall that PNNI is a routing and signaling protocol used to forward SVC setup requests through a network of ATM switches. In filling that role, PNNI uses traditional link state flooding procedures to distribute ATM addressing information around the network. Extensions to PNNI would allow it to distribute IP-layer information (IP address prefixes, router ID, etc.) around the network as well. The first extension is called PNNI Augmented Routing (PAR). This involves operating an instance of PNNI along with any other dynamic IP routing (e.g., OSPF) protocol on ATM-attached routers. PAR allows ATM-attached routers to participate in the PNNI topology distribution process with ATM switches. In fact the ATM-attached routers become nodes in the PNNI topology of one or more peer groups. PAR routers use PNNI flooding mechanisms to distribute and then discover the presence of other non-ATM devices and services attached to the ATM network.

Another PNNI extension is a "lightweight" version of PAR called proxy-PAR. Instead of running PNNI, ATM-attached routers operate a client function that enables them to register their IP and ATM information with a directly attached proxy-PAR server function running on an ATM switch. PNNI is then used to propagate this information throughout the network. Other ATM-attached routers, if they so desire, may retrieve IP information from the nearest proxy-PAR ATM switch. With both PAR and proxy-PAR the objective is to leverage PNNI as a means for IP devices to discover the location and service offered by other IP devices attached to the ATM network. One application of proxy-PAR is to provide ATM-attached routers with the information so that they can establish an initial set of SVCs pursuant to the traditional IP over ATM overlay model.

The third PNNI extension integrates IP and ATM routing into a single routing protocol based on PNNI. Integrated PNNI (I-PNNI) is a single routing protocol that operates on ATM switches, ATM-attached IP routers, and legacy-attached IP routers. Routers and switches become nodes in a peer group and exchange topology information that consists of both ATM and IP address prefixes. The objective of this approach is to support a single routing protocol that would combine both topologies, thereby allowing for a more informed path selection for SVC routing and IP forwarding. It would also reduce to some degree the complexity of the overlay model that requires two routing protocols, one for IP and one for ATM.

14.1 Issues

Network providers that deploy the overlay model, that is IP routing overlayed on top of a switched network, must confront a number of issues.[2] First, these networks require the configuration, maintenance, and operation of two separate routing protocols, two separate address spaces, and two separate topologies. IP routers attached to the switched network run IP routing protocols, exchange routing table updates with neighboring routers, maintain an IP topology database, and forward packets to the next-hop IP router. There is no awareness or knowledge of the underlying switched infrastructure other than whatever familiar logical link (point-to-point, broadcast LAN, etc.) it is emulating. Likewise the ATM switches run an ATM routing protocol, exchange routing table updates with neighboring ATM switches, maintain an ATM switch topology database, and forward SVC requests to destination ATM switches. Even if the switched environment is not supporting SVCs, network providers must still deal with PVC management issues such as configuration, traffic monitoring, and error recovery. It is a simple fact that the duplication of configuration, management, and operation efforts is a necessary requirement for deployment of the overlay model.

Another issue that may arise in the overlay model is the amount of VCs needed to interconnect routers. Figure 14-1 shows a network of three ATM switches and eight routers. A logical topology resulting from a full mesh of VCs connecting each router to every other router is illustrated in Fig. 14-2. This particular logical configuration offers the best performance because traffic flowing between any two routers will traverse a switched path. However, as the number of routers attached to the ATM network increases, so does the number of VCs required to sustain the full mesh configuration and achieve optimal performance. In fact the number of VCs increases by $O(N^2)$ where N is the number of routers. In the modest example shown, a full mesh of VCs interconnecting eight routers requires the provisioning, either dynamically or statically, of 28 point-to-point bidirectional VCs.[3] This may seem like a manageable number, and it probably is. For relatively small and even for some medium-size networks, this is standard practice. But if N becomes large, the number of VCs needed grows dramatically. If N is 100, 4950 VCs must be provisioned. If N is 1000, the number of VCs needed is 499,500. In addition, those networks supporting multiple services may require more than one VC between any two routers.

[2]In the discussion to follow, ATM will be used as an example of a switched network, but the same holds true for Frame Relay or any other NBMA-type network.

[3]If $N = 8$, then $(N^*(N-1))/2 = 28$.

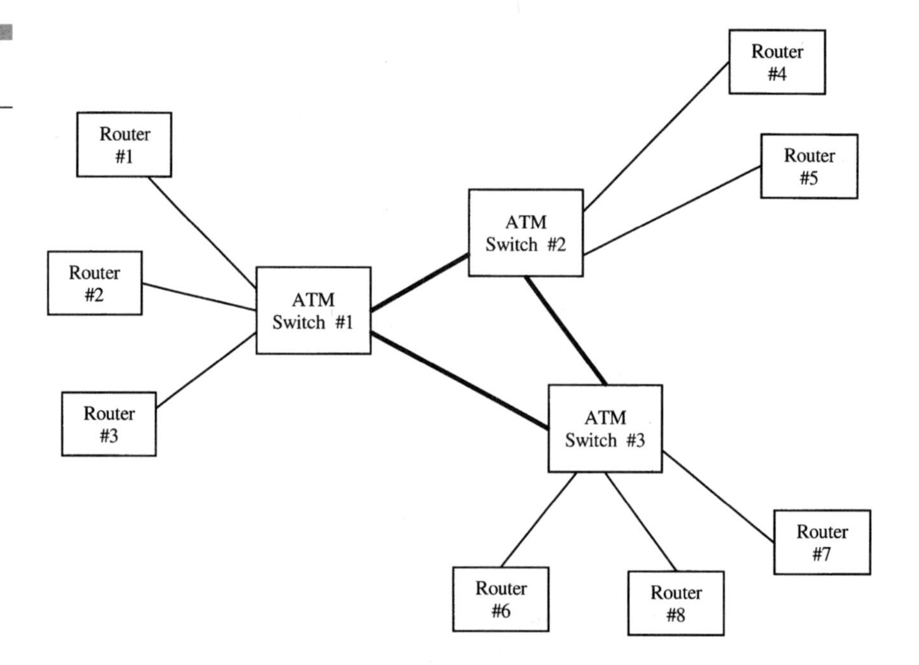

Figure 14-1 Physical topology.

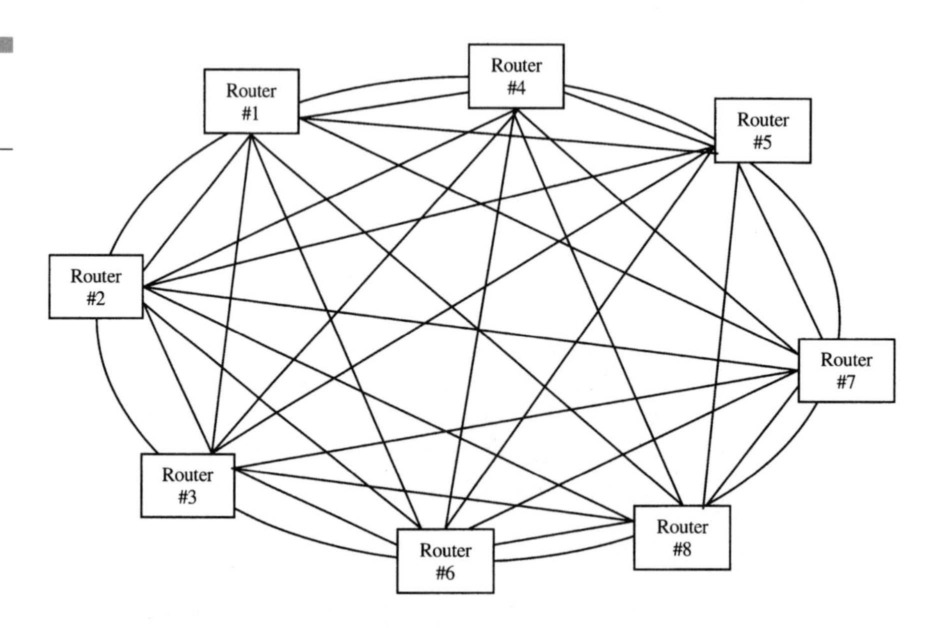

Figure 14-2 Full-mesh logical topology.

As the number of routers grows, the corresponding growth in VCs may impose operational constraints in VC provisioning and capacity constraints on the switched infrastructure. The former has to do with the tasks needed to provision and manage a large number of VCs between multiple edge routers. In the majority of environments, the VCs are manually configured. There are also interior backbone switched links to worry about because they will typically carry a large number of VCs. Therefore, there is a substantial amount of skills, person-power, connection polling, management tools, and so on, needed to ensure that the VC environment is stable. The issues with the switch capacity constraints have to do with the fact that ATM switches can only support a finite number of switched connections. This is because resources in the form of labels, memory, control blocks, buffers, and so on, are allocated for each connection.

Another important issue to consider when deploying the overlay model for IP routing over ATM is the scalability of the IP routing protocol. Specifically the number of separate links represented by individual VCs connecting routers can lead to a large amount of information that must be advertised and then processed by routers. Consider the full mesh logical topology in Fig. 14-2. Each router maintains a dual-purpose adjacency with every other router attached to the network. A routing adjacency enables routers to exchange routing table updates. A forwarding adjacency serves as a next hop toward particular destination networks. If the routers are running a distance vector routing protocol such as RIP or BGP, the amount of information that a router must process is proportional to the number of routing adjacencies and routing table entries. In other words, it is along the lines of O(# of routes * # of routing adjacencies). If the routers are running a link state routing protocol, LSAs will be generated for each link and flooded throughout the network. A large number of VCs results in a large number of links and LSAs that must be advertised by each ATM-attached router. With OSPF the size of the link state database and the overhead of computing routes through the network are proportional to the number of links (VCs in this case) in the network. Moreover, most of the LSAs may contain duplicate information, thereby placing an additional processing load on each router. The duplication occurs because every router, being a neighbor of every other router, propagates LSAs that it receives to every other router.

These issues can be addressed in a number of different ways. First, routing protocol and VC scalability can be addressed by reducing the number of interrouter VCs. Fewer VCs obviously lowers the consumption of switched resources in the network and reduces the number of routing adjacencies. This can be accomplished by establishing a partial mesh of VCs to interconnect routers or by carving up the switched cloud into a number of separate subnets and interconnecting those subnets with routers. But although these approaches enhance scalability, they will almost certainly introduce additional router hops along the data forwarding path, thus contributing to additional latency and less than optimal performance. The performance of the subnet model can be enhanced somewhat through the use of NHRP as a means to establish intersubnet shortcut paths. However, this approach depends upon the addition of an address resolution control protocol and the availability of NHRP servers along the routed path.

Another approach that can address the VC scalability issue is the use of MPT-PT connections, first introduced by the ARIS protocol described in Chap. 12. The use of MPT-PT connections does not exclusively fall under the control of ARIS. An overlay model consisting of separate IP and ATM control and data planes can also benefit from this capability. Rather than support $O(N^2)$ point-to-point connections, a network supporting MPT-PT connections would only require $O(N)$ connections where N is the number of routers. The advantage in terms of fewer VC resources needed to provide complete interrouter switched service is substantial. The graph in Fig. 14-3 illustrates the savings in particular as the number of routers grows.

ARIS is a peer-based IP switching control protocol that supports the establishment of MPT-PT connections. The overlay model requires that the connection setup control protocol support this capability. In the ATM

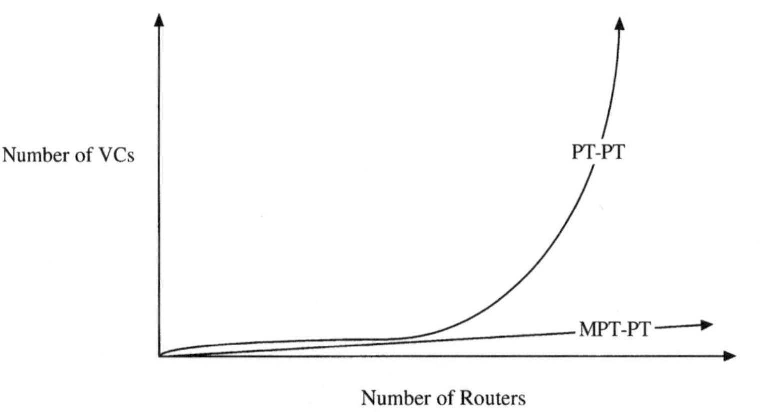

Figure 14-3 VC consumption for MPT-PT and PT-PT environments.

environment it is a combination of UNI and PNNI signaling and routing that performs the connection setup process, and currently only point-to-point and point-to-multipoint switched connections are supported. However, it may be possible to extend UNI/PNNI signaling to support the establishment of MPT-PT connections. One possible technique, as illustrated in Fig. 14-4 would reuse standard root-initiated point-to-multipoint signaling to build the connection but would indicate during connection setup that the polarity of the data flow is leaf to root—not root to leaf as is normally the case. Thus ADD PARTY messages would still originate from the root and would be used to add leaves to a point-to-multipoint VC. However, actual data would flow from the leaf nodes upstream toward the root. This assumes the capability of the ATM switches to support VC merging and assumes that MPT-PT connections will only support UBR connections. Given the obvious benefits of scaling, the ATM Forum may enhance standard UNI/PNNI signaling and routing to support MPT-PT connections.

Relying on suboptimal network configurations or enhanced VC management capabilities is not a complete solution for a scalable and efficient IP/ATM overlay network. Revenue-generating services such as VPNs and bandwidth on demand require a rapid and reliable allocation of network resources. The SVC capabilities of the ATM infrastructure need to be used to dynamically build connections to carry IP packets across the network. However, building SVCs between routers to carry IP packets requires that the destination IP and associated ATM addresses are bound prior to call setup. Currently, server-based technologies such as NHRP provide this binding. However, because the environment

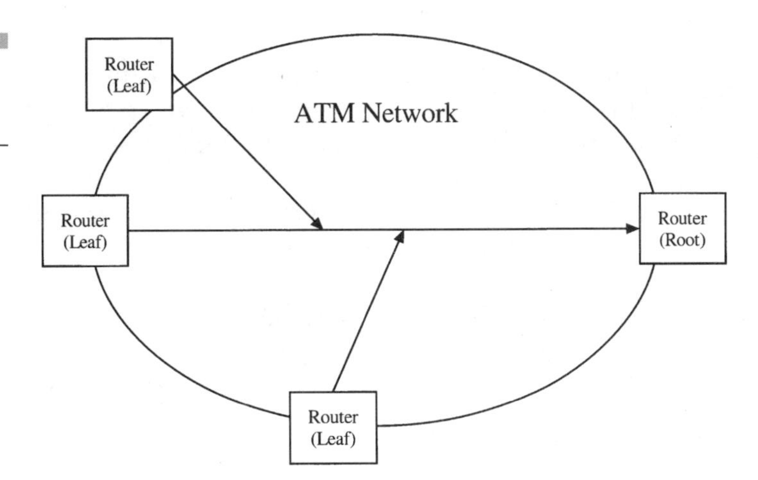

Figure 14-4 ATM MPT-PT connection.

under discussion consists solely of routers and switches, it might be useful to consider incorporating IP and ATM mappings into existing routing protocols. Two possibilities exist: integrate ATM into IP routing or integrate IP into ATM routing.

14.2 OSPF Extensions

OSPF is a link state routing protocol used by routers to build and maintain a database of the network topology inside of an autonomous system (AS). It is from this database that routers compute a routing table used to forward packets to a destination. OSPF is scalable, propagates topology changes quickly, and minimizes control traffic overhead. Another important attribute of OSPF is that it is extensible. Introducing new functions and features that provide an improved network layer service while remaining backward-compatible is very desirable. One possibility to extend OSPF in the ATM environment is to enable it to propagate link- and IP-layer information throughout the network. In other words, OSPF, an IP routing protocol, is extended to distribute IP and associated ATM addressing information throughout a network of ATM-attached routers. Therefore the OSPF routing protocol is the vehicle for delivering IP/ATM address mappings to interested ATM-attached routers. With this information in hand, interrouter SVCs can be built when needed.

14.2.1 Opaque LSA

Opaque LSA is a new class of link state advertisements described in RFC 2370.[4] It is a generalized mechanism or "placeholder" that can easily accommodate any proposed or future extensions to OSPF. The opaque LSA enables new information to be propagated throughout an OSPF domain, using standard OSPF flooding procedures, and then stored in the topology database of routers. The information contained in the opaque LSA can be used directly by OSPF or indirectly by some application that wishes to use OSPF to distribute information throughout an OSPF routing domain.

---

[4]More information on the OSPF opaque LSA and ARA can be found in the appropriate Internet drafts or RFCs located at *http://www.ietf.org/html.charters/ospf-charter.html.*

OSPF routers indicate their ability to support opaque LSAs with a bit in the Options field that is included in OSPF Hello and Database Description packets as well as all LSAs. This enables OSPF routers to determine if their neighbors support the ability to handle opaque LSAs. A flooding scope function in also included that defines the range of topological distribution for a specific opaque LSA. This is used to limit the information carried by opaque LSAs to a specific portion of the OSPF topology (e.g., area only).

The format of an opaque LSA consists of a standard 20-byte LSA header followed by a variable amount of information, as illustrated in Fig. 14-5. The opaque LSA supports three different link state types that define essentially the flooding scope:

- *Link state type 9.* This denotes a link-local scope. This type can only be flooded between routers attached to the same subnet.
- *Link state type 10.* This denotes an area-local scope. This type can only be flooded between routers in the same area.
- *Link state type 11.* This denotes an AS-wide scope. This type can be flooded throughout the entire OSPF AS.

The link state ID portion of the opaque LSA header is divided into an 8-bit opaque type and 24-bit opaque ID. This means that up to 256 different opaque types can be defined. The values of 0 through 127 are reserved for definition and use by the IETF/OSPF community, and the values of 128 through 255 are reserved for private or experimental use.

Figure 14-5
OSPF opaque LSA format.

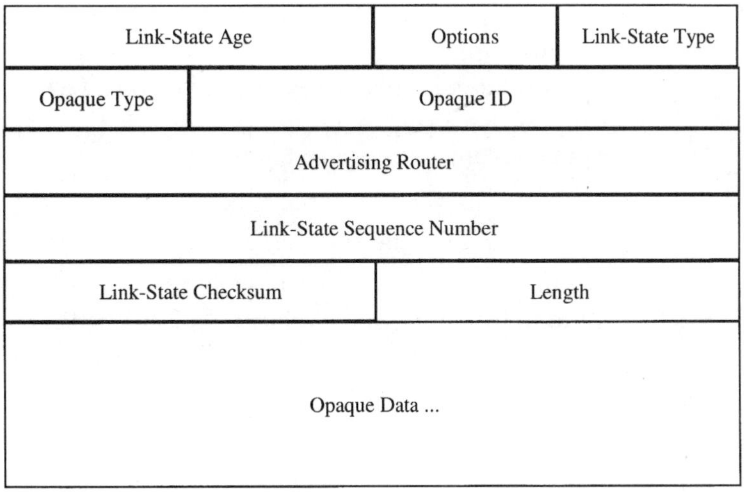

Link-State Age | Options | Link-State Type
Opaque Type | Opaque ID
Advertising Router
Link-State Sequence Number
Link-State Checksum | Length
Opaque Data ...

14.2.2 Address Resolution Advertisements

Two things are needed to enable OSPF to distribute ATM addressing information. One is the opaque LSA that serves as a port by which new information can be inserted and distributed across the OSPF network topology. The second is the actual IP/ATM address resolution information or mappings consisting of either the router's IP and ATM addresses or one or more IP address prefixes and an associated ATM address. The ARA defines the type and format of the address mappings that will be carried in opaque LSAs and distributed to ATM-attached routers. More specifically the ARA is a mechanism that enables routers to distribute IP/ATM address mappings (called link-layer associations) to other routers in the network and enables routers to determine the link-layer association that is closest to a particular destination network within the OSPF routing domain. ARAs can be used to distribute the link-layer associations of routers (called R-ARA) or of the networks that can be reached through routers (called N-ARA) contained within an area or remote to an area.

The basic objective of the ARA mechanism is to utilize OSPF to quickly and reliably provide routers with IP/ATM address mappings so that they can establish both intra- and interarea shortcut SVCs across an ATM network.

The OSPF ARA option has the following characteristics:

- ARA information is transported in link state type 10 opaque LSAs (area-local scope). There are four different opaque types (specified in the opaque type field in the opaque LSA header) that the ARA option supports: intra area router, intra area network, interarea network, and interarea ASBR link-layer associations.
- Only ATM-attached OSPF routers participate in the ARA process.
- Because the ARA is distributed within an OSPF area using reliable flooding mechanisms, it is not subject to packet loss. The same cannot be said for query-to-server mechanisms such as NHRP.
- Supports the establishment of point-to-point, point-to-multipoint, and multipoint-to-point shortcut paths (SVC).
- Can interoperate with existing address resolutions mechanisms such as NHRP/MPOA.
- Provides a capability to associate groups of routers into a logical network. This enables multiple disjoint OSPF networks (e.g., VPN)

to coexist over a single ATM network. It is accomplished by including a logical network ID in the ARA. OSPF will exclude from the routing table any link-layer associations whose logical network ID does not match a preconfigured value on the router.

An example of the OSPF ARA process is illustrated in Fig. 14-6 for routers within a single area. Routers 1 through 4 support the ARA option and are attached to the ATM network through a partial mesh of preconfigured PVCs as shown. Router 5 is also attached but only supports the opaque LSA function.[5] And router 6 is attached to router 4 via a non-ATM serial link and does not support the opaque LSA function. In this example the ARA-capable routers will generate and distribute link-layer associations for their directly attached or reachable networks. The link-layer associations contained in the ARAs and delivered to the ARA-capable routers enable those routers to initiate SVCs to other routers across the ATM network. For example, if router 1 needs to forward packets to network N2, it can establish a direct SVC to router 2. Or if router 1 needs to forward packets to network N6, it can establish a direct SVC to router 4. In both cases router 1 is able to do so because it received ARAs containing link-layer associations from router 2 and router 4.

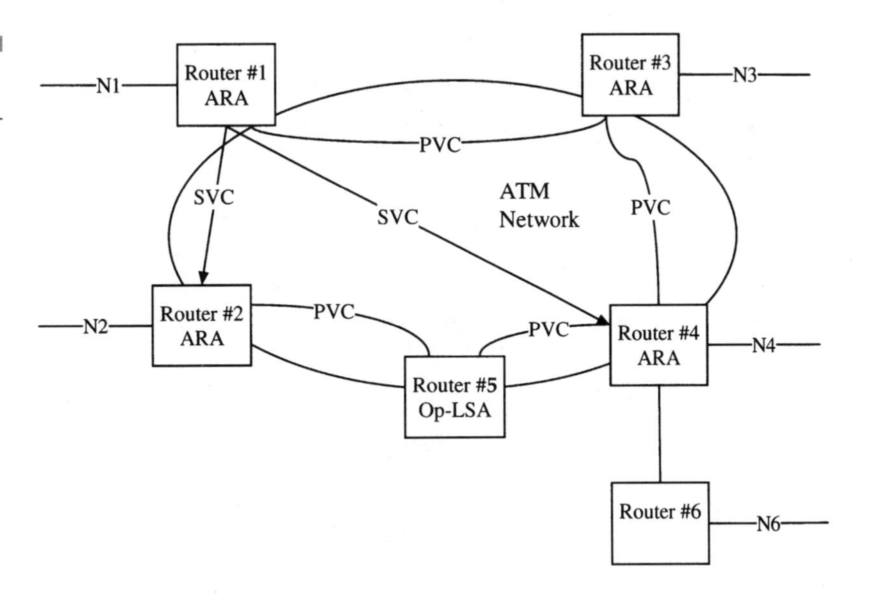

Figure 14-6
OSPF ARA function.

[5]This is required so that router 5 can propagate ARA information.

Figure 14-7
OSPF ARA format.

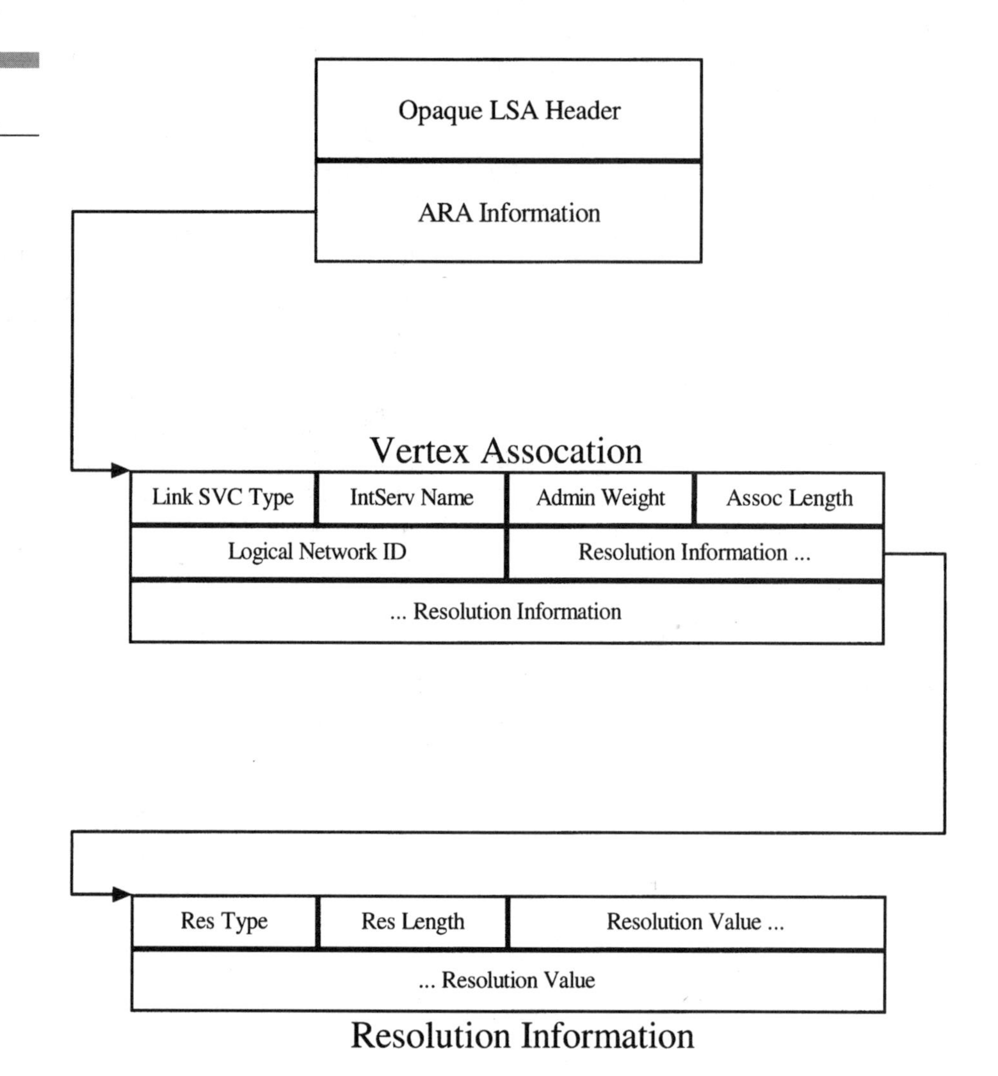

The structure of the OSPF ARA packet is illustrated in Fig. 14-7. It consists of an opaque LSA header followed by one of the four different ARA options. Contained within each of the four ARA options types is a vertex association. The vertex association consists of a link service type (ATM point-to-point, point-to-multipoint, multipoint-to-point), IntServ service name (for mapping IntServ to ATM), administrative weight, length, logical network ID (for VPNs), and the resolution information.

The resolution information in turn defines either an ATM address or an LIJ Call Identifier (for joining existing point-to-multipoint VCs).

The OSPF ARA is designed for OSPF routing domain(s) that for the most part operate on top of an ATM network and that can benefit from establishing shortcut paths between routers. It reuses existing reliable OSPF topology distribution mechanisms to provide all ARA-capable routers with IP/ATM address mappings. An ingress router is therefore not required to query an address resolution server (the case with NHRP) to retrieve this information. Because routers can quickly establish an SVC with other routers in the OSPF network, it is not incumbent upon the network provider to establish a full mesh of VCs. A partial mesh of preconfigured PVCs will do, and if more VCs are needed for enhanced connectivity or to support greater traffic loads, SVCs can be set up. Supporting fewer routing adjacencies also reduces the overhead of routing protocol traffic.

Note that the OSPF ARA function only exists on routers. It does not run on ATM switches. The real ATM network topology is transparent to the routers. In fact the underlying ATM infrastructure still appears as a virtual point-to-point link (e.g., RFC1483), broadcast LAN (e.g., LANE) or whatever, depending of course on the network configuration. This solution still maintains the traditional IP over ATM overlay model except that OSPF has been extended to provide better VC management.

14.3 PNNI Extensions

Just as it is quite feasible to extend OSPF to carry ATM information, it is also quite possible to extend PNNI to carry IP-layer information. Recall that PNNI Phase I is a hierarchical link state routing protocol that is used in a network of private ATM switches. The basic functions of PNNI are to distribute topology information to other ATM switches, compute a source-routed path for SVC setup requests, and move the SVC setup request through the network to the destination. It can also select paths based on the QoS characteristics of the offered traffic load and available capacity of the network and provide a dynamic alternate routing function during connection setup. Because PNNI is being increasingly deployed in ATM networks (enterprise intranets, ISP, etc.) that will transport large volumes of IP traffic, it makes sense to examine several possible ways to integrate IP into the PNNI routing protocol.

14.3.1 PNNI Augmented Routing

PNNI augmented routing (PAR) is the first method that extends the capabilities of PNNI routing to support IP.[6] In a PAR network all IP routers run a traditional IP routing protocol such as OSPF to build and manage IP forwarding tables. All ATM switches run the PNNI routing protocol. PAR enables ATM-attached routers to also participate in PNNI topology distribution operations. Specifically a PAR router operates a full instance of PNNI routing to advertise its existence and capabilities (routing protocol, router ID, IP address, etc.) and in turn learn about locations and services supported on other ATM-attached routers. By filtering through information received from the PNNI topology distribution process, a PAR router extracts the IP and ATM information needed to establish SVCs with other ATM-attached routers. For example, an OSPF router may learn through PAR of other routers attached to the ATM network that belong to the same OSPF area. PAR enables routers with common IP configuration parameters to build and maintain a set of SVCs with each other, thus simplifying network configuration and management. Indeed one of the major benefits of PAR is that it significantly reduces the need to configure interrouter PVCs.

PAR also provides a capability to support multiple VPNs on top of a single ATM network. This is accomplished by tagging all PAR-related information with a VPN id value. If the value in the PAR updates does not match the VPN id of the router, it will be filtered out. PAR routers can also filter IP-layer information such as IP addresses and protocol flags. This ensures that PAR routers will only process the information that is necessary for their operation over an ATM network.

A PAR network is illustrated in Fig. 14-8. Routers 1 through 10 support OSPF. Routers 4 through 7, which are attached to the ATM network, also run PNNI, as do the ATM switches. PAR routers then participate in the PNNI topology management process. They exchange PNNI Hello messages and join peer groups. They exchange PNNI topology state elements (PTSE) with neighboring PNNI nodes and build and maintain a PNNI topology database. They will also insert router-specific non-ATM information into new PTSEs and flood those throughout the peer group. ATM switches running standard PNNI will ignore any router-specific information and transparently pass it on to other nodes

---

[6]PAR is a work item in the PNNI subworking group of the ATM Forum.

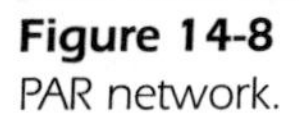

Figure 14-8
PAR network.

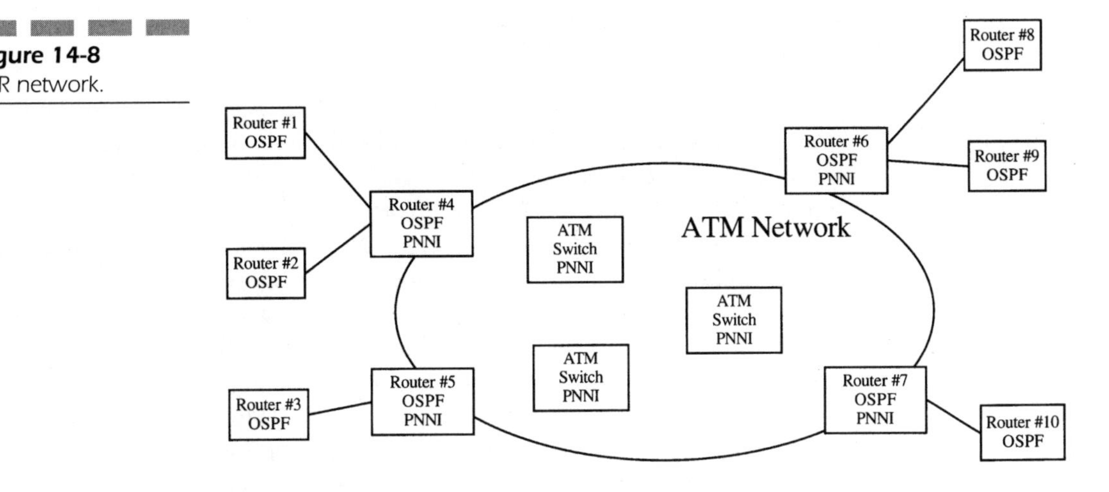

in the peer group. This enables ATM-attached routers to use PNNI to distribute IP-specific information to other routers. And because PAR routers maintain a PNNI database, they have an accurate picture of the real ATM network topology. PAR updates can also be distributed to the different levels of hierarchy in a PNNI network.

The PNNI topology database maintained by a PAR router also contains the addresses and services supported on other PAR routers. Those routers that share common IP configuration information can establish or bootstrap a partial mesh of SVCs at network initialization. This obviates the requirement to preconfigure a set of interrouter PVCs (as is the case with OSPF ARA). For example, routers belonging to the same OSPF area can use PAR to discover each other and then establish a full mesh of SVCs.

PAR requires no change to ATM switches running PNNI with the exception that they are able to flood PNNI PTSEs transparently. A device supporting PAR must be capable of generating and interpreting PAR-related information. Note that PNNI operating on the routers and more generally the PAR solution itself does not specify what a router should do once it receives PAR updates. That is a policy configuration question that can be addressed by the network provider. And finally it is still the IP routing protocol on the routers that determines the best next hop toward a particular destination. PAR just helps out by providing routers attached to the ATM network with a better idea about the functions and services supported on other routers.

14.3.2 Proxy-PAR

One of the drawbacks of PAR is that PNNI is required to run on each router. To get around this less then trivial implementation requirement while delivering similar autodetection and configuration functions, a lightweight version of PAR called proxy-PAR is defined.[7] Proxy-PAR is a minimal version of PAR that enables ATM-attached routers to exploit PNNI topology distribution to discover the presence of services supported by other ATM-attached routers.[8] Proxy-PAR consists of a client component that operates on the router, a server component that operates on the directly attached ATM switch, and series of protocol message formats and exchanges. The client registers IP-layer information with the server. The registered information is distributed throughout the PNNI domain. A client may query a server for related IP-layer information that it may wish to use. Like PAR, the objective of proxy-PAR is to allow non-ATM devices to use PNNI flooding mechanisms for registration and discovery of services.

An example of a network running proxy-PAR is illustrated in Fig. 14-9. Routers 1 through 5 are connected to an ATM network running

Figure 14-9 Proxy-PAR network.

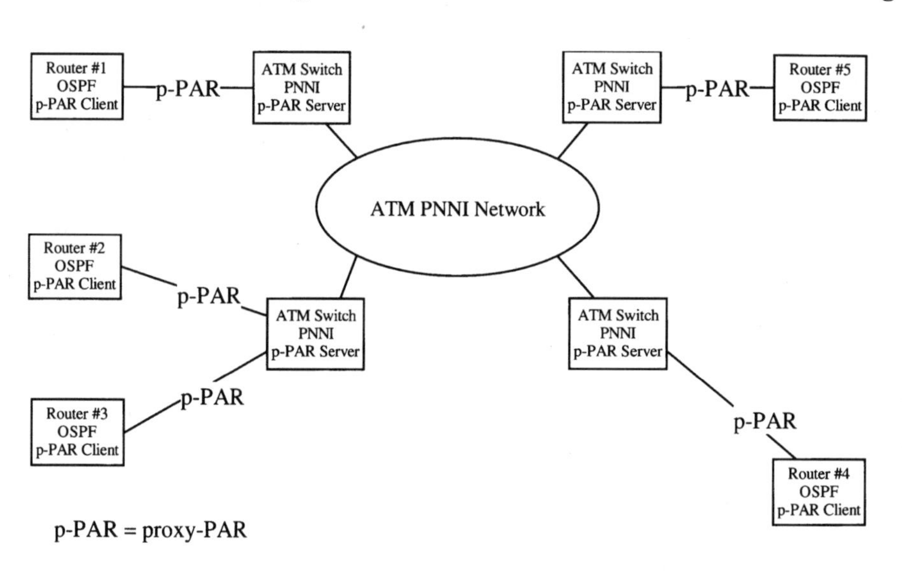

[7]Proxy-PAR is part of the PAR specification that the ATM Forum is working on.

[8]Droz, P., and Przygienda, T., "Proxy PNNI Augmented Routing (Proxy PAR)," *Proceedings from the 1998 First International Conference on ATM ICATM'98,* June 1998, Colmar, France.

PNNI on the switches. Each router has been individually configured to be part of the same OSPF area. Each router is also running the proxy-PAR client. In addition, those switches directly connected to routers are running the proxy-PAR server. At network initialization, the proxy-PAR clients on the routers will contact their server counterparts and register IP-layer information such as OSPF area ID, router ID, and so on. This information in turn is propagated throughout the ATM network using PNNI flooding mechanisms. It is then up to each proxy-PAR client on the router to query the proxy-PAR server in the switch for any IP-layer information that the router may need to continue operations. In this case the queries provide each router with information (including ATM addresses) about the other routers belonging to the same OSPF area. It is then a simple process for the routers to establish a series of SVCs with each other. The result is an automatically configured SVC overlay that reflects the logical topology of the OSPF area. Observe in this case that no preconfigured PVCs were required and that the detection of same-area OSPF routers and subsequent SVC establishment were automatic. In addition note that each router had to unilaterally query the server for the relevant PAR-related information—it was not automatically flooded to the router as is the case with PAR.

The proxy-PAR specification defines three specific functional protocols. The first is the Hello protocol that is essentially a lightweight version of the PNNI Hello protocol defined in the PNNI Phase I specification. The Hello protocol establishes an adjacency between the proxy PAR client and server so that subsequent registration and query messages can flow. A server also informs a client via the Hello protocol of the lifetime that the server assigns to data that was registered by the client.

The second protocol supported by proxy-PAR is the registration protocol. A proxy-PAR client uses this protocol to register the specific IP protocols and services that it supports. Associated with this information is an ATM address and a membership scope within the PNNI hierarchy. The latter enables specific PAR-related information to be distributed to certain levels within the PNNI hierarchy. For example, a router may register OSPF services for distribution to other OSPF routers in the same PNNI peer group and BGP services for distribution to other BGP routers in a higher-level peer group.

The format of a typical proxy-PAR Service Registration packet is illustrated in Fig. 14-10. It consists of a PNNI header, ATM address, sequence number, and membership scope. One or more possible Service Definition Information Groups (IGs) may follow. In this example the

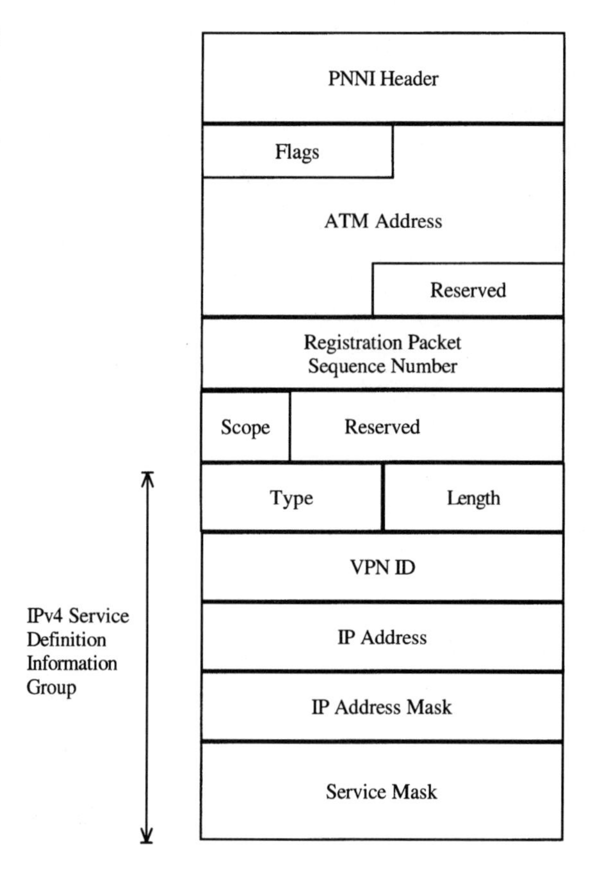

Figure 14-10 Proxy-PAR service registration packet.

IPv4 Service Definition IG that is specific to IPv4 services is shown. It consists of an IPv4 address, address mask, and service mask. The service mask indicates which particular service (OSPF, BGP, PIM, etc.) is being registered. Each registration packet sent by the client is acknowledged by the server. The registered information is then flooded throughout the PNNI network based on the membership scope. It is not directly delivered to other clients—they must explicitly query for it. If a client wants to register new information, any previously registered information is unconditionally overwritten. No state is maintained between the client and server regarding information that was previously registered.

The third protocol supported by proxy-PAR is the query protocol. A proxy-PAR client uses the query protocol to retrieve PAR-related information from the server. Rather than retrieve everything, a client can query for information based on membership scope, IP address, address mask, VPN ID, and service. This enables the proxy-PAR client to request and receive only the information that is relevant to its proper configuration and operation over an ATM network. It is up to the client to periodically query the server to refresh any information that might have become outdated. The general flow of proxy-PAR messages between a client and server is illustrated in Fig. 14-11.

The extensibility of PNNI enables PAR and proxy-PAR routers to register and subsequently access information on just about any IP-layer service (or protocol) for use over an ATM network. Some of the most popular services supported by PAR and proxy-PAR are shown in Table 14-1.

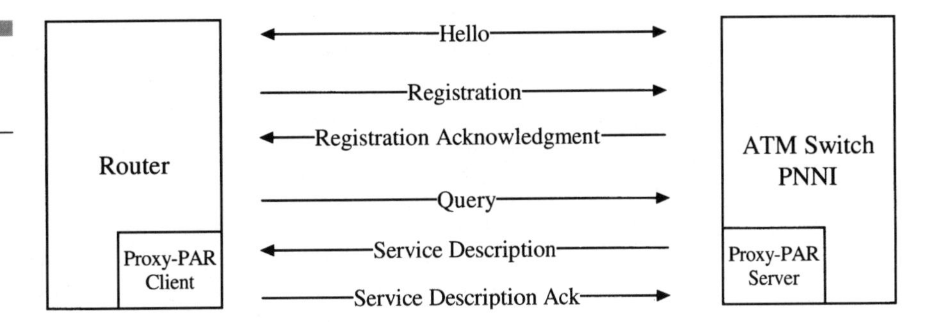

Figure 14-11 Proxy-PAR message flows.

TABLE 14-1 PAR/Proxy-PAR Services

Service	Additional Information
OSPF	Yes
RIP	No
BGP4	Yes
MOSPF	Yes
DVMRP	No
PIM-SM	Yes
PIM-DM	No
MARS	No
NHRP	No
ATMARP	No
DHCP	No
DNS	Yes

The service information may be accompanied in some cases by additional protocol-specific information that is required to complete the configuration. For example, a router advertising OSPF services will include a separate IG that describes the Area ID, designated router (DR) election priority, and the OSPF interface type. Another example is BGP. An additional IG is needed to convey the AS number, BGP id, and router reflector information. In addition there is a systems capability IG that can be used to register proprietary and experimental information.

Proxy-PAR imposes relatively modest processing and memory requirements on the client side. This was a conscious design decision on the part of the proxy-PAR developers. However, this implies that the client will not store and request large amounts of data. The purpose behind proxy-PAR is to use PNNI flooding to distribute service information among a small number of stable clients, in this case routers (or address resolution servers). It is not designed to operate in an environment where changes are frequent. Nevertheless, it provides an efficient, serverless, broadcastfree mechanism for advertising and discovering the existence of ATM-attached devices with shared IP configuration information. Proxy-PAR and PAR in general enable these components to bootstrap a working overlay configuration of SVCs. It is fully compatible with existing and proposed IP/ATM address resolution mechanisms (e.g., ATMARP, NHRP, etc.) that operate in the overlay model. It is even quite possible to use proxy-PAR and the OSPF ARA functions together. The former automatically discovers and connects routers belonging to the same area, and the latter is used to distribute IP/ATM address mappings.

14.3.3 Integrated PNNI

The third possible extension to PNNI that integrates both IP and ATM routing is called Integrated PNNI (I-PNNI). I-PNNI can be viewed as the reverse of the OSPF ARA approach. Instead of using an IP routing protocol to distribute ATM addresses, I-PNNI makes use of an ATM routing protocol (PNNI) to distribute IP addressing. Specifically, I-PNNI consolidates both IP and ATM routing protocol functions into a single protocol based on PNNI. Indeed it really is just PNNI with additional IGs defined to carry IP routing information. I-PNNI is designed to operate on ATM switches, ATM-attached routers, non-ATM routers, and possibly hosts. I-PNNI is fully compatible with ATM switches operating

PNNI. For routers attached to legacy media such as Ethernet or Frame Relay, I-PNNI is used instead of dynamic IP routing protocols such as OSPF or BGP.

From an IP perspective, I-PNNI is just another link state routing protocol that is used to distribute IP address prefixes around the network and compute hop-by-hop paths. As far as ATM is concerned, I-PNNI is just basic PNNI with some extensions used to propagate non-ATM data that the switches will simply ignore. Routers and switches running I-PNNI maintain a single topology database of the entire network consisting of routers and switches. The primary advantage of running a single routing protocol in a combined router and switch network as compared with the traditional dual protocol approach of the overlay model is that it eliminates the routing protocol duplication and vastly reduces the number of routing adjacencies. The latter is a result of the fact that any node in the network, either a router or switch, forms a routing adjacency with its directly attached neighbors to exchange routing table updates. Another important advantage is that IP routing can be improved because routers have a "view" of the real network topology consisting of other routers, ATM switches, and the connecting links.

A two-level hierarchical network consisting of routers and switches running I-PNNI is illustrated in Fig. 4-12. Observe in peer group C the presence of a route server and edge device. A typical route server (e.g., MPOA server) runs an IP routing protocol, so in this case it is I-PNNI. The edge device does not run I-PNNI but rather a standard IP/ATM client protocol such as LANE or MPOA.

The functional characteristics of I-PNNI consist of the following:

- All routers, ATM switches, and route servers run a single routing protocol, I-PNNI.
- Router and switches join peer groups. A peer group can contain all ATM switches, all legacy routers, or a mix of the two. In a mixed environment, the peer group leader (PGL) must support I-PNNI so that IP reachability can be advertised outside of the peer group.
- PNNI topology advertisements (PTSEs) are exchanged between routers, between switches, and between routers and switches.
- IP and ATM addresses are advertised in separate PNNI PTSEs. Therefore, although I-PNNI is a solution for integrated IP and ATM routing, two separate address spaces are still supported and maintained.

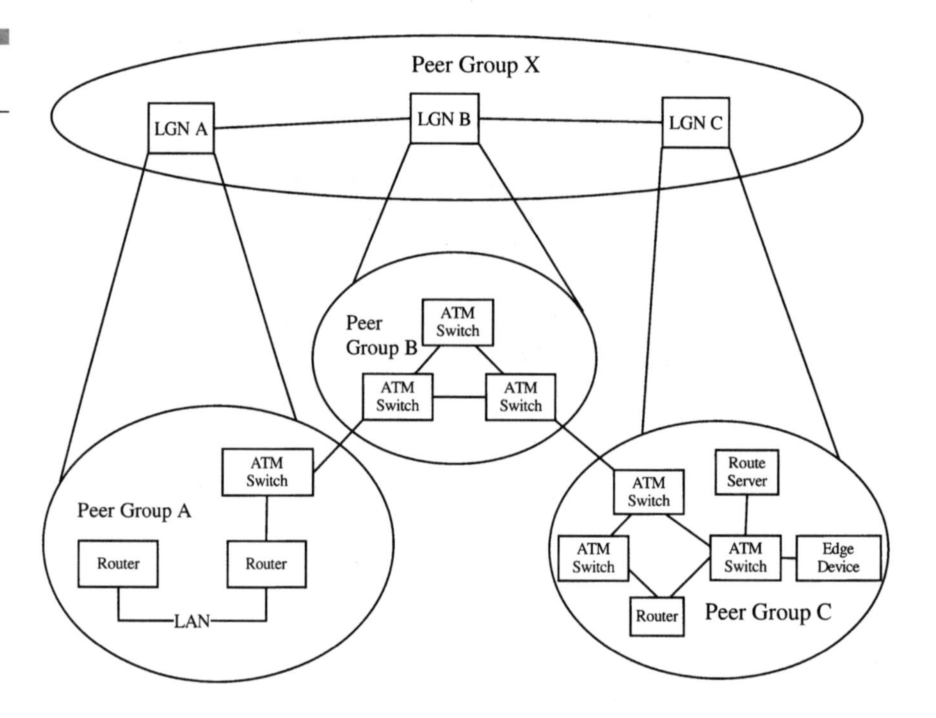

Figure 14-12
I-PNNI network.

- Reachability to an IP address can be advertised in one of three ways:

Direct. The IP address is directly reachable through the advertising router.

Query. The IP address is reachable through the advertising router, but a better way may be found if the requesting IP entity queried (e.g., NHRP) for the address.

In-care-of. The IP address belongs to a third-party. This is used by a router to advertise the IP address and associated ATM address of a well-known host or subnet. This eliminates the need for an NHRP query.

- A single, common topology database is built and maintained by the members (routers and switches) of a peer group.
- Routers are announced as "restricted-transit" nodes because they are not capable of forwarding SVC requests. However, routers can initiate and terminate SVC requests and will utilize standard

PNNI routing techniques (e.g., building the DTL) to accomplish this.

- Routers running I-PNNI that are attached to broadcast-capable LANs such as Ethernet borrow the concept of a designated router and pseudonode from OSPF to represent the LAN in the topology database.
- A multilevel routing hierarchy can be generated.
- NHRP is needed by I-PNNI nodes to resolve IP/ATM address mapping when there is incomplete information. For example, a routing hierarchy will only provide summarized ATM addressing information between peer groups. Or a route server running I-PNNI will not advertise individual host routes of all of the ATM hosts it is supporting.

MORE INFORMATION

R. Callon et al., "ATM Routing Options," *Connexions,* March 1996.

http://www.zurich.ibm.com/%7Edro/proxy/.

CHAPTER 15

Multiprotocol Label Switching

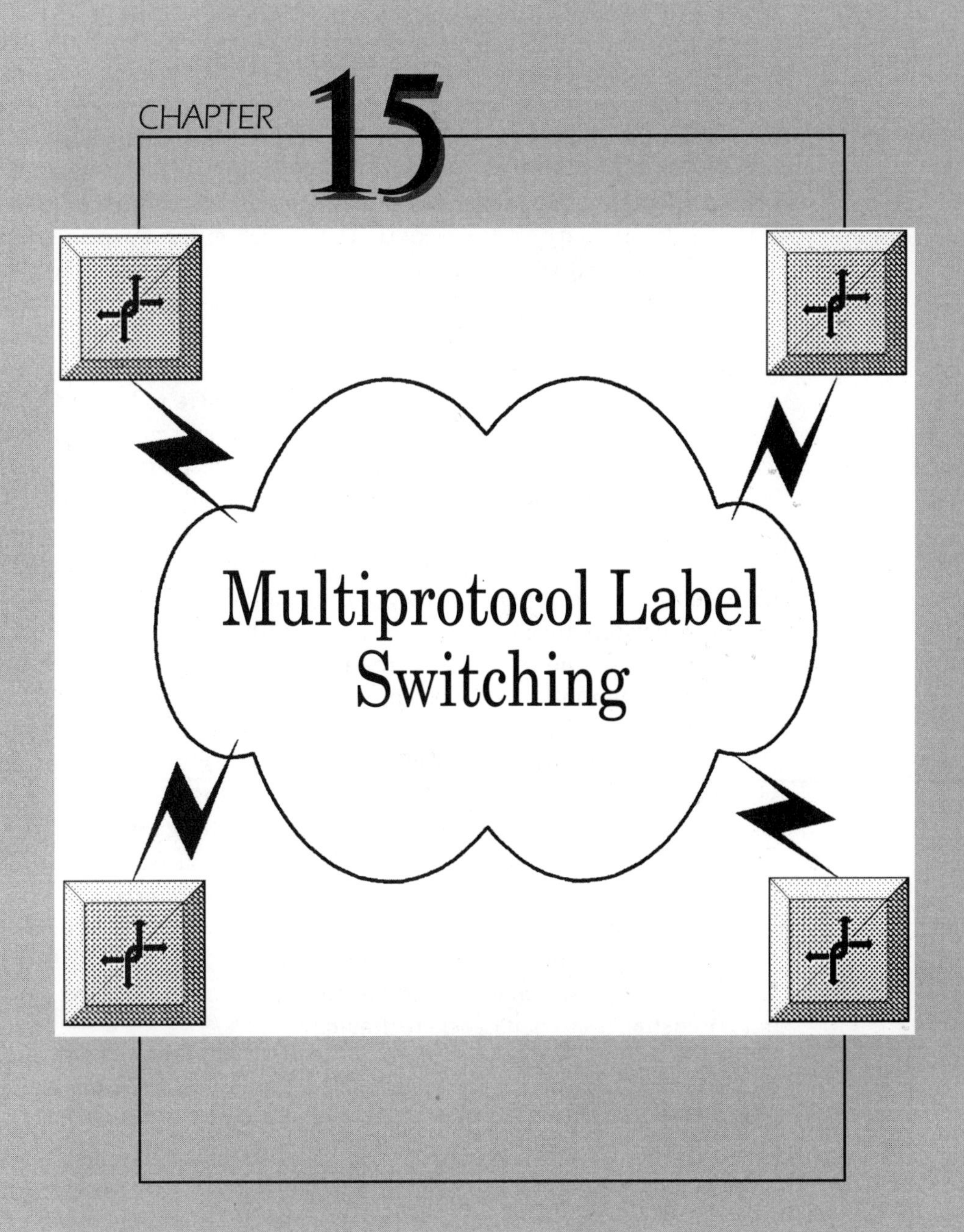

It was evident from the industry attention focused on the IP switching solutions from Ipsilon, Cisco, IBM, and others that this was an important technology of great interest to many. Vendors hoped to develop new platforms (and reuse others) that would dramatically improve the performance and flexibility of IP-based networking. Customers viewed this technology with great interest because it promised to relieve their networks of the capacity bottlenecks that were beginning to spring up as a result of overtaxed router technology and persistent exponential traffic growth. It was also viewed favorably by many network providers as a means of making more efficient use of their frame relay and ATM networks. Industry pundits examined IP switching and expressed guarded optimism about its ability to finds its way and evolve to become a mainstream technology.[1] From a cynical point of view, one could say that the IP-centrists had come up with a way to use ATM without calling it ATM.

Nevertheless, by the fall of 1996 there was a plethora of different IP switching schemes proposed from start-ups (e.g., Ipsilon) and computer industry heavyweights (e.g., Cisco, IBM). Some embraced the flow-driven approach to exploit hardware-based ATM switching and provide improved performance and greater bandwidth for end-to-end host and application traffic flows. Others subscribed to the topology-driven model that is based on routers and routing protocols to progress traffic through a network more efficiently and with better control. In common, these different approaches used a simple label-swapping paradigm based on ATM switching. It was agreed by all concerned that there were significant advantages for integrating routing and switching into a single solution. Routing of course is what the Internet is built on, and switching provides inexpensive, high-capacity, packet/cell-forwarding hardware. But these competing and vendor-proprietary approaches, while generating much interesting press and passionate electronic arm-waiving on various mailing lists, made it difficult for the vendors to move forward and for customers to deploy the technology.

For the good of the industry it was decided, following a BOF session on tag switching held at MIT in December 1996 to form a working group in the IETF to focus on the issue of integrated routing and switching.[2] In fact it was born out of the necessity, technically, politically, or otherwise, to agree on a single standard for integrated routing and switching. Sev-

---

[1]Cautious, yet well-articulated noncommitment to anything new or challenging to the status quo is standard industry pundit behavior.

[2]It was not just altruism that brought these separate interested parties together. Perhaps it was more fear that one solution might dominate over all of the others.

eral names were suggested, but in the end, it was decided to christen the effort Multiprotocol Label Switching (MPLS).[3] The name is based on the fact that simple label switching (swapping) is used as the underlying forwarding mechanism. The use of the word *multiprotocol* in the name suggests that it will support multiple network-layer protocols in addition to IP. Architecturally this is true, but in practice and implementation network-layer support will be limited to IPv4 and IPv6.

The primary objective of the MPLS working group is to develop a standard that integrates routing and switching. Specifically the MPLS effort promises to incorporate network-layer routing and the label-swapping paradigm into a single solution that will provide the following advantages:

- ✓ Improvement in the price and performance of routing
- ✓ Improved scalability of routing vis-à-vis the traditional overlay model
- ✓ More flexibility in the introduction and deployment of new networking services

The last element is of particular importance to network providers who will over time want and need to deliver new and improved services as a means of attracting more subscribers and generating more revenues. And given the affinity of IP switching in general with ATM, it is a solution that can, and most likely will, benefit from the latter's acknowledged traffic management and QoS capabilities.

Given the industrywide interest and participation in the MPLS effort as well as the work already completed (and implemented in some cases) in the several proposed solutions, it is not surprising that MPLS incorporates many of the familiar IP switching concepts. An MPLS network is capable of forwarding packets at layer 3 using standard per-packet processing and at layer 2 using label-swapping as the forwarding technique. It is based on the peer model—that is, each MPLS device runs a single IP routing protocol, exchanges routing table updates with its neighbors, and maintains just one topology and one address space. It maps forwarding equivalence classes (FECs) to labels. A Label Distribution Protocol (LDP) is defined and its function is to allocate and distribute FEC/label bindings among participating MPLS devices called label switch routers (LSRs). Once LDP has completed its job, a label switched path (LSP) is built from the network ingress to the egress. When pack-

---

[3] *http://www.ietf.org/html.charters/mpls-charter.html*

ets enter the network, the ingress LSR examines multiple fields in the packet header to determine what FEC the packet belongs to. If an FEC/label association (as a result of a downstream-to-upstream LDP message or other means) is present, the ingress LSR affixes a label to the packet and directs it to the appropriate outbound interface. The packet is then label-swapped through the network until it reaches the egress LSR, where the label is removed and the packet is processed at layer 3. The performance gain comes from the fact layer-3 processing is pushed to the edge (ingress) and performed only once rather than at each intermediate hop. Intermediate processing consists of matching a short, fixed-length label in the packet with a corresponding entry in the LSR connection table and then swapping the labels—a process performed typically in high-performance hardware-based switches.

Although performance and efficiency are important, they are not the only benefits offered by MPLS. In the eyes of those who operate large router-based networks for a living, it is its ability to perform advanced traffic engineering with no performance penalty that is of particular interest. Recall from the chapters on tag switching and ARIS that it is possible and fairly straightforward to create a nondefault explicit route. It involves identifying the appropriate FEC (of the packets to be placed on the explicit route) and then allocating the associated labels in the devices along the nondefault path.[4] In the case of MPLS, the forwarding paradigm for default and nondefault traffic is exactly the same. One possible use of explicit routing is to distribute different FEC traffic over different paths to balance the load across a network backbone. Another use might be to direct traffic over different LSPs based on some configured policy such as the location of network ingress. Moreover, just about any kind of FEC that one can imagine can be bound to a set of labels and forwarded through an MPLS network. One other possible and very interesting use of MPLS is to map intra-VPN traffic to labels, thereby creating high-performance LSPs between sites of the same VPN.

From the perspective of ATM, MPLS should be viewed as another control plane that is extremely partial to IP traffic and routing. Loosely speaking it is another way to build an ATM VC, except MPLS calls it an LSP. When running on ATM hardware, both MPLS and ATM Forum protocols use the same packet formats (53-byte cells), the same labels (VPI/VCI), the same label-swapping cell switching forwarding mecha-

---

[4]Observe that the overhead of processing the entire sequence of nodes in the path is only performed at LSP setup. Contrast that with normal IP routing in which each packet must carry a copy of all of the nodes in the path.

nisms, and the same ingress and egress functions.[5] Both require a connection setup protocol (e.g., LDP for MPLS and UNI/PNNI signaling for ATM).

The fundamental differences lie in the fact that MPLS has no use for ATM addressing, ATM routing, and ATM Forum protocols. Instead, MPLS uses IP addressing, dynamic IP routing, and an additional control protocol (LDP) to map FECs to labels and thus form LSPs. In general MPLS is concerned solely with creating and distributing FEC/label mappings so that IP traffic can be better and more efficiently forwarded through a network over a default or nondefault path. When operating on ATM hardware, MPLS is just another ATM control plane, albeit one driven primarily by IP routing protocols and designed exclusively for the benefit of routed IP traffic. Of course in practice MPLS will most likely coexist with native ATM in a ships-in-the-night (SIN) configuration where the two operate in a mutually exclusive manner, or MPLS nodes may communicate through native ATM switches (integrated). Finally it should be pointed out that the MPLS architecture is capable of operating over any data link technology—not just ATM. Therefore a network provider may configure and run MPLS in a single domain consisting of Point-to-Point Protocol (PPP), frame relay, ATM, and broadcast LAN data links.

The formation of the MPLS working group and the standardization effort that resulted was and is a positive development for all who operate and use IP-based networks (just about everyone). It served as a forum to discuss important issues related to network scalability, router design and performance, protocol behavior, and the need to be able to cost-effectively introduce new network services. It brought to light the need and desire of many network providers who have deployed ATM to develop a solution that better utilizes their infrastructures to transport large volumes of IP traffic. Many of the best engineers and minds in the industry have participated in and contributed to the MPLS effort. Indeed some of the more clever ideas first conceived in the proprietary predecessors to MPLS have found their way into the MPLS discussions. Examples are the notion of penultimate hop popping and the label stack (Tag), the VCID for associating an FEC with an ATM Forum built VC (Toshiba), and loop prevention and VC merging (IBM). Many of the contentious issues that first arose have been settled through compromise or

---

[5]The ingress and egress functions are the same to the extent that per-packet layer-3 processing is performed for both MPLS and traditional IP over ATM overlay solutions.

group consensus. And to be sure that we all do not take our work too seriously, it was an opportunity for humor.[6]

15.1 Architecture

MPLS owes its existence to the initial suite of IP switching solutions, so its architecture is based on notions, concepts, and components that have previously been presented. Thus one of its fundamental objectives, as with the previous solutions, is to simplify the forwarding of IP packets through a network. And it seems that there is ample opportunity to achieve this. Recall that with conventional IP forwarding, information contained in the header of each packet is analyzed at each router hop.

---

[6]Bala R.'s "Guide to MPLS Terminology," Aug. 3, 1997, submission to MPLS Mailing List:

MPLS: My Product Line Needs a New Story.

Label: Stuff you stick to the front of your box for product identification.

Label Swapping: The process of swapping a "router" label with a "switch" label on a box. Label swapping typically requires an engineer or two but can be executed by an entire marketing dept.

Label Information Base: A database containing all the labels so far used in the industry, so that a new one may be selected by a company entering the fray. Taken labels include "IP switching," "Cell Switch Router," and "Tag switching."

Flow: The influx of employees leaving a big corporation to join an IP switching startup.

Stream: The aggregate of many flows; many employees leaving many big corporations in droves to join many IP switching startups.

Layers: OSI has defined three MPLS layers. Layer 3 is the marketing hype. Layer 2 is the hand-waving logic to prop up the marketing hype. Layer 1 is the set of dubious performance numbers supporting layer 2. Some IP switching products also rely on a layer 0, which is the glib dismissal of anything ATM.

VC: Very Confounding—indicating arguments in favor of standardizing MPLS.

Shortcut VC: A VC argument which skips unnecessary details such as performance results and directly jumps to the conclusion that MPLS is the salvation.

Loop: A circular strategy whereby a vendor uses doubts on conventional router performance to sell IP switching products and skepticism about IP switching to sell more of its conventional routers.

Loop Prevention: A drastic step whereby some customers stick to SNA over frame-relay to avoid the whole nasty business altogether.

Stack: Steadily accumulating drafts and white papers on IP switching.

LDP: Let's Do Packets—New slogan for ATM vendors.

LSR: Low-Selling Router—A device being converted to an "IP switch" by a router vendor.

This analysis and subsequent per-packet processing consists of parsing the header, extracting the destination address, performing a routing table lookup, determining the next-hop address, calculating the header checksum, decrementing the TTL, applying the appropriate outbound link-layer encapsulation, and then transmitting the packet. Put in less rigorous terms, the function performed at each router hop for each packet is to analyze the network-layer header, assign the packet to an FEC (based on a destination address prefix), and then map that FEC to a next-hop router.

MPLS aims to simplify this process by performing the network-layer header analysis and FEC assignment functions once at the network ingress. Rather than map the FEC to a next-hop router, the ingress router appends a label to the packet that represents the FEC that the packet belongs to. At subsequent router hops, it is not necessary to examine the network-layer header because the packet has already been associated with an FEC. The label is used to index a connection table that contains an outbound port and a new label. The old label is replaced by the new label, and the packet is then forwarded out the outbound port to the next-hop device. The simplified forwarding process of MPLS as compared with conventional IP forwarding is illustrated in Fig. 15-1.

The MPLS portion of the figure shows that it is possible at the ingress router to map the packet(s) to any number of different FECs. For example, an FEC may be based on the destination network address, a group of destination addresses, a source/destination address pair, a source address only, or even the physical point of entry into the network. An FEC can also represent all packets that are to traverse an explicit nondefault path. Independent of whatever complex policy is invoked to assign a packet to an FEC, the forwarding of the packet through the network is still based on label-swapping. Thus MPLS facilitates the use of policy-based routing in a much simpler and more straightforward manner than would otherwise be possible using traditional IP forwarding.

In developing the MPLS specification a number of specific functional requirements arose that for the most part represent a composite of those that drove the development of earlier IP switching solutions. Nevertheless, raising these important issues ensures that MPLS will address the primary challenges of integrating routing and switching in a scalable manner. The basic requirements that the MPLS specification must address are to

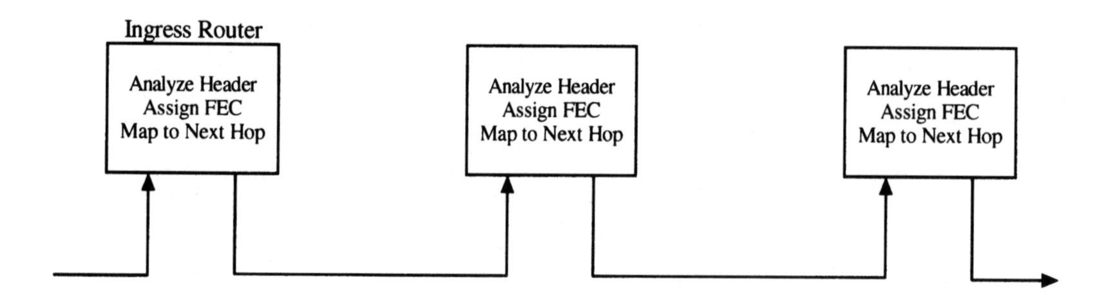

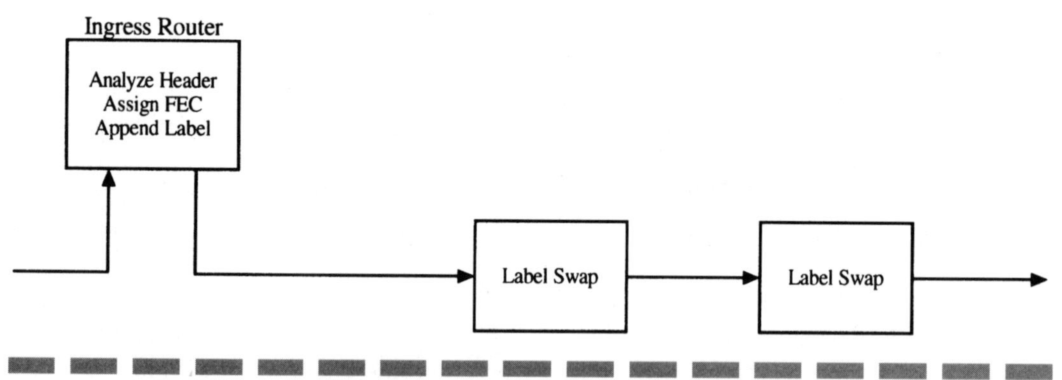

Figure 15-1 *MPLS and conventional IP forwarding.*

- Simplify packet forwarding to reduce costs and improve performance.
- Operate independently of specific data link technologies; that is, it must work over frame and cell media.
- Be compatible with but yet independent of the operation of existing and future network-layer routing protocols.
- Support loop prevention and detection. The former prevents loops from ever forming; the latter is to detect the presence of a loop.
- Allow aggregate forwarding of multiple traffic flows over a single LSP.
- Be compatible with existing IP network management tools.
- Support hierarchical operation.

- ✓ Support unicast and multicast traffic flows.
- ✓ Support O(N) switched paths for best effort traffic where N is the number of MPLS nodes in the network.
- ✓ Support interoperation with non-MPLS switches.
- ✓ Support SIN operation with non-MPLS switching technologies.
- ✓ Support both topology- and flow-driven IP switching models.

There are several core technologies or components that comprise the MPLS solution. First, the MPLS term for an IP switch is a *label switch router* (LSR). Like a generic IP switch, it is capable of forwarding packets at layer 3 and switching packets at layer 2. It also operates traditional IP routing protocols and may run a specialized control protocol to coordinate FEC/label bindings with neighboring LSRs. Next are the labels. A label is a short, fixed-length value that is contained in each packet and is used to forward a packet through the network. A pair of LSRs must agree up front on the value and meaning of the label. For example, the downstream LSR will tell the upstream LSR that a particular label X will be used to represent a certain FEC called A. Thus a label only has significance between a pair of communicating LSRs and is taken to represent the packets belonging to a particular FEC that are flowing from the upstream LSR to the downstream LSR. MPLS can support labels that are appended to existing frame or packet structures (e.g., Ethernet, PPP), or it can use label structures that are contained in the data link layer (e.g., frame relay, ATM).

The third important core technology used in MPLS is the forwarding mechanism, which in the case of MPLS is label-swapping. As can be seen from the performance and capacity of typical frame relay and ATM switches, implementations that support label-swapping, it is a fast and simple forwarding procedure. Unlike conventional IP routing, there is no need to analyze variable-length portions of the header's contents. The label as a whole (and possibly additional fields in the label field such as TTL and COS) is handled by the switch component. And even if a packet contains a stack of labels, the MPLS device is only required to process the top label in the stack.

Figure 15-2 illustrates the process of forwarding a packet along an LSP. At the ingress of the LSP, a label is pushed onto the packet, creating a stack of labels with a depth M. Intermediate MPLS nodes along the LSP receive and process the packet with a stack of labels of depth M. Only the top label in the stack is acted upon and that involves swapping the label with a new label that corresponds to the next-hop downstream

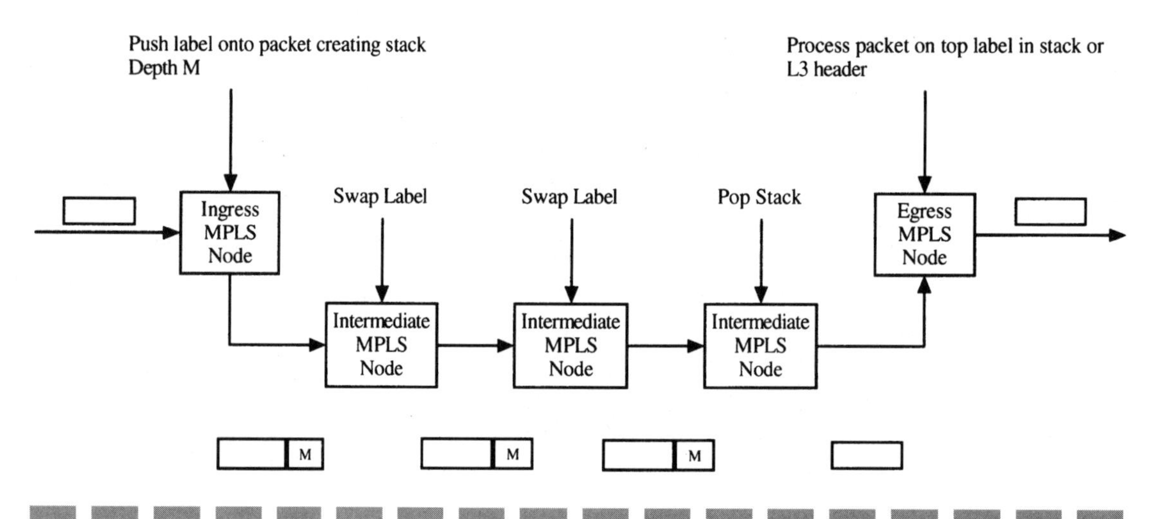

Figure 15-2 *MPLS forwarding along an LSP.*

LSR that will receive the packet. The LSR at the egress of the LSP will make a forwarding decision based on the contents of the next lower label in the stack.[7] This means that the egress LSR will need to pop the stack to get to the next label in the stack. A subtle optimization to this process, called penultimate hop popping, can be achieved if the egress node tells the next to last LSR in the LSP to pop the stack.[8] The packet then arrives at the egress MPLS device with the label that will be used to forward the packet already at the top of the stack. This saves the egress device from having to do a table lookup on a label that was not going to be used to forward the packet anyway.

The third core MPLS technology is label distribution. Label distribution is the process of distributing FEC/label bindings among participating MPLS devices to form an LSP for the purpose of label swapping the packets belonging to the specified FEC. This can be accomplished by a separate Label Distribution Protocol (LDP), or the FEC/label bindings can be transported (piggy-backed) in existing control protocols.[9] The basic operation is for the downstream LSR to allocate the label and then

[7] If there is only one label, the forwarding decision is based on the contents of the network layer header.

[8] A special label, the Implicit Null label. is used to tell the upstream MPLS node to pop the stack.

[9] Piggy-backing labels in existing control protocols such as BGP and RSVP requires a change to those protocols.

distribute the FEC/label binding to the adjacent upstream LSR. In the case where the FEC corresponds to an address prefix distributed by a dynamic routing protocol, the formation of an LSP (using either LDP or another control protocol) can be done in an independent or ordered manner. Independent LSP control is the case where an LSR makes an independent decision to bind a label to an FEC and communicates this binding to its LSR neighbor(s). Ordered LSP control occurs when the LSR will bind a label to an FEC only if it is the egress of the LSP or if it has already received a binding for the specified FEC from the next hop LSR for that FEC. Ordered LSP control is used to build a path with nondefault properties. Paths that pass through a specified sequence of nodes or paths that must be loop-free are two examples of paths with nondefault properties. The MPLS architecture is capable of supporting both independent and ordered LSP control.

MPLS LSPs can be built based on the arrival of specific data flows (flow-driven), reservation setup messages (e.g., RSVP) or routing table update messages (topology-driven). Given that one of the major requirements is scalability and the fact that MPLS is for the most part designed for very large IP networks, it is the topology-driven approach that will most commonly be deployed. Figure 15-3 illustrates conceptually how the presence of routing table updates initiates the exchange of FEC/label bindings between a pair of LSRs. Routing table updates received by LSR #2 trigger new path calculations that result in new or modified entries in the routing table (FIB). LSR #2 recognizes this, and

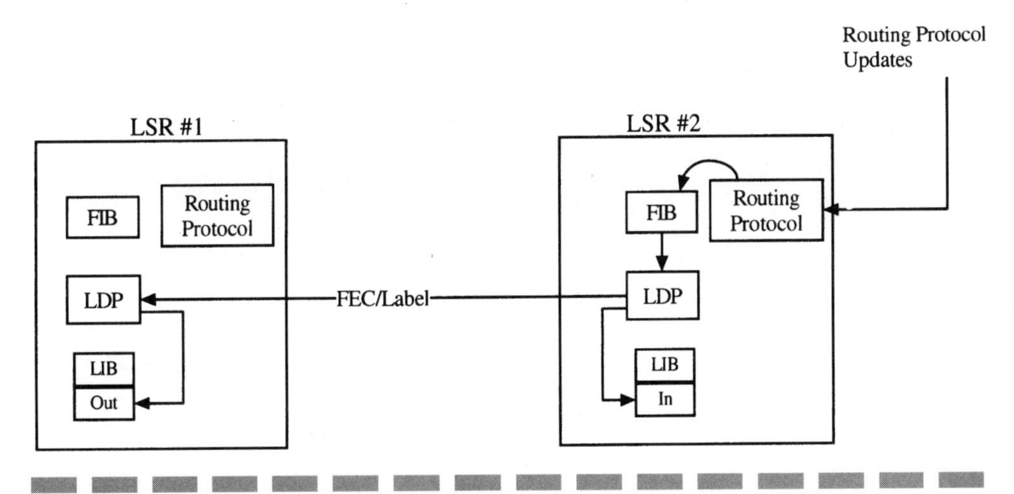

Figure 15-3 Topology-driven label assignment.

for each FEC (address prefix in this case) a label is allocated on the inbound port and placed in the Label Information Base (LIB). The label and the associated FEC are communicated upstream to LSR #1 (using LDP), which places the label in a corresponding outbound port entry in its LIB.

15.2 Components

In an attempt to create some distance from its proprietary ancestors as well as to prevent one contributing body from claiming too much credit for its creation, a lot of thought went into creating the new terminology that defines the following components and functions of MPLS:

- ✓ *Multiprotocol Label Switching (MPLS).* Name given to the IETF effort to create a standard for integrated routing and switching.
- ✓ *Label switch router (LSR).* A device that supports MPLS and is capable of forwarding packets at layer 3 and label swapping packets at layer 2. More specifically an LSR can be a traditional switch (e.g., ATM) augmented with IP routing and MPLS support, or it can be a traditional router that has been upgraded to support MPLS. The latter is the case where the router forwards the packets based on the contents of an explicit label that is contained in each packet.
- ✓ *Label edge router (LER).* A traditional router that is capable of forwarding packets to and from an MPLS domain. It can also communicate with interior MPLS LSRs to exchange FEC/label bindings.
- ✓ *Label Distribution Protocol (LDP).* MPLS control protocol that is used to communicate FEC/label bindings between LSRs.
- ✓ *Label switched path (LSP).* An ingress-to-egress switched path built by MPLS nodes to forward the packets of a particular FEC using a label-swapping forwarding mechanism.
- ✓ *Label information base (LIB).* The connection table maintained in an LSR (LER) that contains FEC/label bindings and associated port and media encapsulation information. A conceptual sample LIB is illustrated in Table 15-1; the value of an individual entry in the LIB consists of the following:

TABLE 15-1

Sample Label Information Base

Input Port	Input Label	Next-Hop LSR	FEC Identifier	Label Operation	Outbound Label	Outbound Port	Link-Layer Encap
1	1	LSR A	Net A	Replace	2	2	ATM
1	4	LSR D	Net D	Replace	5	3	ATM
2	7	LSR A	Net A	Replace	2	2	ATM
3	3	LSR C	Net C	Push	3	2	ATM
4	2	LSR A	LSR A	Pop			

- Inbound port
- Inbound label
- Next-hop LSR
- FEC identifier (e.g., address prefix)
- Operation to perform on the label. Could *replace* the inbound label with an outbound label, *pop* the label stack, or replace the inbound label with an outbound label and *push* a new label on top of that.
- Outbound label
- Outbound port
- Outbound link-layer encapsulation.

✔ *Stream.* A stream of packets that are forwarded over the same path and treated in the same manner. A stream may consist of one or more flows. In the MPLS architecture a stream is identified by a stream member descriptor (SMD).

✔ *Forwarding equivalence class (FEC).* A group of IP packets that are forwarded over the same path and treated in the same manner and therefore can be mapped to a single label by an LSR.[10] An FEC

[10]An exception to the FEC-to-one-label mapping occurs if the LSR does not support a stream merge capability.

can also be defined as an operator that maps packets to a particular stream.

- *Label.* A short fixed-length physically contiguous value that is used to identify a stream. The format of a label depends on the media that the packet is encapsulated in. For example, ATM-encapsulated packets (cells) use the VPI and/or VCI values as the label, and frame relay PDUs use the DLCI as the label. For those media encapsulations that have no native label structure, a special shim value is used. The format of the 4-byte shim label is illustrated in Fig. 15-4 and contains a 20-bit label value, a 3-bit COS value, a 1-bit stack indicator, and an 8-bit TTL value. In addition, if the shim value is inserted into a PPP or Ethernet frame, a protocol ID (or ethertype) contained in the respective frame headers indicates that the frame is either an MPLS unicast or MPLS multicast frame.
- *Label stack.* An ordered set of labels appended to a packet that enables it to implicitly carry information about more than one FEC that the packet belongs to and the corresponding LSP that the packet may traverse. A label stack enables MPLS to support a routing hierarchy (one label for the EGP and one label for the IGP) and to aggregate multiple LSPs onto a single "trunk" LSP.
- *Stream merge.* The case where several smaller streams are merged into a single larger stream. Media-specific examples of this capability are VP merge and VC merge under ATM. The use and support of a stream merge capability enables an MPLS network to support O(*N*) switched paths for all best-effort traffic.

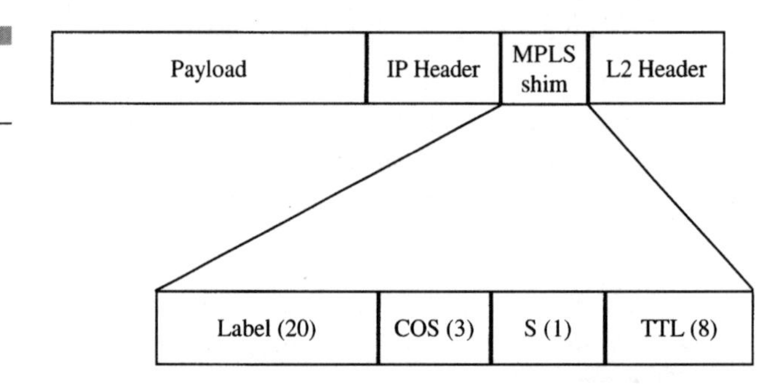

Figure 15-4
MPLS label format.

15.3 Label Distribution Protocol

The MPLS LDP is a separate control protocol that LSRs use to exchange and coordinate FEC/label bindings.[11] More specifically LDP is the sequence of message exchanges and message formats that enable peer LSRs to agree on the value of a specific label to use for packets belonging to a specific FEC. A TCP connection is established between peer LSRs to ensure that LDP messages are reliably delivered in the right sequence. LDP mapping messages can originate locally from any LSR (independent LSP control) or from the egress LSR (ordered LSP control) and flow from the downstream LSR to its upstream neighbor. The exchange of LDP messages can be triggered by the arrival of a specific data flow, a reservation setup (RSVP) message, or routing protocol updates. Once individual pairs of LSRs have exchanged LDP messages for a particular FEC, an ingress-to-egress LSP is formed after each LSR "splices" the incoming label to a corresponding outgoing label in the LSR's LIB.

15.3.1 Message Flow

LDP messages are divided into three classes: discovery, adjacency, and mapping. Discovery class messages announce and maintain the existence of an LSR in the network. This is accomplished by multicasting an LDP Link Hello message to the "all routers" group address to advertise one's existence to another LSR on the same link or by directing a targeted LDP Hello message to a specific IP address for LSRs that are not directly connected to each other.[12] Adjacency class messages are used to establish, maintain, and terminate adjacencies between LSR peers. This involves establishing a TCP connection and then exchanging session negotiation messages. The parameters that are negotiated include LDP protocol version, label distribution method (downstream or downstream-on-demand), timer values, VPI/VCI ranges for label controlled ATM, and so on. Advertisement class messages are used to create, modify, and

---

[11]To remain consistent with MPLS terminology, the term *stream/label mapping* will be used instead of the FEC/label binding for the rest of the chapter.

[12]One might need to do this to exchange stream/label mappings between BGP peers that are separated by multiple intermediate LSRs.

delete stream/label mappings between LSR peers. A typical advertisement class message is an LDP mapping message that is used by an LSR to communicate a stream/label mapping to neighboring LSRs. This message will contain a stream identifier and an associated label along with possibly several additional objects including a COS value, LSR ID vector (used for loop prevention), hop count (number of LSR hops in the LSP), and MTU size.

Figure 15-5 illustrates the general flow of LDP messages between three adjacent LSRs. Each LSR discovers the presence of a neighbor LSR on the same data link by transmitting and receiving Hello messages. Next a TCP connection is built and Initialization messages are exchanged. After that, stream/label mappings are generated for stream = A by the downstream LSRs and conveyed to the upstream LSR neighbor in LDP mapping messages.

15.3.2 Stream Member Descriptors

The MPLS LDP specifies a number of stream member descriptors that define the type and granularity of the traffic that is mapped to an LSP. The SMDs that LDP supports are

- *Wildcard.* This specifies that any SMD associated with a given label is processed
- *Network address.* Variable-length address prefix
- *BGP Next_Hop.*
- *OSPF router ID.*
- *OSPF ABR router ID.*
- *Aggregation list.* Contains a list of address prefixes that are being aggregated over a single LSP

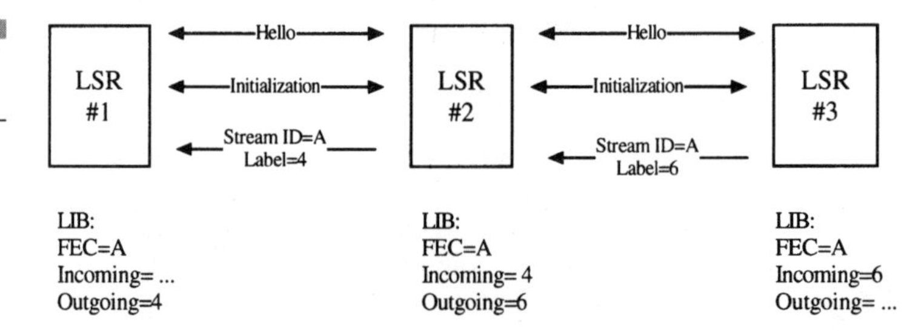

Figure 15-5 LDP Message Flow.

- *Opaque tunnel.* Used to identify packets that will pass through an explicit route LSP (ERLSP)
- *Flow.* Source/destination address pair.

Additional SMDs can be added in the future, including those that specify a multicast flow.

15.4 MPLS Services

An MPLS network is capable of supporting a number of different networking services. The most elementary service that it offers is a faster way of forwarding packets through a network. But although this is an important function and one that any network provider will always strive to achieve, it is not the only reason to choose to implement MPLS. One may choose to deploy MPLS because it offers a more scalable and efficient way to support routed IP traffic over ATM networks than does the traditional IP over ATM overlay model. Certainly performance and scalability are absolute prerequisites for any network provider. But the primary reason to implement MPLS is because it is a simple, fixed-length, self-contained label, and not network-layer information, that is used to forward a packet through a network. It is the flexibility to take any customer's traffic, apply a default or nondefault service, associate it with a set of labels, and then use the same, high-performance, high-capacity label-swapping forwarding procedures to move the traffic through the network. It enables network providers to develop and offer new network-layer services that are visible to the customer base while internally supporting a singular high-performance transport network. This is possible because the data forwarding path is completely autonomous from the network layer service. Thus MPLS will serve as an enabler for network providers who wish to develop and offer interesting and attractive network-layer services (IP specific) that, presumably, customers and subscribers will be willing to pay for.

15.4.1 ATM

Although it is possible to operate MPLS over any data link technology, it is ATM, for many reasons, that receives the most attention and focus. First MPLS has its roots in the previously discussed IP switching technologies that for the most part all attempt to exploit the performance and capacity of hardware-based ATM cell switching. It makes absolute sense to carry this work forward to better take advantage of *N*th-generation ATM switching technology. Second ATM switches can be found at the core of many ISP/carrier-type networks that currently are using the traditional overlay method. As these networks grow in size and number of routers, scaling issues in terms of the number of VCs become a major concern. Providing a more efficient and scalable IP routing capability that utilizes the existing ATM infrastructure would be a useful undertaking. At the same time it is not practical to assume that the existing mechanisms used to build and maintain VCs will be immediately replaced with MPLS. In fact it may not even be possible to upgrade some existing ATM switches to support MPLS. Additionally some of these networks may be quite comfortable with and depend on native ATM forum protocols to support a particular service that is being offered. Therefore the mechanisms employed by MPLS over an ATM network must coexist and/or interoperate with existing ATM protocols. And finally ATM is the only data link technology that currently supports data, voice, and video with QoS. Indeed it is this particular attribute of ATM that makes it such an attractive technology for multiservice network providers. And one of those services can be advanced, high-performance IP switching.

MPLS can be supported over an ATM network in a number of different ways. The first example is an invocation of the peer model. The MPLS device is this case is an ATM switch that is converted into an LSR, or more specifically to an ATM-LSR. An ATM-LSR supports normal IP routing protocols, may be capable of forwarding packets at layer 3, is able to label-swap cells at layer 2 and runs an instance of LDP. ATM-LSRs are directly connected to each other over serial links and exchange routing table updates as well as LDP messages. An ATM-LSR in this example has been stripped of all ATM forum protocols and instead uses LDP to build and establish LSPs through the network.

The peer MPLS/ATM implementation places much less of a routing protocol burden on the system than the IP/ATM overlay model because each ATM-LSR only forms routing adjacencies with its neighbor ATM-LSR—not each router at the other end of a VC. The network topology that is maintained is based on the number of actual inter-LSR physical links, so it is most certainly not as "link dense" as the VC-rich overlay model. However, what may be lost here are some of the attractive ATM features such as loop-free QoS-based path selection and more generally

QoS support. In addition if one is not exploiting the native QoS features of ATM, it might be hard to justify the overhead of the cell tax that is introduced into this solution.

Another example is the integrated approach in which ATM-LSRs communicate with each other through native ATM switches. Neighboring ATM-LSRs may establish PVCs or utilize SVC signaling to establish VCs with each other. Control traffic, default IP traffic, LDP messages, and per-stream LSP traffic flow over a number of VCs established between neighboring ATM-LSRs. When a stream is mapped to a VC, a special identifier, the VCID is needed to associate that stream with a VC that passes through some intermediate ATM switches. Recall that the reason behind this is that the label in use (VPI/VCI) may be swapped as the cells traverse native ATM switches. Other techniques such as setting up a preestablished VP are also supported. An ingress-to-egress LSP is created when the ATM-LSRs along a path concatenate the inbound VC for a specific stream/label mapping with the corresponding outbound VC.

One of the nice features of the integrated approach is that it enables MPLS to be introduced in an incremental fashion to an existing ATM network. For example, a core set of MPLS LSRs could be initially deposited in the middle of a large ATM cloud to serve as "VC aggregation" hubs. Interhub traffic would consist of LSP "trunks" that carry the aggregate traffic flows delivered by VCs from the edge routers. From a routing adjacency perspective the edge routers would simply peer with the nearest core LSR. Although this might introduce additional router hops in the core, this scenario could be viewed as a temporary relief from the "*n*-squared" dilemma.[13] In the future, the MPLS function could be distributed closer to the edge of the network, thus enabling the establishment of edge-to-edge LSPs. Another important feature is that existing ATM VC management techniques remain in place. This may be beneficial from an operations sense (network procedures and operations may revolve around ATM VC management) as well as from the fact that some ATM switches may not be MPLS-upgradeable. The integrated approach might be the easiest migration path to take if moving from the IP/ATM overlay model to MPLS. Although it should be noted that depending on the configuration and implementation of the ATM-LSR, ATM VC operations would play only a supporting role—that being to

---

[13]It is possible that the one or both of the core LSR router hops could be avoided by aggregating inbound VCs into a VP trunk at the ingress LSR and then deaggregating at the egress LSR.

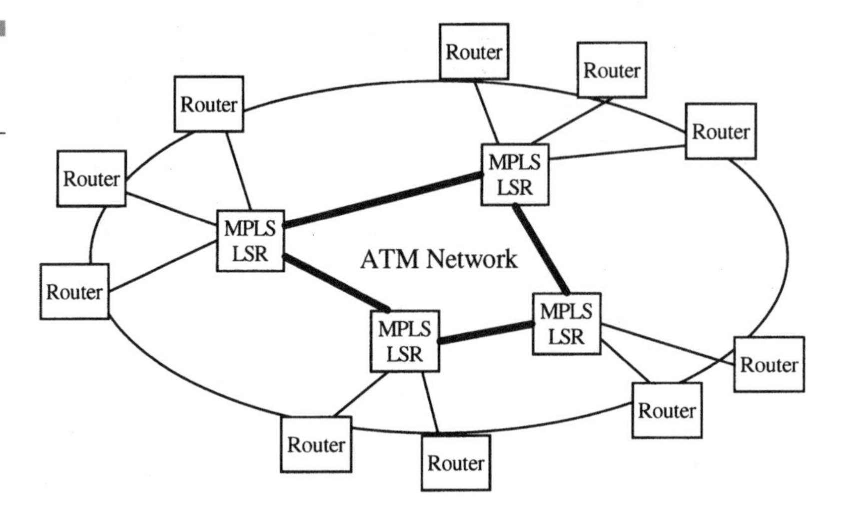

Figure 15-6 MPLS over ATM integrated approach.

provide inter-LSR connectivity. An example of the MPLS over ATM integrated approach is illustrated in Fig. 15-6.

The third approach to supporting MPLS over ATM is the so-called SIN configuration. In this model, the physical ATM network is divided into two disjointed topologies: one supported and maintained by PNNI and the other by IP and MPLS. Figure 15-7 illustrates this concept showing an MPLS LSR that is capable of providing intermediate switching services for MPLS LSPs, ATM VCs, and possibly IP packets. Although it is really a switch-dependent implementation issue, it typically requires the partitioning of the VPI/VCI label space into MPLS and ATM portions. Additionally, a particular implementation may require that some ports be physically configured for MPLS and others for ATM. Switch resources are then under the control of coresident but mutually exclusive MPLS and ATM switch control points. The switch component may also run a VC routing protocol such as PNNI. Thus a network of ATM-LSRs employing the integrated approach would support concurrent MPLS and ATM topologies, routing, and VC management.

Although this might seem like an overly complex model that is difficult to manage, it should be placed in its proper perspective. First, the common single element in this model is ATM, so it is a single network infrastructure that is being utilized. With that comes a single set of operational procedures, management tasks, management tools, and hardware. Second, for all its perceived faults and complexity, ATM does indeed do what it was designed to do—provide QoS over virtual connections for data, voice, and video mediums. A network provider may

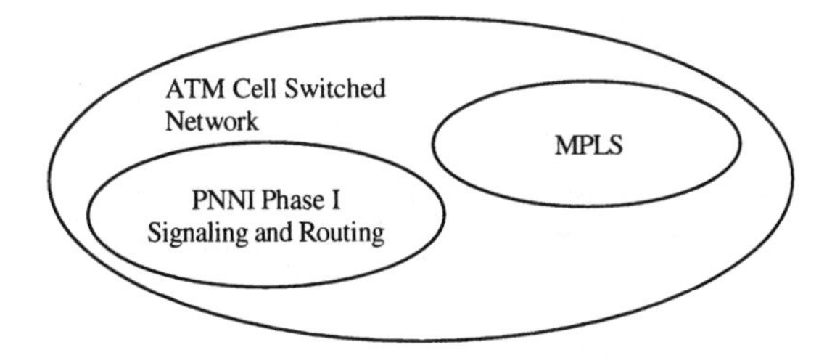

Figure 15-7 MPLS over ATM ships-in-the-night approach.

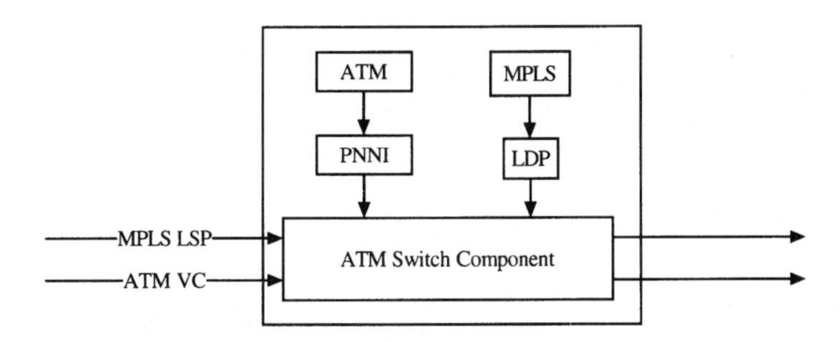

depend on this particular technology to support residential broadband services (xDSL), high-quality video services (MPEG2 over ATM), circuit emulation services, and so on. Rather than implement the overlay model that could deplete VC resources, a controlled deployment of the SIN-capable ATM-LSRs might be a suitable option. Doing so allows an ATM-centric network provider to continue to provide ATM services while at the same time introducing an additional control plane that is optimized for scalable IP routing.

Because ATM does not offer any loop detection in the forwarding path, methods for preventing or at least containing the damage caused by loops must be taken into consideration. In practice dealing with the possibility of ATM loops depends on several factors including the particular switch implementation, the desire to prevent loops, and configuration options. Building on techniques first described in the ARIS and tag switching proposals, MPLS can address the loop problem in one of several ways:

- Don't worry about it and hope it does not happen. This is the extreme optimist's approach.
- Depend on per-VC queuing and queue management schemes (e.g., EPD, RED) in the ATM-LSR to contain the damage caused by

loops. This is a reasonable approach but probably would require new switch hardware.

- Use egress-oriented ordered LSP control to build a loop-free path. This method prevents an LDP path setup message from passing through a switch more than once. This is the safest technique but takes the most time in LSP path setup.
- Employ the diffusion process to systematically prune upstream links that cause a loop when appended to a new downstream LSP after a topology change.
- Implement faster, simpler, but as yet undiscovered loop detection algorithms.

Stream merge is a generalization of the ATM-specific VP merge and VC merge functions first mentioned in the ARIS specification. It enables an MPLS network to scale using only $O(N)$ switched paths. With respect to operation over an ATM network, several factors related to MPLS functions must be taken into account:

- The VC-merge function on a switch requires a per-VC buffering and/or queuing mechanism to prevent cells from different sources from interleaving on the merged downstream VC.
- If only VP-merge is supported, VCI collisions on the last VP to the egress ATM-LSR must be prevented by assigning a nonoverlapping VCI range to each ingress ATM-LSR. This is probably most easily accomplished by a configuration process, although it is not inconceivable that other more dynamic means (e.g., single node assigns VCI ranges) could accomplish this.
- ATM-LSRs advertise their ability to support VP- or VC-merge.
- An ATM-LSR that must merge the cells of a stream onto a single downstream path but is not capable of performing VC-merge must request multiple labels for a particular FEC from the downstream ATM-LSR. The number of labels is determined by the requirements of the requesting ATM-LSR itself (it may be an ingress node for a particular FEC) as well as any labels that might be needed for upstream ATM-LSRs that need an outbound label for an FEC. In general a VP-merge node requests one VPI per FEC, and a VC-merge node requests one VPI/VCI per FEC. A nonmerge node may request multiple VPI/VCI labels for the upstream nodes as well as one for itself on a per-FEC basis.

15.4.2 Traffic Engineering

One of the most useful applications of MPLS is the ability to perform traffic engineering. Traffic engineering, loosely defined, is the ability to direct streams of traffic over separate, nondefault paths. Instead of depending on a dynamic IP routing protocol to compute a path based on a single metric (number of hops or lowest link cost) for all streams sharing a common destination or egress point, MPLS can be used to build separate LSPs for separate streams that may share a common ingress and egress node. In addition, the forwarding path for both default and nondefault LSPs is the same, so there is no performance penalty.

There are a number of reasons why a network provider may want to perform traffic engineering. One is to evenly distribute the aggregate network traffic load across the entire network. This may be required to provide more capacity based on changing network usage patterns or to avoid network bottlenecks. Another basic reason is to attempt to use all available network resources to support the customer's traffic. It simply does not make sense to have available resources standing idle (e.g., links, switches) while others may not have sufficient capacity or are being overused. Another reason might be to offer a value-added service with more bandwidth and lower delays by directing a particular customer's traffic over a separate path.

The ability to direct traffic over specific paths can be accomplished using several techniques, depending upon the network implementation. In a pure router environment it is possible to use loose or strict source-routing with IPv4 to forward packets over a specific path. This is not a particularly attractive solution because each packet is required to contain the entire path as a separate IPv4 option header. In addition there is some amount of overhead in processing IPv4 options versus having to process just the IPv4 header. The routing header in IPv6 is a more efficient implementation of per-packet source-routing, but of course that requires IPv6-capable routers in the network.

In an IP overlay environment network providers may provision additional VCs (e.g., frame relay, ATM) through a specific set of links and nodes to achieve better network utilization or offer higher capacity and better performance for a customer's traffic flows. With respect to traffic engineering, VCs have a number of useful capabilities:

- ✓ QoS attributes such as bandwidth and delay can be applied on a per-VC basis.

- Admission control performed at each node along a path ensures that there are sufficient resources to support the traffic flowing over the path.
- Unless a link or node along the path is affected, the path itself is not affected by changing topological conditions. Thus there is some level of stability to the VC approach.
- VCs can be manually provisioned or dynamically established using explicit signaling.
- Traffic management functions (e.g., policing, shaping) can be applied on a per-VC basis.

The attractiveness of provisioning VCs over a specific path does come at a price. Unless SVC signaling and routing are used, there may be some administrative delay in the provisioning of a VC. The delay may be such that when the newly provisioned VC is available to handle data, it may no longer be of any use. There is also the fact that it is the overlay model and its associated overhead that are deployed in this environment and that enable VCs to be provisioned as needed. Thus a solution that provides the traffic engineering functionality of the overlay model but without its overhead or that associated with per-packet source-routing is desirable.

MPLS is considered an ideal solution to the problem of efficient traffic engineering because specific nondefault explicit paths for a specific stream can be easily provisioned over any data link technology. Specifically, MPLS possesses the following attributes suitable for traffic engineering:

- Provisioning of per-stream LSPs .
- Stability of nondefault LSP is maintained during changes in topology that may result in new default paths computed by dynamic IP routing.
- Ability to apply QoS attributes to an explicit path.
- No performance penalty for packets traversing explicit path.
- Explicit paths can be manually provisioned or dynamically computed from topological information maintained by routing protocols.
- Actual sequence of LSP hops comprising the explicit path is only carried during path setup versus the case with IP in which the entire path sequence is carried in each packet.

- Supports explicit paths over ATM networks in a more scalable and efficient manner than the overlay model.
- Enables the function and properties of explicit paths to be supported in a traditional connection-oriented environment (e.g., frame relay, ATM) as well as in a packet-based connectionless environment.

It is possible to provision explicit paths in an MPLS environment using one of two techniques. In the first technique, LDP is used to establish an explicitly routed LSP (ER-LSP) through a network of LSRs. The LDP ER-LSP message is initiated at the egress LSR of the LSP and is forwarded upstream, visiting each LSR in the specified sequence of LSRs until it reaches the ingress LSR . The message contains a list of LSRs that the ER-LSP message must pass through, a label, an SMD that identifies the traffic that will traverse the LSP, and, optionally, a COS value that can be applied to the label. The other technique makes use of RSVP in conjunction with an explicit route object (ERO) and a label request object. In this scenario the ingress LSR initiates an RSVP PATH message containing an ERO and label request object. The label request object indicates that a stream/label mapping is requested and the type of network-layer protocol that will flow over the explicit path. The ERO contains a set of LSRs that the PATH message must visit. Upon receiving the PATH message, the egress LSR (of the LSP) sends an RESV message upstream toward the sender (the ingress LSR of the LSP in this case). The RESV message contains a label that is allocated by the downstream LSR for incoming traffic and passed to the upstream LSR that uses the same label for outgoing traffic. Additionally, the RESV message may contain QoS parameters (e.g., TSPEC, RSPEC) that were derived from information transferred from the ingress LSR, thus enabling QoS state to be installed over the ER-LSP as well. A conceptual depiction of both approaches is illustrated in Fig. 15-8.

15.4.3 QoS/COS

Another important service that an MPLS network can and must support is COS. This is important because it enables individual customer streams to benefit from specific COS attributes that can be applied to the LSP. Providing this service is something that network providers can do for a premium fee and therefore generate additional revenues. MPLS

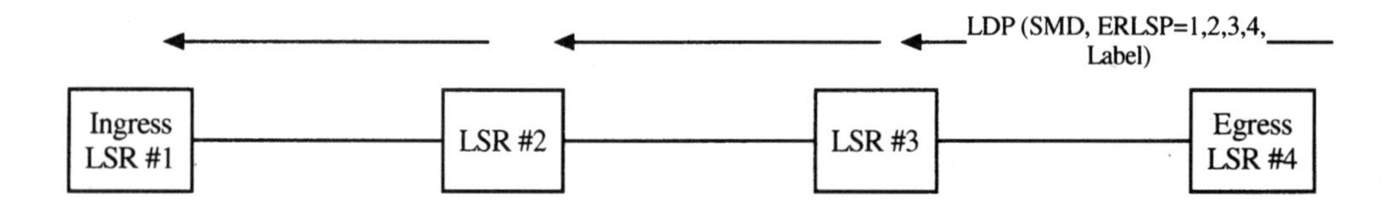

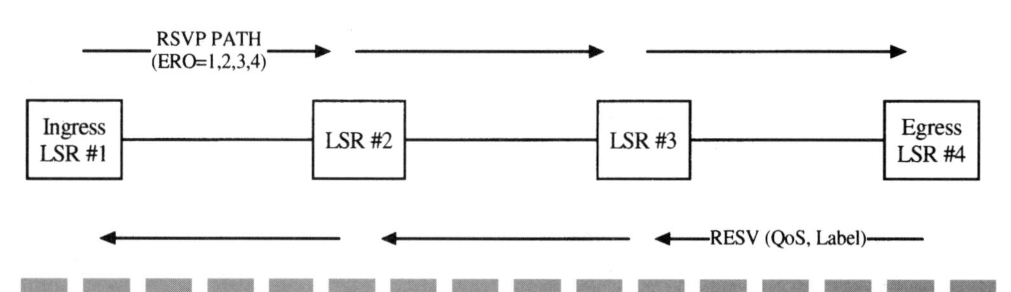

Figure 15-8 MPLS ER-LSP establishment using LDP and RSVP.

actually facilitates this process by first identifying the customer streams at the edge of the network (by examining multiple fields in the packet header) and then placing those streams on a specific LSP that may have some COS or QoS attributes attached to it. MPLS can support a nondefault service over an LSP using one of several possible techniques:

- A COS indicator can be explicitly transported in the label that is affixed to each packet. In addition to swapping the label at each LSR hop, the packet can be serviced on the outbound link based on its specified COS attribute. The MPLS shim header contains a COS field. The COS indicator can be derived from a differentiated service policy installed at the ingress LSR that might say something like "all packets with source address equal to A should be treated as high-priority."
- A COS value can be implicitly associated with a particular LSP. This would require LDP or possibly RSVP to assign a nondefault COS value to the LSP so that the packets of the stream are handled appropriately. This is required if the label does not contain an explicit COS indicator.

- ✓ A particular LSP that is built using underlying ATM signaling and routing (integrated approach) can exploit the QoS mechanisms inherent in ATM.

15.4.4 Virtual Private Networks

A VPN is a logical network overlay on top of a public network facility. Use of the public Internet as the network transport for a VPN is becoming quite common. Links between VPN member sites consist of tunnels built by encapsulating intra-VPN packets in externally routable IP packets. This means though that intra-VPN traffic that is traversing a public network is subject to the performance and capacity constraints of existing IP routing.

MPLS can associate any abstract network-layer entity with a set of labels that are used to forward the traffic independent of the entity. Thus it is quite possible to associate the members of VPN (and their respective network address prefixes) with a set of labels so that they may erect intra-VPN LSPs and exchange traffic in a secure and predictable manner. Because it is likely that a network provider may support more than one VPN (or rather provide transport services between members of multiple VPNs), there needs to be some way of disambiguating the routing table entries for these different networks. This can be accomplished by assigning a VPN Identifier (VPN ID) to each VPN. This is used to associate routing table entries that are specific to a particular VPN. Furthermore, it then becomes a straightforward matter to advertise stream/label mappings (the SMDs is this case are the address prefixes unique to the specific VPN) to the corresponding members of a VPN. This can be accomplished using one of several techniques such as piggybacking labels in BGP update messages or by using LDP. The notion of using MPLS to form VPNs is conceptually depicted in Fig. 15-9.

15.4.5 Multicast

IP multicast traffic flows are typically sustained for some period of time and function best with adequate bandwidth and low delay. Examples are multiparty videoconferencing applications and shared whiteboard tools. MPLS can facilitate the efficient transport of multicast packets by mapping a layer-3 delivery tree to an LSP tree. In addition to the added

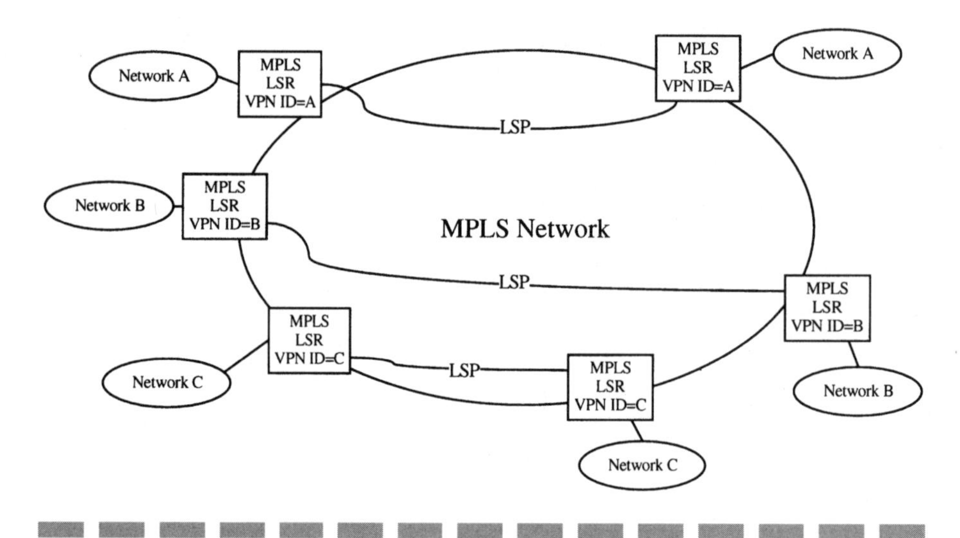

Figure 15-9 MPLS VPN support.

benefit of a fast performance and lower delays, multicast packets flowing on their own LSPs are separated from other traffic types flowing on other LSPs. This is important because multicast traffic is completely ignorant about available network capacity. In other words it does not react to network congestion like TCP or TCP-friendly applications. This means that a multicast application streaming large doses of packets into a network could introduce congestion and cause the TCP contingent to react and reduce their information transfer rates, thus affecting performance.

MPLS assigns a label to the branch of a multicast delivery tree identified by either a (source, group address) for source-rooted trees or by just a (*, group address) for shared delivery trees. The assignment of the labels can be based on the presence of new entries in a multicast forwarding cache (as driven by the arrival of multicast data packets) or by the presence of explicit multicast routing control messages (e.g., PIM-join). Other considerations that may arise in the support of multicast over an MPLS network follow:

- Because there is no loop detection in ATM, the formation of a looping LSP on top of an ATM network must be avoided.
- If the physical link of a particular multicast LSP is broadcast-capable (e.g., Ethernet), it will be necessary to partition the overall available label space and assign nonoverlapping ranges to partici-

pating LSRs. This is so that separate LSRs transmitting to separate multicast groups do not use the same label.

15.5 MPLS Network Examples

MPLS for the most part is a topology-driven IP switching solution that is deployed in service provider networks and any other large IP networks that utilize both ATM and packet-based infrastructures. In the former it offers a more efficient way of supporting IP over ATM switched paths than does the existing overlay model. Indeed, whereas the latter requires $O(N^2)$ links and places an inordinate demand for routing protocol processing on a network with a large number of routers (N), MPLS with the use of stream merge only requires $O(N)$ links. With respect to operating over a packet-based infrastructure, MPLS provides an efficient and straightforward technique for building and maintaining stable nondefault explicit paths. This ability to establish "VCs" without requiring VC-specific data link technology is a useful mechanism for the engineering of traffic through a packet-based router network of arbitrary size, complexity, and capacity.

Figure 15-10 is a simple illustration of two possible situations where MPLS is able to build separate per-stream LSPs between the same ingress-egress pair of LERs. In the first example one LSP carries all packets to networks A, B, or C and another one carries all packets to network D. Even though the ingress and egress points to the MPLS domain are the same, separate LSPs can be established that carry varying granularities of traffic. The granularity of traffic placed on the LSP could even descend to the host-to-host or even application-to-application level if the network provider so wished. In the second example a separate ER-LSP is established to transport all packets to network D. This may be done for performance or security reasons or just as means of balancing backbone traffic across multiple paths.

15.6 MPLS Comparison with MPOA

The primary contributors to the MPLS specification consist of many who initially worked on the various proprietary IP switching schemes, and as

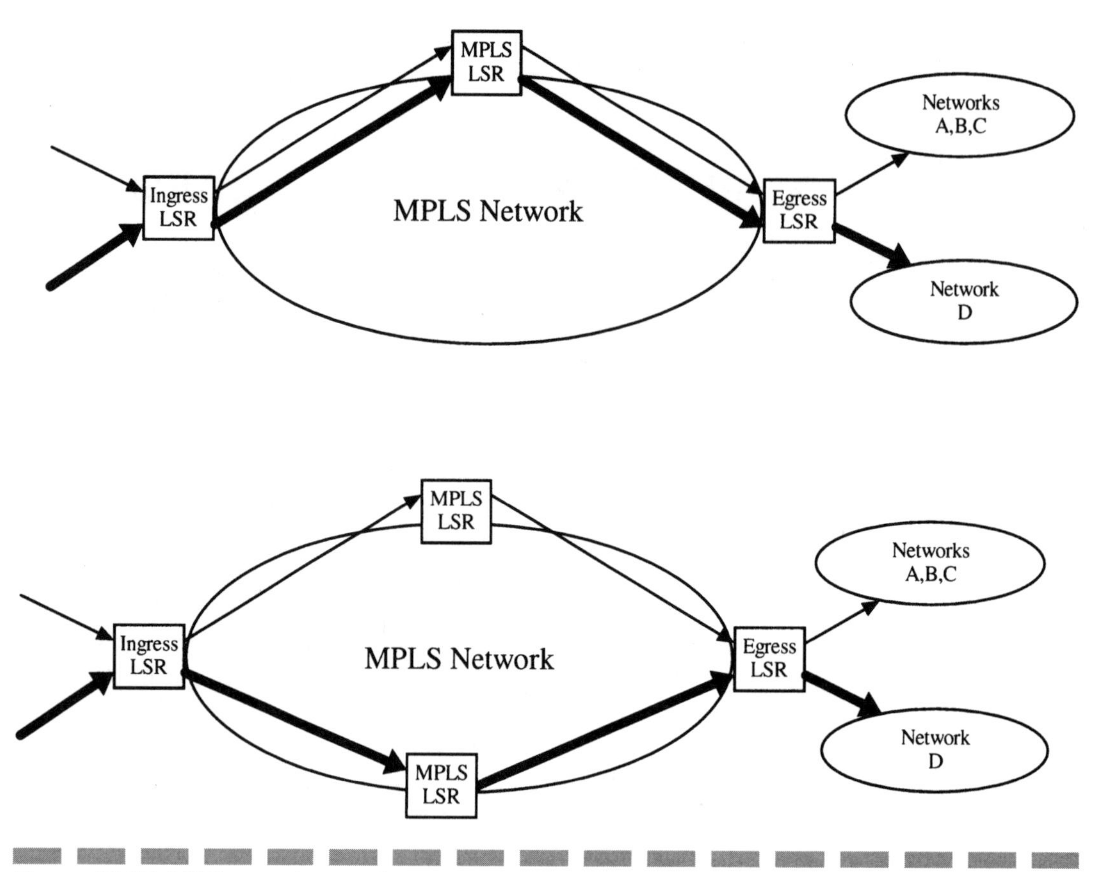

Figure 15-10 MPLS network examples.

a result MPLS incorporates many of the more useful features from those solutions. But as with any standard there were some compromises made to accommodate a "choice" of possible functions, but nevertheless it is for the most part a sound and stable effort.

Rather than compare MPLS with its ancestors, it might be useful to compare it with another well-known IP over ATM (switching) solution, MPOA. Perhaps the most elementary difference is the environments that the solutions were designed for. MPOA is basically a campus-based solution that enables hosts and edge devices to set up shortcut VCs through an ATM network. It offers a virtual router service that can be overlayed on top of an ATM network. Its usefulness, though, depends on the density of ATM switches in the campus. On the other hand, MPLS is really designed for use in a WAN consisting of not just ATM switches but other data link technologies as well. It provides a simple mechanism for engineering traffic flows across the network and in some cases enhances the scaling properties of the underlying infrastructures. Both options provide better performance than traditional IP routing and both can either dynamically use backbone resources or at least facilitate their optimal use. MPOA accomplishes this with PNNI routing; MPLS does it with traffic engineering.

Table 15-2 compares some of the functional attributes of MPOA and MPLS.

TABLE 15-2

MPOA and MPLS Comparison

Attribute	MPOA	MPLS
Environment	Campus, WAN	WAN
Router entity	Virtual router	Label switch router
Model	Overlay	Peer
Addressing	Separated, IP and ATM	IP only
Routing protocols	IP unicast and multicast plus ATM routing (PNNI)	IP unicast and multicast
Switched path establishment	Flow-driven	Topology-driven (can also support flow- and reservation-driven)
Control protocols	IP, MPOA, NHRP, ATM Forum protocols	IP and LDP
Devices	Hosts, edge devices, routers	Routers and switches
Native ATM support	Yes	No but can coexist
Data path selection	PNNI Phase I	Dynamic IP routing or explicit routes
Standard	ATM Forum	IETF
Non-ATM data links	Via edge devices	Yes
Single point of failure	Yes, MPOA Route Server	No

MORE INFORMATION

Feldman et al., "Evolution of Multiprotocol Label Switching," *IEEE Communications,* vol. 36, no. 5, May 1998.

Hagard, G., et al., "Multiprotocol Label Switching in ATM Networks," *Ericsson Review,* no. 1, 1998.

http://www.employees.org/~mpls/

http://dcn.soongsil.ac.kr/~jinsuh/home-mpls.html

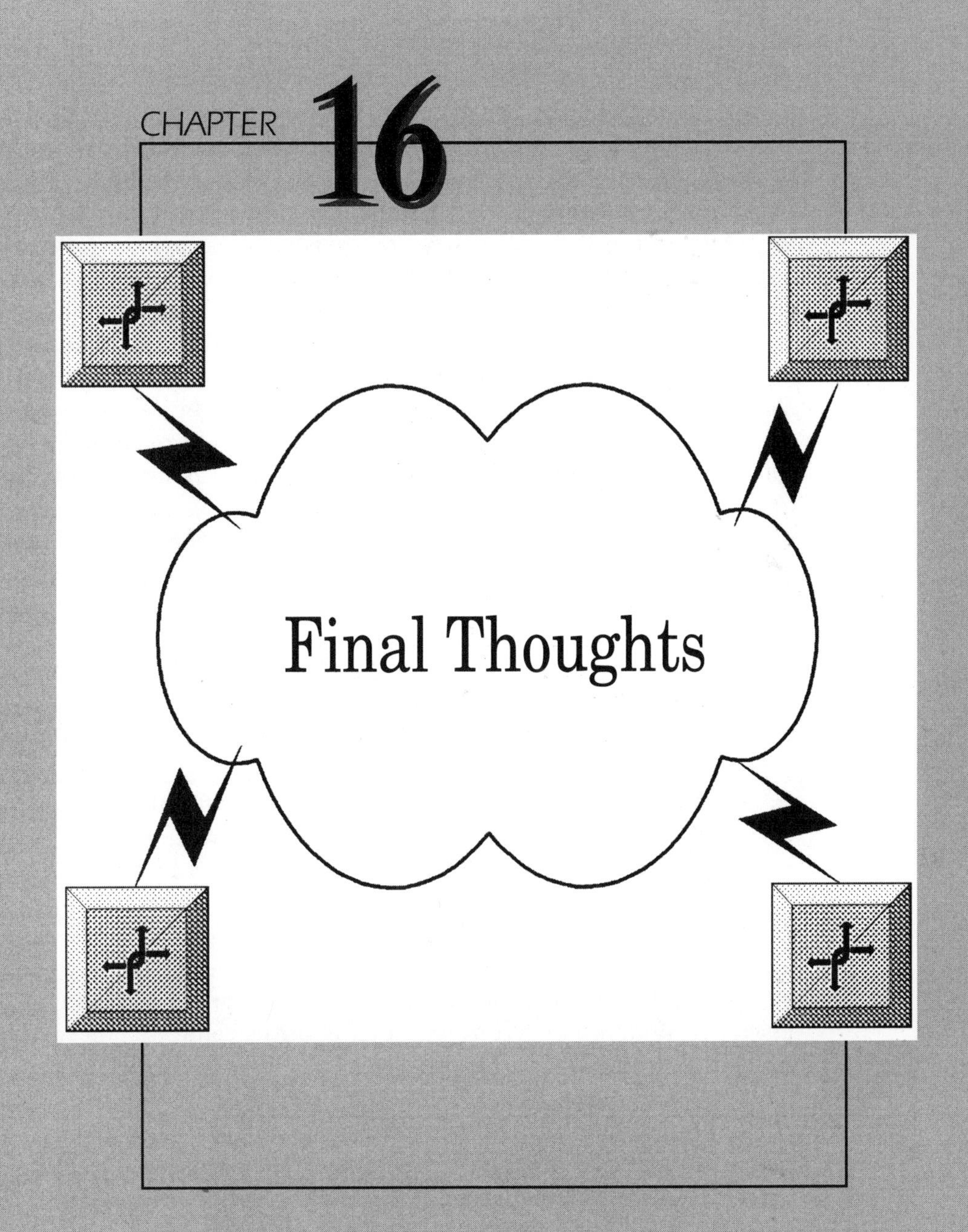
CHAPTER 16
Final Thoughts

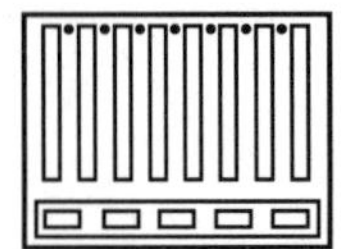

IP switching is a technology that first arose out of a need to provide greater capacity and better performance for IP traffic flows. It was built on the notion that the per-packet processing performed in each router could and should be circumvented as a means of improving performance. Rather than build faster routers, the design decision at the time was to reuse already existing multigigabit forwarding devices, with ATM switches as the high-capacity, high-performance packet forwarding engines. But ATM switches are not packet-forwarding engines per se; they are cell switches—that is, they switch small cells from an input port to an output port. There is no information in the cell header that a switch can use to make an IP forwarding decision. The most important field in the cell header is a connection identifier—a label. It binds the contents of a cell to an end-to-end virtual connection. Thus, to turn an ATM switch into a router (or part of a routing system), it is necessary to introduce a control component that will associate the forwarding path for aggregate or individual IP traffic flows with an ATM virtual connection. Once the control protocol does its job, routed traffic that was formerly processed on a per-packet basis at each router along the forwarding path can instead enjoy uninterrupted switched service through the network.

In fact all of the IP switching solutions, including the MPLS effort, are simply control mechanisms that bind IP traffic flows to ATM virtual connections. The concept of mapping flows of packets to a small, fixed-size label can be, as first suggested in tag switching and later incorporated into MPLS, generalized to associate IP traffic flows with labels over any media. A label switched path, analogous to an ATM VC, can be established through a network of packet routers to provide, if not the performance of traditional layer-2 switching, at least the option of directing traffic over nondefault paths with no performance penalty. There is of course a price for achieving this binding between IP and switching—the overhead and complexity of mapping a connectionless packet relaying architecture on top of a static, connection-based infrastructure. Even the soft-state IP-centric approach embodied by the peer model still requires that the binding process occur in each segment along the path between the network ingress and egress. It doesn't really matter whether it is executed in an independent or ordered fashion—the process must be completed end to end. The same holds true for the so-called heavyweight signaling mechanisms found in the ATM Forum signaling protocol suite. The price many indeed be well worth it because the return that one can expect to receive is better raw throughput, or a better utilization of available network resources.

But other simpler solutions are rapidly emerging that do, or at least promise to, solve some of the same problems that IP switching addresses. Gigabit Ethernet, the next turn of the Ethernet crank, supports a 1000 mb transfer rate. Layer-3 switches in the campus offer an inexpensive way to forward millions of packets per second from one Ethernet to another. Gigabit and terabit routers in the WAN, designed for the next-generation IP carrier network, will support on the order of tens and hundreds of gigabits per second of forwarding capacity. They will attach directly to fiber networks and through sheer speed promise to provide service differentiation. Not only will data be supported, but many claim that IP will carry voice and video—all of it.

16.1 Changing Network Requirements

The role of information networks and what is expected from them is rapidly changing from the perspective of the end user and applications and from those who operate and provide the service. The very public and well-documented evolution of the Internet from underground research experiment to global mass media outlet has instilled a new set of expectations that in some cases may exceed the reality. Recall that it was not too long ago that to just connect to another host to download a file was the task at hand. And to dash off email to a colleague in another part of the country, not 100 percent sure that he or she would even receive it, was a quaint demonstration of one's computer literacy. And corporations owned and operated their own networks, consisting of their own equipment and private communications links, to interconnect their own people with their own applications. Outside communications involved face-to-face meetings, phone calls, or hard-copy exchanges.

Now with the global Internet and the URL replacing phone numbers, business cards, and television commercials, that is all expected to change. It is no longer enough just to connect to another computer. It must be done quickly, transparently, and securely. Now instead of downloading files, the end user or the application on behalf of the end user will execute network programs and present networked content. Some of the content may be on disparate computers located in different corners of the world—that is of no consequence to the end user who is expecting instantaneous gratification from the content presentation. Having to wait for information to be delivered is just not tolerated.[1]

Email communication is expected to change as well. Whereas before it was presented in a primitive but useful text-based format, it now comes in a number of jazzed up presentation formats supporting a myriad of different attachments. These include voice and video attachments that enable the correspondents to include both verbal and visual nuances that are so important to human interaction. Real-time voice and video communications over IP-based networks, while still in their nascent stage, are expected to become as common as email is today in the near term.

Corporations are breaking down the hard physical barriers that separated their information networks from the rest of the world. They are rapidly developing a presence on the Internet as a means of announcing and advertising their goods to the millions of "eyes" that regularly navigate the Web. Electronic commerce is viewed by many as a "can't miss" opportunity to win new customers, transact more business, and ultimately increase market share. Corporations are also looking at the Internet as an inexpensive and simple way to develop and foster "electronic relationships" with business partners and suppliers. In addition, efficient and secure communications with tele-employees is seen as an important requirement for conducting business on a global scale.

And the network providers who collectively form the interconnected mesh of networks that comprise the global Internet must now put in the infrastructure that will support these new applications and services. They are now faced with an unprecedented growth in traffic volumes coupled with the desires of the customer and subscriber bases to receive a better and more reliable service. Simply put, the network providers must offer new and competitively priced services that customers will be willing to pay extra for, or they face extinction.

Investment in a new technology, even if it is fully compatible with the status quo, is now based on the changing set of network requirements that can be summarized by the following:

- ✔ *Service differentiation.* Customers want a choice of varying levels of service that they can choose from and that will be priced accordingly. It is no longer the case that simple connectivity or even bandwidth will do.

---

[1]It is understood that waiting is part of the game at the moment, but it is becoming increasingly intolerable. Growing expectations will further erode this tolerance.

- *Simplicity and scalability.* New technology deployed in the interior of networks (the part that the network provider deals with) must be simple, scalable, and manageable. It must be easy to configure so that the network provider can offer the special services that customers desire (and will pay for). Externally, the customer should have varying levels of service to choose from. It is the dichotomy of interior simplicity and external complexity.
- *IP-based.* This goes without saying—it is an IP world now, and all information communications will begin to travel with a 20-byte IP header appended to them.

16.2 Popular Alternatives

One of the major advantages of IP switching, in particular when ATM hardware is used, is performance. There is no question that generic ATM switching technology is extremely fast. It has for some time enjoyed price/performance superiority over its packet-based router cousins. However, a number of emerging data link technologies and a significantly faster and less-expensive packaging of traditional router-based functions into hardware have been introduced for the campus and WAN environments. The notion that switching (IP or otherwise) is cheaper and faster may be challenged—in fact it is being challenged. Indeed if one can deploy faster pipes with IP forwarding devices that can fill those pipes and maintain cost parity with switching, are there any reasons why one might choose to implement an IP switching solution?

Upon first examination Gigabit Ethernet is just that—regular old Ethernet operating at 1 gigabit per second or 1000 MB per second.[2] It provides a tenfold increase in raw bandwidth over Fast Ethernet and is 100 times faster than traditional 10-MB Ethernet. It operates over single-mode fiber (SMF) for extended distances of up to 5 km and over multimode fiber for shorter distances (500 m) at a lower cost. Coupled with matching increases in workstation and server CPU performances and the inevitable decrease in costs as the industry begins to ship in mass quantities, it would seem that Gigabit Ethernet solves all outstanding problems for next-generation campus networks. The speed is there, the

[2] *http://www.gigabit-ethernet.org/*

cost is there, and it is simple. One could not really ask for more.

However, one needs to carefully position Gigabit Ethernet in the context of a campus network. It is a faster LAN segment—that's all. It gives an individual workstation or server a means for sending and receiving Ethernet frames at a higher speed. A campus network is not built by extending an individual LAN segment to all end-user desktops. In reality multiples of individual LAN segments are interconnected by bridges or routers, and these segments can be Gigabit Ethernet, Fast Ethernet, Token Ring, and so on. Furthermore, if a campus network is to offer some level of robustness, it should be resilient to link or node failures. This would imply that there are alternate paths available and that the network of interconnected LAN segments is capable of dynamically rerouting Ethernet frames over an active and available alternate path.

Figure 16-1 illustrates a network of Gigabit Ethernet LAN segments connected into a backbone of generic layer-2 switches. These switches behave like any old LAN switch except they offer nonblocking segment-to-segment forward rates at media speeds. Observe that the spanning tree algorithm forms a single loop-free path between any two segments. Also note that this particular path is passing through a subset of the entire topology. In other words the backbone path is concentrated on a specific few links and switches in the network. Even though other links and nodes are available to be used, and in fact would be if a topology change calculates a new spanning tree, they are currently not being

Figure 16-1
Gigabit Ethernet in a backbone of layer-2 switches.

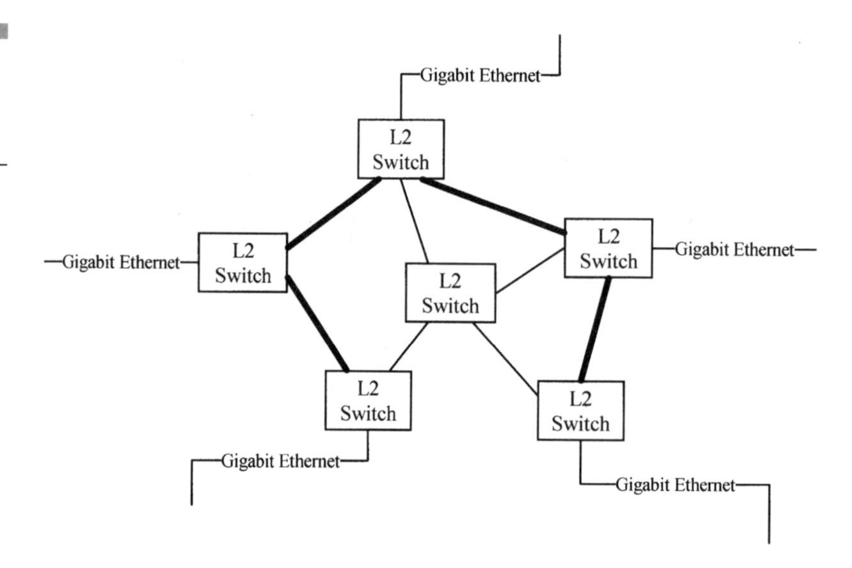

used nor can they be (except in the case of certain broadcast traffic). They are in a blocking state. The switch ports and links that form the spanning tree are in an active or forwarding state, and they are the only ones that can forward Ethernet frames. Therefore, this particular topology, despite the presence of Gigabit Ethernet segments and nonblocking wire-speed layer-2 switches, only uses a subset of the entire network topology. A portion of the network is pruned from consideration and cannot be used unless a new spanning tree calculation selects it. If there is congestion on a particular link or switch, performance over the backbone may suffer. This is not the fault of Gigabit Ethernet. After all, devices on the same segment can exchange frames at gigabit speeds. Nor is it the fault of the layer-2 switches. But rather it is the fault of the spanning tree path selection algorithm. The price for necessarily providing loop freedom is a suboptimal use of network resources. Of course one can get around this limitation by designing a topology that only has one possible path between any two segments.

Another highly touted and popular solution for evolving campus networks is the so-called layer-3 switch. This new genre of internetworking devices comes in many different flavors. Essentially these devices can be classified as inexpensive, wire-speed hardware-based routers. They are typically Ethernet port dense and support switched 10/100 Ethernet and Gigabit Ethernet. They for the most part run traditional IP unicast and multicast routing protocols. And they possess forwarding rates on the order of millions of packets per second. Now that the campus network can be outfitted with inexpensive high-performance routers, it would seem that all problems are solved.

Again the focus is on the behavior of the backbone because that is the key resource that gigabit-capable end users and applications will be vying for. Figure 16-2 illustrates the same topology as the previous example except that layer-3 switches are employed in the backbone. In this scenario, dynamic IP routing protocols select a path based on the fewest number of hops (e.g., RIP) or the lowest link cost (e.g., OSPF) for packets destined for a particular destination address. Packets sent from Net A to Net B travel through L3 switches 1, 2, and 3 toward Net B. But what happens if there is congestion at L3 switch 2 or on the link connecting L3 switches 2 and 3 and packets are being dropped? Nothing happens except that performance for the Net A-to-Net B traffic flows will suffer. Even though alternate paths through the network are available, the dynamic IP routing protocol has no way of directing packets away from the congested path. Even though an alternate path to the

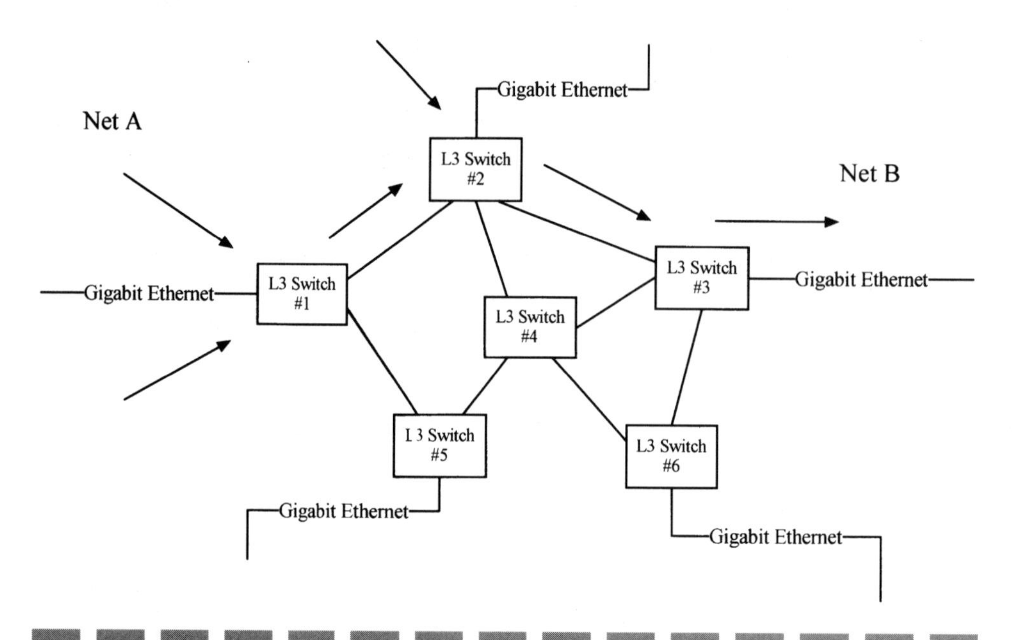

Figure 16-2 Gigabit Ethernet in a backbone of layer-3 switches.

destination is congestion-free, dynamic IP routing cannot redirect new or existing flows over the alternate path and it does not even know that one even exists.

This particular scenario illustrates once again that, despite the presence of Gigabit Ethernet at the desktop and wire-speed forwarding devices interconnecting Gigabit Ethernet segments, network resources (e.g., bandwidth, backbone forwarding capacity) are not utilized to the fullest. Although a portion of the network is overutilized as it attempts to forward more packets than it is capable of handling, another portion of the network is standing by, possibly in an idle state. In this case the culprit is the dynamic IP routing protocol that uses just a single static metric to compute a path to the destination. It does not take into account available capacity along a path or whether congestion is present along the path, and it does not know or even care what the offered traffic load is.[3] Therefore in a meshed backbone configuration consisting of Gigabit Ethernet segments interconnected by layer-3 switches running

[3]RFC2386, A Framework for QoS-based Routing in the Internet, looks at this problem and possible solutions in more detail.

dynamic IP routing protocols, traffic may flow over a congested path. Another little subtle behavior that one can experience in this scenario is the situation where a topology change causes dynamic IP routing to "flip" all traffic onto a new fewer-hops lower-cost path that may be congested.

Can an IP switching solution offer any means for avoiding these seemingly unavoidable situations in a meshed campus backbone? This depends on the choice of IP switching solution. Something along the lines of IFMP or FANP still depends on dynamic IP routing for path selection and also requires an additional control protocol. Tag switching and ARIS are really designed for large ISP-type networks. In a campus one needs to stay with a solution that is inexpensive and open to choice from multiple vendor suppliers. If it is a given that Ethernet will be installed to the desktop, the fundamental requirement that arises is for a robust and dynamic backbone with multigigabit forwarding capacity and one that is capable of utilizing the entire backbone topology.

An IP switching solution that fits the bill quite nicely is an MPOA virtual router. First, it provides an aggregate multigigabit forwarding capacity that is scaled by adding an additional MPOA client at the edge or ATM switching capacity in the core. It is an open standard that is supported by a number of different vendors. It is Ethernet-port-dense by virtue of Ethernet switches with ATM uplinks (as well as other media), runs standard IP unicast and multicast routing protocols, and is fully interoperable with legacy routers. But perhaps the most important feature of the MPOA virtual router is that IP routed traffic is exposed to PNNI Phase I path selection. Recall that when PNNI computes a path, it factors in reachability to a particular destination, the available capacity and QoS characteristics of the path, and the traffic and QoS requirements of the actual data that will flow over the path. Once the path is computed, an implicit check on the validity of the path is performed as the connection setup request traverses the network and connection admission control (CAC) is performed at each switch along the path. Once the path through the backbone has been established, traffic will flow over that path for the duration of the flow. It will not be subject to the "flip" problem described above.

Of course more efficient backbone path selection does come at a price. First, one needs to deploy ATM switches in the backbone and convert frames to cells at the ingress and cells to frames at the egress. Second, there is a bit more work to be done in the control plane with the presence of not only IP routing protocols but also ATM Forum protocols and a separate address resolution protocol (MPOA/NHRP) that is needed to resolve IP and ATM addresses. But MPOA is a stable, open standard, and the approach of querying an address resolution server and then setting up an SVC has been done in various flavors (e.g., Classical IP, LANE) for a number of years now.

Table 16-1 contrasts the attributes of layer-3 switching and IP switching (MPOA) in their support of meshed campus backbones. However, as it the case with layer-2 switching, if the topology is relatively simple enough, a layer-3 switching solution would be sufficient.

In the WAN environment a whole host of new gigabit and even terabit routers are beginning to appear. These devices possess orders of magnitude greater forwarding capacities over existing high-speed routers and even switches. They of course run the same dynamic IP routing protocols that traditional routers use except these devices provide a rich assortment of different high-speed interfaces including ATM, Gigabit Ethernet, fiber channel, packet SONET, and more. With the introduction of more fiber and the deployment of dense wave division multiplexing (DWDM) equipment, bandwidth may not be the issue that it once was. And gigabit routers will be used to direct IP packets over these marvelously fast new data link technologies. But it must not be forgotten that an IP world means two things. First, our networks and individual hosts on those networks are addressable using an IP address—not some wavelength on the light spectrum. Second, one can only forward

TABLE 16-1

Layer-3 Switching versus IP Switching

Attribute	Layer-3 Switching	IP Switching (MPOA)
Full IP routing	Yes	Yes
Media	Ethernet (10/100/1000)	Ethernet, ATM, and others
Environment	Campus	Campus, WAN
QoS	No	Yes
Applications	IP	IP, ATM
Data path calculation	Based on IP routing and topology	Based on ATM routing and topology
Additional control protocols	No	Yes, ATM Forum protocols plus MPOA/NHRP
Single point of failure	No	Yes, MPOA Server

packets to those IP addressable entities using routers.

Does IP switching have any role in a network consisting of "photonic" IP routers with gigabit/terabit capacities? Maybe not so much for any performance gain but more so as a means of directing packets over non-default paths. Consider the mesh network of gigabit routers interconnected by packet SONET links illustrated in Fig. 16-3. With multi-gigabit forwarding rates, 90 + percent payload efficiency (of PPP over SONET) and none of the overhead of the overlay model, it would seem that all problems have been solved. A network provider can provision differentiated services policies at the edge of the network and service the packets (based on the DS field) in the core. This provides the internal simplicity and external service differentiation that network providers are seeking.

But as is the case with layer-3 switching in the campus, it is dynamic IP routing that is selecting the path through the network. If a transient high-volume burst suddenly arrives, there is the possibility that some packets will be dropped (or downgraded in service). For security or other policy-specific reasons it might be preferable to override dynamic IP routing and direct a particular customer's traffic over a specific nondefault path. Or maybe as a result of an external event, traffic patterns suddenly change and a topological subset is inundated with an excessive amount of network traffic. Clearly a means of supporting a nondefault explicit routing would be useful. As discussed in the previous chapter, it is very difficult to direct packets over a nondefault path using native IP techniques such as per-packet source-routing or per-router filters/static routes. MPLS even in this scenario with no ATM or frame relay enables the construction of nondefault explicit routes. Therefore, it is quite rea-

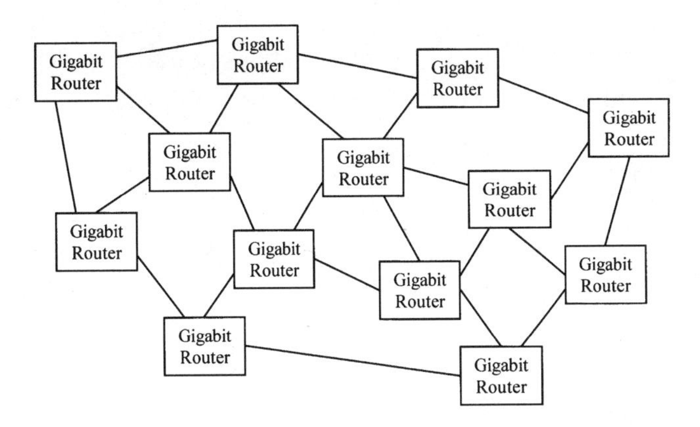

Figure 16-3 Gigabit routers in a packet Sonet mesh.

sonable to expect that MPLS would be utilized to support fast and efficient nondefault routing in such a network.

16.3 What Has IP Switching Done for Us?

There have been and are so many wonderful, creative, and brilliant ideas that the networking industry is exposed to on a not too infrequent basis. Some, like a whale, surface and through their sheer size and beauty move us to gape in awe. But after a few spellbound moments, a stream of mist emerges from a blowhole, and what follows is a loud crashing splash followed by calm seas. It is as if nothing ever happened. Others, like an alien cuisine, initially repulse us, but over time we grow to appreciate the taste and even love it. And still others, like a piece of contentious legislation, only appear in a watered down version after much backroom machinations, compromise, arm twisting, and implied threats.

What metaphor best fits IP switching? The whale, the alien dish, or the legislation? It's hard to say, but one could suggest that there are elements of all three in IP switching. Certainly it made a big splash when it first appeared in 1996. But the flow-driven approaches initially suggested by Ipsilon and Toshiba have since lost some allure due to scalability concerns. Some of the solutions, namely, MPOA and elements of both tag switching and ARIS, could be viewed as alien dishes, but many now have grown to handle the taste and even appreciate them. Of course the legislation bit is a direct analogy to the standardization process that is ongoing in the IETF with respect to MPLS.

In the final analysis IP switching is a very positive contribution to the evolution of internetworking. In particular it has given us the following:

- ✓ Proof that IP is the current method of networking and its future
- ✓ Explicit recognition that past and current IP routing technologies are slow
- ✓ Demonstration that ATM is and can be a very useful technology for forwarding IP packets
- ✓ An explicit means for labeling customer packets with "special service" indicators

- Lightweight control plane for building and managing ATM virtual connections
- More efficient and scalable mechanism for supporting routed IP traffic over ATM networks
- Better mechanism for explicit routing
- Indirect "push" for vendors to develop faster, cheaper routers

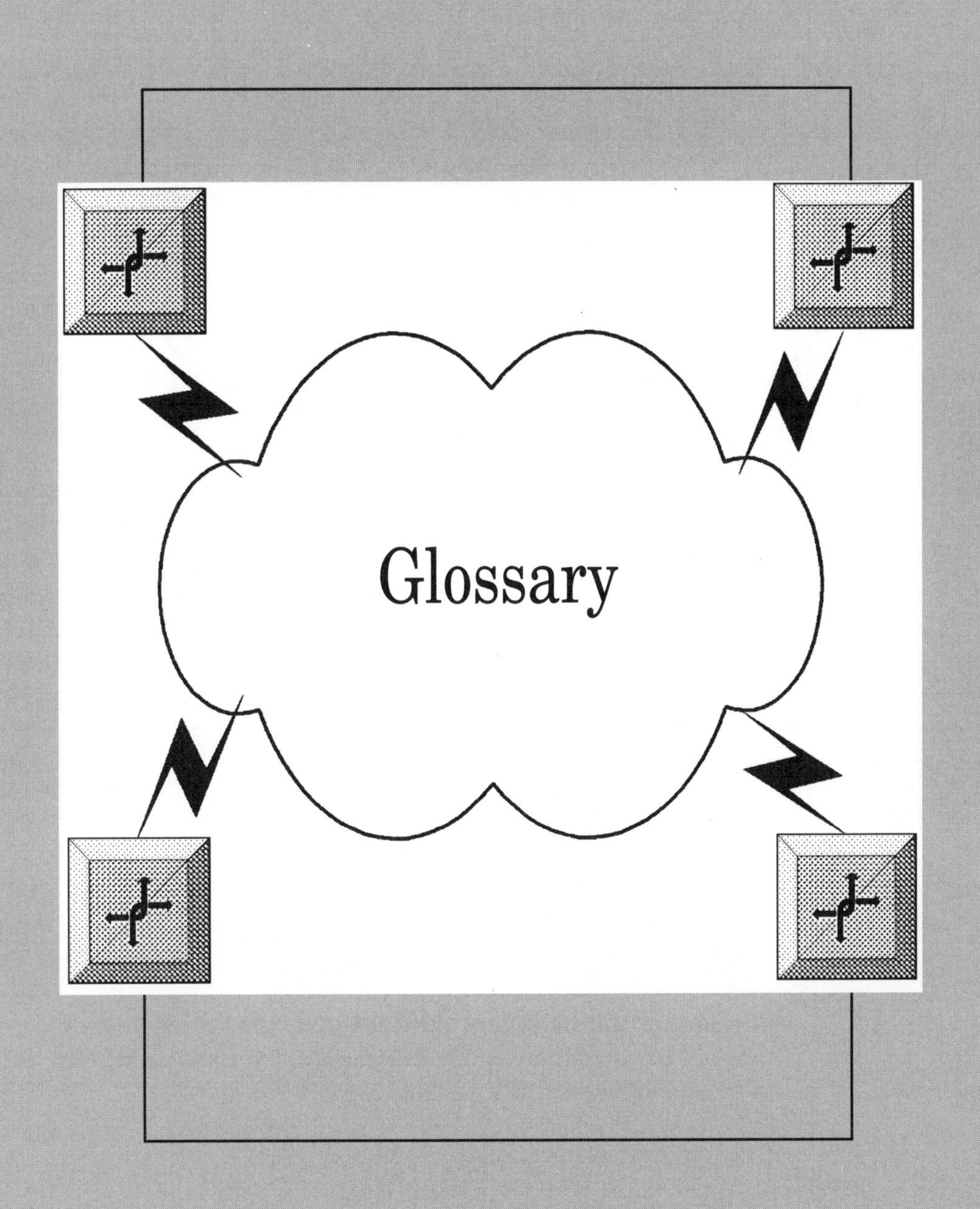
Glossary

802.1P. IEEE standard that defines techniques for supporting dynamic group multicast filtering and traffic prioritization in 802 LANs. The latter is accomplished by the use of a 3-bit user priority field contained in the 802.1Q VLAN tag.

802.1Q. IEEE standard that defines the method for supporting virtual LANs (VLANs) across 802 LANs. It works by inserting a 4- or 10-byte VLAN tag in each frame as it traverses one or more VLAN-capable bridges (switches).

AAL (ATM ADAPTATION LAYER). A collection of standardized protocols that provide services to higher layers by adapting user traffic to a cell format.

AAL1 (AAL TYPE 1). An AAL used for the transport of constant bit rate (CBR) traffic (i.e., audio and video) and for emulating TDM-based circuits (i.e., DS1, E1).

AAL2 (AAL TYPE 2). An AAL used for supporting time-dependent variable bit rate (VBR-RT) connection-oriented traffic (i.e., packetized video and audio).

AAL3/4 (AAL TYPES 3 AND 4). An AAL used for supporting both connectionless and connection-oriented variable bit rate (VBR) traffic. It is also used to support SMDS.

AAL5 (AAL TYPE 5). The most common AAL type used for the transport of data packets.

ABR (AVAILABLE BIT RATE). One of the two best-effort service types (the other is UBR), where the network makes no absolute guarantee of cell delivery; however it guarantees a minimum bit rate for user transmission. An effort is also made to keep cell loss as low as possible.

ACCESS RATE. The bit per second (bps) rate at which a user can transmit over the network's lines.

ACK (ACKNOWLEDGMENT). A message that acknowledges the reception of a transmitted packet. ACKs can be separate packets or piggybacked on reverse traffic packets.

ACR (ALLOWED, OR AVAILABLE, CELL RATE). The available bandwidth, in cells per second, for a given QoS class, which is dynamically controlled by the network.

AH (AUTHENTICATION HEADER). IPsec object that is inserted in each IP packet that provides data origin authentication and integrity.

ANSI (AMERICAN NATIONAL STANDARDS INSTITUTE). A U.S. technology standards organization.

ARIS (AGGREGATE ROUTE-BASED IP SWITCHING). A topology-driven IP switching solution proposed by IBM.

ARP (ADDRESS RESOLUTION PROTOCOL). A TCP/IP protocol used for resolving local network addresses by mapping a physical address (i.e., a MAC address) to an IP address.

ARPANET. One of the original Internet backbones.

ARS (ADDRESS RESOLUTION SERVER). A general term that describes a server function that (1) maintains a table/cache of LAN or network-layer addresses and associated ATM addresses and (2) responds to queries for this information from associated clients.

ASYNCHRONOUS (ASYNCHRONOUS TRANSMISSION). An approach for acquiring synchronization on a per-byte basis. Start and stop bits are used as delimiters.

ASYNCHRONOUS TRANSFER. An efficient approach for transmitting information where time slots are used on a demand basis (ATDM, ATM) rather than on a periodical one (TDM, STM).

ATM (ASYNCHRONOUS TRANSFER MODE). A broadband switching and multiplexing, connection-oriented, high-performance, and cost-effective integrated technology for supporting B-ISDN services (i.e., multimedia). Because no clock control is necessary, it is called asynchronous. (*See also* STM.) Information is transmitted at very high rates (up to hundreds of megabits per second) in fixed-size format packets called cells. Traffic streams are distinguished and supported according to different QoS classes.

ATM FORUM. Originally founded by a group of vendors and telecommunication companies, this formal standards body comprises various committees responsible for making recommendations and producing implementation specifications.

ATM LAYER. The second layer of the ATM protocol stack model that constructs and processes the ATM cells. Its functions also include Usage Parameter Control (UPC) and support of QoS classes.

ATMARP. Protocol and message formats that enable a client to request and receive resolution of a destination's IP address with an ATM address from an ATMARP server so that the client may establish an SVC to the destination. It only resolves addresses for devices attached to the same LIS. ATMARP is defined in RFC1577/2225 and collectively known as Classical IP and ARP over ATM.

ATMARP CLIENT. Function present on host or router that queries ATMARP server for address mappings and caches responses. It will establish SVCs to other devices on the same LIS.

ATMARP SERVER. Server component that maintains a table of IP/ATM mappings. It responds to queries from ATMARP clients and can run on a standalone device or in a route server or router.

BEC (BACKWARD ERROR CORRECTION). An error-correction scheme where the sender retransmits any data found to be in error, based on the feedback from the receiver.

BEST EFFORT. A QoS class in which no specific traffic parameters and no absolute guarantees are provided. Traditional IP offers a best-effort service, as do ATM ABR and UBR.

BT (BURST TOLERANCE). Proportional to the MBS, burst tolerance is used as a way (leaky bucket parameter) to check the conformance of the SCR.

BURSTINESS. A source traffic characteristic that is defined as the ratio of the peak cell rate (PCR) to the average cell rate. It is a measure of the intercell spacing. (*See also* MBS.)

BUS (BROADCAST AND UNKNOWN SERVER). A server that forwards multicast, broadcast, and unknown-destination address traffic to the attached LECs.

CAC (CONNECTION ADMISSION CONTROL). An ATM function that determines whether a virtual circuit (VC) connection request should be accepted or rejected.

CAT-3 (CATEGORY 3 UNSHIELDED TWISTED PAIR). A type of UTP commonly used with ATM interfaces for cell transmission at low speeds, 25 to 50 Mbps, and at distances up to 100 m.

CAT-5 (CATEGORY 5 UNSHIELDED TWISTED PAIR). A type of UTP commonly used with ATM interfaces for higher-speed cell transmission (more than 50 Mbps).

CBR (CONSTANT, OR CONTINUOUS, BIT RATE). One of the five ATM classes of service, which supports the transmission of a continuous bit stream of information where traffic, such as voice and video, needs to meet certain QoS requirements. (*See also* QoS classes.)

CBT (CORE BASED TREES). Multicast routing algorithm in which a single group-specific multicast delivery tree is rooted at a core router(s), and all senders for that group must forward packets through the core. PIM-SM and CBT are examples of multicast routing protocols that support the CBT algorithm.

CDV (CELL DELAY VARIATION). A QoS parameter that measures the difference between a single cell's transfer delay (CTD) and the expected transfer delay. It gives a measure of how closely cells are spaced in a virtual Circuit (VC). CDV can be introduced by ATM multiplexers (MUXs) or switches.

CDVT (CELL DELAY VARIATION TOLERANCE). Used in CBR traffic, it specifies the acceptable tolerance of the CDV (jitter).

CELL. A 53-byte packet, comprising a 5-byte header and a 48-byte payload. User traffic is segmented into cells at the source and reassembled at the destination.

CELL HEADER. The 5-byte ATM cell header contains control information regarding the destination path and flow control. More specifically it contains the following fields: GFC, VPI, VCI, PT, CLP, and HEC.

CER (CELL ERROR RATE). A QoS parameter that measures the fraction of transmitted cells that are erroneous (they have errors when they arrive at the destination).

CES (CIRCUIT EMULATION SERVICE). An ATM-provided class of service, where TDM-type constant-bit-rate (CBR) circuits are emulated by the AAL1.

CI (CONGESTION INDICATION). A bit in the RM cell to indicate congestion (it is set by the destination if the last cell received was marked).

CIR (COMMITTED INFORMATION RATE). A term used in Frame Relay; it defines the information rate the network is committed to provide the user with, under any network conditions.

CIRCUIT EMULATION. A virtual-circuit (VC) service offered to end users in which the characteristics of an actual, digital bit-stream (i.e., video traffic) line are emulated (i.e., a 2- or 45-Mbps signal).

CLASSICAL IP. IETF-defined protocols for developing IP over ATM networks (i.e., IP support for the QoS classes, ARP over SVC and PVC networks) so that common applications (i.e., FTP, Telnet, SMTP, SNMP) can be supported in an ATM environment. The main issues in the transport of IP over ATM are the packet encapsulation and the address resolution.

CLP (CELL LOSS PRIORITY). A 1-bit field in the ATM cell header that corresponds to the loss priority of a cell. Lower-priority (CLP = 1) cells can be discarded under congestion situations.

CLR (CELL LOSS RATIO). A QoS parameter that gives the ratio of the lost cells to the total number of transmitted cells.

CONGESTION CONTROL. A resource and traffic management mechanism to avoid and/or prevent excessive situations (buffer overflow, insufficient bandwidth) that can cause the network to collapse.

CONNECTIONLESS NETWORK. A communications service where packets are transferred from source to destination without the need for a preestablished connection. Examples are IP and SMDS.

CONNECTION-ORIENTED NETWORK. Communications service where an initial connection between the endpoints (source and destination) has to be set up. Examples are ATM and Frame Relay.

COS (CLASS OF SERVICE). The ability for a network to classify packets into one of several service classes and then handle those packets on a per-class priority basis.

CPCS (COMMON PART CONVERGENCE SUBLAYER). Part of the AAL convergence sublayer (CS). It always has to be present in the AAL implementation. Its task is to pass primitives to the other AAL sublayers (SAR, SSCS). It supports the functions of the standardized Common Part AALs: AAL1, AAL3/4, and AAL5.

CPE (CUSTOMER PREMISES EQUIPMENT). Computer and communications equipment (hardware and software) used by a carrier's customer and located at the customer's site. (*See also* DTE.)

CPI (COMMON PART INDICATOR). A 1-byte field in the header of the CPCS-PDU in AAL3/4 that indicates the number of bits the BASize field consists of.

CRC (CYCLIC REDUNDANCY CHECK). A bit-error detection technique that employs a mathematical algorithm, where, based on the transmitted bits, it calculates a value attached to the information bits in the same packet. The receiver, using the same algorithm, recalculates that value and compares it to the one received. If the two values do not agree, the transmitted packet is then considered to be in error.

CS (CONVERGENCE SUBLAYER). The upper half of the AAL. It is divided into two sublayers, the Common Part (CPCS) and the Service Specific (SSCS). It is service-dependent, and its functions include manipulation of cell delay variation (CDV), source clock frequency recovery, and forward error correction (FEC). Although each AAL has its own functions, in general the CS describes the services and functions needed for conversion between ATM and non-ATM protocols. (*See also* SAR.)

CS-PDU (CONVERGENCE SUBLAYER PROTOCOL DATA UNIT). The PDU used at the CS for passing information between the higher layers and the SAR, where they are converted into cells.

CSR (CELL SWITCH ROUTER). A device defined in the CSR/FANP architecture that is capable of forwarding packets at layer 3 and switching ATM cells at layer 2.

CSU (CHANNEL SERVICE UNIT). Equipment at the user end that provides an interface between the user and the communications network. A CSU can be combined with a DSU in the same device (*see* DCE)

CTD (CELL TRANSFER DELAY). A QoS parameter that measures the average time it takes a cell to be transferred from its source to its destination over a VC. It is the sum of any coding, decoding, segmentation, reassembly, processing, and queueing delays.

DATAGRAM. A packet transport mode where packets are routed independently and may follow different paths; thus there is no guarantee of sequence delivery. (*See also* VC.)

DCE (DATA CIRCUIT-TERMINATING EQUIPMENT OR DATA COMMUNICATIONS EQUIPMENT). Device at the user end, typically a modem or other communications device, which acts as an access point to the transmission medium.

DIFFSERV (DIFFERENTIATED SERVICES). IETF effort to define a technique for supporting COS over IP networks. It uses packet marking at the edge of the network and per-priority queuing in network devices to prioritize individual packets as they are forwarded through the network.

DS-0 (DIGITAL SIGNAL 0). Physical interface for digital transmission at the rate of 64 Kbps.

DS-1 (DIGITAL SIGNAL 1). Physical interface for digital transmission at the rate of 1.544 Mbps. Also known as a T-1 standard, it can simultaneously support 24 DS-0 circuits.

DS-3 (DIGITAL SIGNAL 3). Physical interface for digital transmission at the rate of 44.736 Mbps.

DSU (DATA SERVICE UNIT). Equipment at the user end that acts as a telephony-based interface between low-rate (i.e., 56 Kbps) services and higher-rate circuits.

DTE (DATE TERMINAL EQUIPMENT). The host computer (PC or workstation) to provide the end user with access to a communications network. The DTE is connected to a DCE that performs the signaling operation. (*See also* CPE.)

DVMRP (DISTANCE VECTOR MULTICAST ROUTING PROTOCOL). An RPM-based multicast routing protocol, so named because it utilizes distance vector routing mechanisms (e.g., split horizon, distance vector routing table) in its operation. It is also capable of supporting unicast tunnels through networks of routers that are not multicast-capable.

DWDM (DENSE WAVE DIVISION MULTIPLEXING). Capability for a single optical fiber to simultaneously carry a large number of multigigabit traffic streams (e.g., greater than 32) by multiplexing each over a separate light wavelength.

DXI (DATA EXCHANGE INTERFACE). A frame-based ATM interface between a DTE (such as a router or a local switch) and a DCE. DXI interfaces to the ATM UNI and has been chosen by the ATM Forum as an affordable solution for providing ATM capabilities over WAN.

E.164. An ITU-T-defined 8-byte address format. In ATM it is typically used in public networks and is provided by the telecommunication carriers, while 20-byte NSAP format addresses are used within private networks.

E-1 (EUROPEAN DIGITAL SIGNAL 1). European standard for digital physical interface at 2.048 Mbps.

E-3 (EUROPEAN DIGITAL SIGNAL 3). European standard for digital physical interface at 34.368 Mbps. It can simultaneously support 16 E-1 circuits.

EFCI (EXPLICIT FORWARD CONGESTION INDICATION). A 1-bit field in the PTI that contains information about whether congestion at an intermediate node has been experienced. The EFCI bit is set when, for example, a buffer threshold has been exceeded.

EGP (EXTERIOR GATEWAY PROTOCOL). Another name for interdomain or inter-AS routing protocols. Examples are EGP and BGP.

ELAN (EMULATED LAN). ATM Forum LAN Emulation term for a group of LE clients connected to the same LES/BUS.

ER (EXPLICIT RATE). A field in the RM cell header specifying the cell rate a user should use for transmission over a virtual connection (VC), as dictated by the RM.

ESP (ENCAPSULATING SECURITY PAYLOAD). An IPsec object that is inserted in each IP packet that provides data confidentiality, authentication, and integrity.

FANP (FLOW ATTRIBUTE NOTIFICATION PROTOCOL). Control protocol used between Toshiba CSR devices in which an upstream CSR device provides a downstream CSR device with flow/label mapping information. It may also include the exchange of a VCID value if the upstream and downstream CSR devices need to communicate through native ATM switches.

FDDI (FIBER DISTRIBUTED DATA INTERFACE). An ANSI-defined standard for implementing a high-speed (100 Mbps) LAN over fiber.

FEC (FORWARD ERROR CORRECTION). An error-correction technique where there are no retransmissions, and therefore the receiver is responsible for correcting any errors in the packets.

FEC (FORWARDING EQUIVALENCE CLASS). A group of IP packets that are forwarded over the same path, treated in the same manner, and can be mapped to a single label.

FIB (FORWARDING INFORMATION BASE). Another name for a routing table that consists of destinations, next-hop addresses, and interfaces.

FLOW CONTROL. A method used in networking for congestion avoidance and traffic regulation. It has three techniques: window-based control, where a sliding window is used to determine how many cells can be transmitted during a predefined period; rate-based control, where the rate at which the source can transmit is monitored and controlled; and credit-based control, where a source can transmit a cell if there is a credit available. CAC is also part of the flow control.

FLOW LABEL. A 20-bit field in the IPv6 header used to identify packets belonging to a specific flow.

FLOW TYPE. Defines the type of flow (e.g., host-host, appl-appl) that will be mapped to a VC in Ipsilon IP switching.

FLOW-DRIVEN IP SWITCHING. An IP switching solution that builds shortcut paths based on the arrival of individual IP flows. IFMP, FANP, and MPOA are examples of flow-driven IP switching protocols.

FRAME RELAY. A packet-switching technology to provide a very reliable packet delivery over virtual circuits (VC). Some of the concepts used in Frame Relay have been incorporated in ATM networks.

FRTE (FRAME RELAY TERMINATING EQUIPMENT). A function on an internetworking device (e.g., router, FRAD) that terminates a Frame Relay VC.

FTP (FILE TRANSFER PROTOCOL). A protocol used for transferring files between different machines across a network.

GARP (GROUP ADDRESS REGISTRATION PROTOCOL). A receiver-driven protocol for registering attributes across a network of GARP-capable devices. The attributes can consist of multicast addresses or in the case of VLANs, membership information by VLAN. GARP is defined in the IEEE 802.1Q specification.

GBPS (GIGABITS PER SECOND). Transmission speed or rate of a hundred million bits per second.

GCRA (GENERIC CELL RATE ALGORITHM). A reference model proposed by the ATM Forum for defining cell-rate conformance in terms of certain traffic parameters. It is usually referred as the Leaky Bucket algorithm.

GFC (GENERIC FLOW CONTROL). A 4-bit field in the ATM cell header that supports multiplexing functions. Its default value is '0000' when the GFC protocol is not enforced. The GFC mechanism is intended to support simple flow control in ATM connections.

GMRP (GARP MULTICAST REGISTRATION PROTOCOL). GMRP is an application that enables end systems and switches to register and deregister multicast group membership information and for that information to be disseminated throughout the network. GMRP is defined in the IEEE 802.1p specification.

GVRP (GARP VLAN REGISTRATION PROTOCOL). GVRP is used by VLAN-aware devices (e.g., workstations and switches) to dynamically register and distribute VLAN membership information across a network of VLAN-aware devices. This enables a collection of VLAN switches to establish and update which VLANs are active and through which switch ports they are reachable. GVRP is defined in the IEEE 802.1Q specification.

HEC (HEADER ERROR CHECK OR HEADER ERROR CONTROL). A 1-byte field in the cell header used for header error correction and detection. Due to the information contained in the header, HEC is quite significant.

HOL (HEAD-OF-LINE). The head position of a buffer (i.e., inside a switch). A blocking phenomenon is associated with the HOL that refers to the fact that cells in the queue have to wait for the HOL cell to depart first.

IEEE (INSTITUTE OF ELECTRICAL AND ELECTRONIC ENGINEERS). A standards and specification organization that has extensive activities in the areas of computers and electronics.

IETF (INTERNET ENGINEERING TASK FORCE). A body that was initially responsible for developing specifications required for the interoperable implementation of IP.

IFMP (IPSILON FLOW MANAGEMENT PROTOCOL). Control protocol used between Ipsilon IP switches in which the downstream IP switch instructs the upstream IP switch to relabel the cells of a particular flow with a new VPI/VCI label. IFMP is defined in RFC1953 and implemented on Ipsilon IP switches and edge devices.

IGMP (INTERNET GROUP MANAGEMENT PROTOCOL). Group membership protocol that operates on hosts and routers supporting IP multicast.

IGP (INTERIOR GATEWAY PROTOCOL). Another name for intradomain or intra-AS routing protocols. Examples are RIP, OSPF, and IGRP.

IISP (INTERIM INTERSWITCH SIGNALING PROTOCOL). A protocol that uses UNI-based signaling and static address prefix tables to forward SVC requests through a network of ATM switches.

ILMI (INTERIM LOCAL MANAGEMENT INTERFACE). An ATM Forum-defined Network Management System (NMS) based on SNMP that can provide configuration, performance, and fault management information concerning virtual circuit (VC) connections available at its UNI (public and private). It operates over AAL3/4 and AAL5 and will be eventually replaced once it becomes standardized by ITU-T.

INTSERV (INTEGRATED SERVICES). IP model that supports best-effort and real-time traffic flows. Its components consist of setup protocol (e.g., RSVP), traffic control, traffic classes, and flow specification. INTSERV is defined in RFCs 2205 through 2216.

IP (INTERNET PROTOCOL). A networking protocol for providing a connectionless service to the higher transport protocol. It is responsible for discovering and maintaining topology information and for routing packets across homogeneous or heterogeneous networks. Combined with TCP, it is commonly known as the TCP/IP Protocol.

IP MULTICAST. IP network transport service in which packets addressed to a group address are delivered by routers to those networks with group members. A group membership protocol (IGMP) is used by hosts to tell routers which multicast group they wish to join/leave, and the routers run a multicast routing protocol to build a delivery tree from the source's network(s) out to all networks that have group members.

IP SWITCH. A device or system that can forward IP packets at layer 3 and possesses a switching component(s) that enables packets to be switched at layer 2 as well. An IP switch possesses mechanisms to classify which packets will be forwarded at layer 3 and which will be switched at layer 2 and then to redirect some or all packets over a layer-2 switched path. Most IP switches utilize an ATM switching fabric, but other layer-2 switching technologies can be used as well.

IP SWITCHING. General term for a suite of technologies that uses layer-2 switching to accelerate the forwarding of IP packets.

I-PNNI (INTEGRATED PNNI). PNNI extension that integrates IP and ATM routing into a single routing protocol based on PNNI.

IPNG (IP NEXT GENERATION). Another name for IPv6.

IPSEC (IP SECURITY). Defines the protocols and mechanisms that support network-layer security for IPv4 and IPv6. Enables security to be embedded within each individual IP packet. It supports data origin authentication and confidentiality and is based on a technology that enables secure channels to be formed over unsecure networks so that Virtual Private Networks can be built on top of the public Internet.

IPX. Network-layer protocol developed by Novell and used for Novell client/server communications.

ISAKMP/OAKLEY. Internet Key Exchange (IKE) framework that supports the dynamic generation, distribution, and exchange of shared secret keys between communicating IPsec entities.

ISDN (INTEGRATED SERVICES DIGITAL NETWORK). An early CCITT-adopted protocol reference model intended to provide a ubiquitous, end-to-end, interactive, digital service for data, audio, and video.

ISOCHRONOUS. Refers to the fact that a time slot can be divided into equal-size mini slots allocated to different channels for synchronous transmission of information (used in DQDB).

ISSLL (INTEGRATED SERVICES OVER SPECIFIC LINK LAYERS). IETF working group that is developing guidelines on supporting IntServ on top of specific link layers such as ATM, 802 LANs, and low-bit-rate links.

ITU-T (INTERNATIONAL TELECOMMUNICATIONS UNION-TELECOMMUNICATIONS STANDARDS SECTOR). A formal international standards, specifications, and recommendations body, formerly known as CCITT. ITU-T is part of the International Telecommunications Union (ITU) founded in 1948 and sponsored by the U.N. to promote telephone and telegraphy issues.

JITTER. Cell or packet delay variation.

LABEL. A short fixed-length physically contiguous value that is used to identify a stream. The format of a label depends on the media that the packet is encapsulated in. For example, ATM-encapsulated packets (cells) use the VPI and/or VCI values as the label and Frame Relay PDUs use the DLCI as the label. For those media encapsulations with no native label structure, a special 4-byte shim value is used.

LABEL DISTRIBUTION PROTOCOL (LDP). MPLS control protocol that is used to communicate FEC/label bindings between LSRs.

LABEL EDGE ROUTER (LER). A traditional router that is capable of forwarding packets to and from an MPLS domain. It can also communicate with interior MPLS LSRs to exchange FEC/label bindings.

LABEL INFORMATION BASE (LIB). The connection table maintained in an LSR (LER) that contains FEC/label bindings and associated port and media encapsulation information.

LABEL STACK. An ordered set of labels appended to a packet that enables it to implicitly carry information about more than one FEC that the packet belongs to and the corresponding LSP that the packet may traverse. A label stack enables MPLS to support a routing hierarchy (one label for the EGP and one label for the IGP) and to aggregate multiple LSPs onto a single "trunk" LSP.

LABEL SWITCH ROUTER (LSR). A device that supports MPLS and is capable of forwarding packets at layer 3 and label swapping packets at layer 2.

LABEL SWITCHED PATH (LSP). An ingress-to-egress switched path built through a series of LSRs to forward the packets of a particular FEC using a label-swapping forwarding mechanism.

LAG (LOGICAL ADDRESS GROUP). A collection of hosts and routers connected to a physical NBMA network that are capable of establishing shortcut paths with hosts and routers on different subnets.

LAN (LOCAL AREA NETWORK). A network that interconnects PCs, terminals, workstations, servers, printers, and other peripherals at a high speed over short distances (usually within the same floor or building). Various LAN standards have been developed, with Ethernet as the most widely used.

LAN EMULATION. A technique that specifies the interfaces and protocols needed for providing LAN-supported functionality and connectivity in an ATM environment so that legacy protocols can be interoperable with the ATM protocols, interfaces, and devices.

LANE. *See* LAN Emulation.

LEAKY BUCKET. A flow control algorithm in which cells are monitored to check whether they comply with the connection parameters. Nonconforming cells are either tagged (as violators) or dropped from the network. The analogy is taken from a bucket (memory buffer) with a hole in its bottom that allows the fluid (cells) to flow out at a certain rate. (*See also* GCRA; Traffic contract, UPC.)

LE-ARP (LAN EMULATION ARP). The ARP used in LAN Emulation for binding a requested ATM address to the MAC address.

LEC (LAN EMULATION CLIENT). Typically located in an ATM end system (i.e., ATM host, LAN switch); its task is to maintain address resolution tables and forward data traffic. It is uniquely associated with an ATM address.

LECS (LAN EMULATION CONFIGURATION SERVER). A server whose main function is to provide configuration information to an LEC (such as the ELAN it belongs to or its LES).

LES (LAN EMULATION SERVER). A server that provides support for the LAN Emulation address resolution protocol (LE-ARP). The LECs register their own ATM and MAC addresses with the LES. A LES is uniquely identified by an ATM address.

LIS (LOGICAL IP SUBNET). An IP subnet consisting of ATM-attached devices that share a common address prefix and can communicate with each using ATM PVCs or SVCs.

LLC/SNAP ENCAPSULATION. Encapsulation technique supported in RFC1483 in which the AAL frame is encapsulated in an 8-byte LLC/SNAP header that typically contains an identifier of the network layer protocol being transported. It enables packets of multiple protocols to be transported across a single VC.

LNNI (LAN EMULATION NETWORK NODE INTERFACE). Defines protocols for exchange of control information between LANE server components (LES, LECS, BUS).

LUNI (LAN EMULATION USER NETWORK INTERFACE). Specifies the UNI between an LEC and the network providing the LAN Emulation.

MARS (MULTICAST ADDRESS RESOLUTION SERVER). An address resolution protocol that resolves IP multicast group addresses with ATM addresses so that IP multicast can operate on top of an ATM network.

MBPS (MEGABITS PER SECOND). Transmission speed or rate of one million bits per second.

MBS (MAXIMUM BURST SIZE). A traffic parameter that specifies the maximum number of cells that can be transmitted at the peak rate (PCR).

MCR (MINIMUM CELL RATE). A parameter that gives the minimum rate that cells can be transmitted by a source over a virtual connection (VC).

MCS (MULTICAST SERVER). An entity that receives source-initiated multicast traffic and forwards out an MCS-rooted multicast tree.

MIB (MANAGEMENT INFORMATION BASE). A data structure that defines objects for referencing variables such as integers and strings. In general, it contains information regarding a network's management and performance (i.e., traffic parameters).

MOSPF (MULTICAST EXTENSIONS FOR OSPF). Functional extensions to OSPF that support the establishment of intra- and interarea multicast delivery trees.

MPEG (MOTION PICTURE EXPERTS GROUP). A video technology standard that specifies the digital encoding, transmission, and decoding protocols capable of presenting VCR-quality motion video.

MPLS (MULTIPROTOCOL LABEL SWITCHING). IETF effort to define a standardized technique for integrating routing and switching. It works by associating the packet of an FEC with a set of labels. At the network ingress the FEC is identified, the packet is labeled, and then an LSP is transversed to the network egress.

MPOA (MULTIPROTOCOL OVER ATM). A set of standards to support (distributed) routing protocols other than IP, (distributed) routing protocols. Developed on top of LANE and NHRP, it will support switches, route servers, and hosts all attached to an ATM network.

MULTIMEDIA. A way of presenting to the user a combination of different forms of information such as text, data, images, video, audio, graphics (i.e., videoconference).

MULTIPOINT-TO-POINT (MPT-PT) SWITCHED PATH. A name for an ARIS-built switched path that is rooted at the egress ISR and branches out to all ingress ISR devices. Traffic flows in the reverse direction from the leaves of the tree back toward the root, which is the egress ISR. MPT-PT trees make use of VP or VC merging to limit consumption of switched resources.

MUX (MULTIPLEXER). A local networking device that combines multiple streams of information so they can share a common physical medium.

NBMA NETWORK (NONBROADCAST MULTIACCESS NETWORK). A network that consists of devices attached to a common intrastructure but does not have any native broadcast capability. Examples are X.25, Frame Relay, and ATM.

NDIS (NETWORK DRIVER INTERFACE SPECIFICATION). Generic name for a device driver for NIC, which is independent of any hardware or software implementation.

NHC (NEXT HOP CLIENT). A client function on a host or router that queries an NHRP server.

NHRP (NEXT HOP RESOLUTION PROTOCOL). An inter-subnet address resolution protocol.

NHS (NEXT HOP SERVER). An NHRP server function that runs on a route server or a router.

NIC (NETWORK INTERFACE CARD OR CONTROLLER). The hardware communications interface (circuit board) required for the DTE (workstation, PC) to access the network (also called an Adapter Card).

N-ISDN (NARROWBAND INTEGRATED SERVICES DIGITAL NETWORK). Predecessor to the B-ISDN, N-ISDN encompasses the original standards for the ISDN.

NMS (NETWORK MANAGEMENT SYSTEM). Set of OAM&P functions for setting the required hardware and software parameters used in managing a network.

NNI (NETWORK NODE INTERFACE (OR NETWORK-TO-NETWORK INTERFACE)). ITU-T-specified standard interface between nodes within the same network. The ATM Forum distinguishes between two standards, one for private networks called PNNI and one for public networks known as public NNI.

OAM (OPERATIONS AND MAINTENANCE). Set of administrative and supervisory actions regarding network performance monitoring, failure detection, and system protection. Special-type cells are used to carry OAM-related information.

OC-N (OPTICAL CARRIER-N). ITU-T-specified physical interface for transmission over optical fiber at *n* times 51.84 Mbps (i.e., OC-3 is at 155.52 Mbps, OC-12 at 622.08 Mbps, OC-48 at 2.488 Mbps).

OCTET. Eight bits or 1 byte.

OVERLAY MODEL. This occurs when an IP routing topology operates on top of and independently of the underlying layer-2 (e.g., ATM) switched topology. It requires definition and maintenance of separate address spaces, separate routing protocols, and separate topologies.

PAR (PNNI AUGMENTED ROUTING). An extension to PNNI in which an instance of PNNI along with any other dynamic IP routing (e.g., OSPF) protocol operates on ATM-attached routers. PAR allows ATM-attached routers to participate in the PNNI topology distribution process with ATM switches.

PAYLOAD. Part of the ATM cell; it contains the actual information to be carried and occupies 48 bytes. (*See also* PTI.)

PCR (PEAK CELL RATE). A traffic parameter that characterizes the source and gives the maximum rate at which cells can be transmitted. It is calculated as the reciprocal of the minimum intercell interval (time between two cells) over a given virtual connection (VC).

PDU (PROTOCOL DATA UNIT). Term originally used in the OSI model to describe the primitive passed across different layers; it contains header, data, and trailer information and is also known as a message.

PEAK DURATION. A source traffic characteristic that gives the duration of a transmission at the peak cell rate (PCR). It is equivalent to the burst length (in cells).

PEER MODEL. This occurs when the network forwarding nodes operate a single IP topology. This model supports a single IP topology and a single IP address space and runs a single IP routing protocol.

PHY (PHYSICAL LAYER). The bottom layer of the ATM protocol reference model, it is divided into two sublayers, the Transmission Convergence (TC) and the Physical Medium (PM). It transmits the ATM cells over the physical interfaces that interconnect the ATM devices.

PIM (PROTOCOL INDEPENDENT MULTICAST). Umbrella name for two distinct multicast routing protocols: PIM Dense Mode and PIM Sparse Mode. PIM-DM is RPM-based and PIM-SM uses the concept of a rendezvous point (RP) to receive all source-rooted multicast traffic and forward out separate per-group RP-rooted multicast delivery trees.

PNNI (PRIVATE NETWORK NODE INTERFACE). Dynamic signaling and routing protocol used to forward SVC requests through a network of ATM switches.

PRI (PRIMARY RATE INTERFACE). An ISDN specification that provides twenty-three 64-Kbps B-channels and one 64-Kbps D-channel intended for use over a single DS1 or E-1 line.

PROXY PAR. PNNI extension that serves as a registration, query, and discovery mechanism for devices operating internetworking services over an ATM network. ATM-attached routers operate a client function that enables them to register their IP and ATM information with a directly attached proxy-PAR server function running on an ATM switch. PNNI is then used to propagate this information throughout the network and make it available to other ATM-attached routers if they wish to query for it.

PTI (PAYLOAD TYPE IDENTIFIER). A 3-bit cell header field for encoding information regarding the AAL and EFCI.

PVC [PERMANENT (OR PROVISIONED) VIRTUAL CONNECTION]. A virtual connection (VPC/VCC) provisioned for indefinite use in an ATM network, established by the network management system (NMS).

QOS (QUALITY OF SERVICE). The set of ATM performance parameters that characterize the traffic over a given virtual connection (VC). These parameters include the CLR, CER, CMR, CDV, CTD, and the average cell transfer delay.

RFC (REQUEST FOR COMMENT). Draft documents that contain proposed standards and specifications. RFCs are approved or just archived as historical recommendations.

ROUTING. A network management function that is responsible for forwarding packets from the source to their destinations. Numerous algorithms exist that satisfy various network topologies and requirements.

RPF (REVERSE PATH FORWARDING). Multicast forwarding algorithm that forwards packets out all interfaces except on the interface that takes the shortest path back to the network of the source (sender).

RPM (REVERSE PATH MULTICASTING). Functional extension to RPF algorithm that optimizes use of network resources by pruning links and interfaces that have no group members from the delivery tree.

RSVP (RESERVATION PROTOCOL). An IP signaling protocol used for reserving resources (e.g., bandwidth) through a network of routers and hosts. RSVP is a component of the IP Integrated Services model.

RTCP (RTP CONTROL PROTOCOL). Additional control protocol that provides session management information to the participants in a real-time session.

RTP (REAL-TIME TRANSPORT PROTOCOL). Application-layer framing protocol that provides a sequence number, payload-type identifier, time-stamp, and source identifier so that a source can transmit real-time data (e.g., voice, video) to a destination and the destination can play it back in the right sequence with the right timing.

RTT (ROUND-TRIP TIME). Round-trip time between a source and a device, such as a switch; it is usually measured in number of cells (which depends on the buffering capabilities of the device). It is used as a window in flow control.

SAAL (SIGNALING AAL SERVICE). Specific parts of the AAL protocol responsible for signaling. Its specifications, being developed by ITU-T, were adopted from N-ISDN.

SAR (SEGMENTATION AND REASSEMBLY). The lower half of the AAL. It inserts the data from the information frames into the cell. It adds any necessary header or trailer bits to the data and passes the 48-octet to the ATM layer. Each AAL type has its own SAR format. At the destination, the cell payload is extracted and converted to the appropriate PDU.

SAR-PDU (SEGMENTATION AND REASSEMBLY PROTOCOL DATA UNIT). The 48-octet PDU that the SAR sublayer exchanges with the ATM layer. It comprises the SAR-PDU payload and any control information that the SAR sublayer might add.

SCR (SUSTAINABLE CELL RATE). A traffic parameter that characterizes a bursty source and specifies the maximum average rate at which cells can be sent over a given virtual connection (VC). It can be defined as the ratio of the MBS to the minimum burst interarrival time.

SCSP (SERVER CACHE SYNCHRONIZATION PROTOCOL). An interserver cache synchronization and replication protocol. SCSP enables server-based protocols such as LANE, ATMARP, NHRP, MARS, and MPOA to distribute internetworking services across multiple servers.

SDH (SYNCHRONOUS DIGITAL HIERARCHY). A hierarchy that designates signal interfaces for very-high-speed digital transmission over optical fiber links. (*See also* SONET.)

SEAL (SIMPLE EFFICIENT ADAPTATION LAYER). The original name and recommendation for AAL5.

SERVICE TYPES. There are four service types: CBR, VBR, UBR, and ABR. CBR and VBR are guaranteed services, whereas UBR and ABR are described as best-effort services.

SHORTCUT PATH. A layer-2 switched path that is established between entities on different networks that bypasses any intermediate routing hops. The establishment and maintenance of shortcut paths is the primary functional objective of all IP switching protocols.

SMDS (SWITCHED MULTIMEGABIT DIGITAL SERVICE). A connectionless, internetworking service, based on 53-byte packets, that targets the interconnection of different LANs into a switched public network.

SMTP (SIMPLE MAIL TRANSFER PROTOCOL). The protocol standard developed to support electronic mail (email) services.

SNA (SYSTEMS NETWORK ARCHITECTURE). A host-based network architecture introduced by IBM, where logical channels are created between endpoints.

SNMP (SIMPLE NETWORK MANAGEMENT PROTOCOL). An IETF-defined standard for handling management information. It is normally found as an application on top of the user datagram protocol (UDP).

SONET (SYNCHRONOUS OPTICAL NETWORK). An ANSI-defined standard for high-speed, high-quality digital optical transmission. It has been recognized as the North American standard for SDH.

SS7 (SIGNALING SYSTEM NUMBER 7). A common channel signaling standard developed by CCITT. It was designed to provide the internal control and network intelligence needed in ISDNs.

SSCOP (SERVICE SPECIFIC CONNECTION-ORIENTED PROTOCOL). Part of the SSCS portion of the SAAL . SSCOP is an end-to-end protocol that provides error detection and correction by retransmission and status reporting between the sender and the receiver, while it guarantees delivery integrity.

STM (SYNCHRONOUS TRANSFER MODE). A packet-switching approach in which time is divided into time slots assigned to single channels during which users can transmit periodically. Basically, time slots denote allocated (fixed) parts of the total available bandwidth. (*See also* TDM.)

STM-1 (SYNCHRONOUS TRANSPORT MODULE-1). An ITU-T-defined SDH physical interface for digital transmission in ATM at the rate of 155.52 Mbps.

STM-N (SYNCHRONOUS TRANSPORT MODULE-N). An ITU-T-defined SDH physical interface for digital transmission in ATM at *n* times the basic STM-1 rate. There is a direct equivalence between the STM-*n* and the SONET STS-3*n* transmission rates.

STP (SHIELDED TWISTED PAIR). Two insulated copper wires twisted together and wrapped by a protective jacket shield. (*See also* UTP.)

STREAM. A stream of packets that are forwarded over the same path and treated in the same manner. A stream may consist of one or more flows. In the MPLS architecture a stream is identified by a stream member descriptor (SMD).

STS-1 (SYNCHRONOUS TRANSPORT SIGNAL-1). SONET signal standard for optical transmission at 51.84 Mbps. (*See also* OC-1.)

STS-N (SYNCHRONOUS TRANSPORT SIGNAL-N). SONET signal format for transmission at *n* times the basic STS-1 signal (i.e., STS-3 is at 155.52 Mbps).

SVC (SWITCHED VIRTUAL CONNECTION). A connection that is set up and taken down dynamically through signaling.

T1. A TDM digital channel carrier that operates at a rate of 1.544 Mbps.

T3. A TDM digital channel carrier that operates at 44.736 Mbps. It can multiplex 28 T1 signals, and it is often used to refer to DS-3.

TAG SWITCHING. Topology-driven IP switching solution proposed by Cisco.

TAXI (TRANSPARENT ASYNCHRONOUS TRANSMITTER/RECEIVER INTERFACE). An interface that provides connectivity over multimode fiber links at a speed of 100 Mbps.

TCP (TRANSMISSION CONTROL PROTOCOL). A reliable, connection-oriented transport protocol used on IP-based devices.

TCP/IP. The internetworking protocol suite used to interconnect hosts and networks on the Internet and corporate intranets.

TDM (TIME-DIVISION MULTIPLEXING). A technique for splitting the total bandwidth (link capacity) into several channels to allow bit streams to be combined (multiplexed). The bandwidth is allocated by dividing the time axis into fixed-length slots, and a particular channel can then transmit only during a specific time slot.

TDP (TAG DISTRIBUTION PROTOCOL). Control protocol used between Tag Switching devices that distributes route/tag bindings

TM (TRAFFIC MANAGEMENT). Means for providing connection admission (CAC) and congestion and flow control (i.e., UPC, traffic shaping).

TOPOLOGY-DRIVEN IP SWITCHING. An IP switching solution that builds shortcut paths based on the presence of entries in a routing table. ARIS and Tag Switching are examples.

TOS (TYPE OF SERVICE). Eight-bit field contained in the IPv4 header that will be redefined by the IETF DiffServ working group to specify an individual packet's service priority. It has been renamed the DS byte (or field).

TRAFFIC CLASS. IntServ term for Gauranteed Service, Controlled Load, and best-effort traffic classes.

TRAFFIC CONTRACT. An agreement between the user and the network management agent regarding the expected QoS provided by the network and the user's compliance with the predetermined traffic parameters (i.e., PCR, MBS, burstiness, average cell rate).

TRAFFIC DESCRIPTORS. A set of parameters that characterize the source traffic. These are the PCR, MBS, CDV, and SCR in ATM and the Tspec/Rspec in IP IntServ.

TRAFFIC SHAPING. A method for regulating noncomplying traffic (i.e., violates the traffic parameters, such as PCR, CDV, and MBS as specified by the traffic contract).

UBR (UNSPECIFIED BIT RATE). Best-effort service in which hosts may transmit up to a peak cell rate if available. Makes no guarantees on bandwidth, delay, or other QoS parameters.

UDP (USER DATAGRAM PROTOCOL). A connectionless transport protocol without any guarantee of packet sequence or delivery. It functions directly on top of IP.

UNI (USER-NETWORK INTERFACE). The interface, defined as a set of protocols and traffic characteristics (i.e., cell structure), between the CPE (user) and the ATM network (ATM switch). The ATM Forum specifications refer to two standards being developed, one between a user and a public ATM network, called public UNI, and one between a user and a private ATM network, called P-UNI.

UPC (USAGE PARAMETER CONTROL). A form of traffic control that checks and enforces the user's conformance with the traffic contract and the

QoS parameters. Commonly known as traffic policing, it is performed at the UNI level.

UTP (UNSHIELDED TWISTED PAIR). A twisted pair (copper) wire without any protective sheathing, used for short distances wiring (i.e., building).

VBR-NRT (VARIABLE BIT RATE-NONREAL TIME). One of the service types for transmitting traffic when timing information is not critical and which is characterized by the average and peak cell rates. It is well suited for long data packet transfers.

VBR-RT (VARIABLE BIT RATE-REAL TIME). One of the service types for transmitting traffic that depends on timing information and control and which is characterized by the average and peak cell rates. It is suitable for carrying traffic such as packetized (compressed) video and audio.

VC (VIRTUAL CHANNEL). Describes unidirectional flow of ATM cells between connecting (switching or end-user) points that share a common identifier number (VCI).

VC MERGE. Layer-2 switch function that enables multiple upstream VCs to merge into a single downstream VC without interleaving cells from different frames. A VC merge-capable switch must collect all the cells of a frame on the inbound VC and then transmit the frame's cells in a contiguous fashion with a new VCI value on the outbound link. The cells transmitted on the new VC must be grouped contiguously by frame so that the destination device can distinguish cells from different sources. VC merging may require special hardware or a per-VC buffering capability in the switch.

VC MULTIPLEXING ENCAPSULATION. Encapsulation technique supported in RFC1483 in which an AAL frame is transported natively between VC endpoints that originate or terminate in a network-layer instance. It requires a separate VC for each network-layer protocol.

VCI (VIRTUAL CHANNEL IDENTIFIER). A 16-bit value in the ATM cell header that provides a unique identifier for the virtual channel (VC) that carries that particular cell. It can be used as a label in the IP switching technologies.

VCID (VC IDENTIFIER). Unique value that is used to associate a specific VC with a flow. The VCID is needed if intermediate ATM switches operating between a pair of IP switching devices swap the VPI/VCI label.

VIRTUAL CIRCUIT. A connection set up across the network between a source and a destination where a fixed route is chosen for the entire session and bandwidth is dynamically allocated. (*See also* datagram.)

VIRTUAL CONNECTION. A connection established between end users (source and destination), where packets are forwarded along the same path, and bandwidth is not permanently allocated until it is used.

VIRTUAL IP SWITCH. A collection of switches augmented with an IP routing service. Its components consist of ingress and egress edge devices, a route server that runs routing protocols and provides default IP forwarding, and the switches, typically ATM. An example of a virtual IP switch is an MPOA virtual router.

VLAN (VIRTUAL LAN). A networking environment where users on physically independent LANs are interconnected in such a way that it appears as if they are on the same LAN workgroup.

VLSM (VARIABLE LENGTH SUBNET MASK). The ability for a single routing system to support subnet masks for variable lengths to best match the host address space capacity with the actual number of hosts on a subnet. RIPv2, OSPF, and E-IGRP are routing protocols that support VLSM.

VP (VIRTUAL PATH). A set of virtual channels (VCs) grouped together, between switches.

VP MERGE. Standard ATM switch function that enables multiple upstream Virtual Paths (VPs) to merge into a single downstream VP. The inbound and outbound VCI values remain unique and unchanged, so no special VC merging function is required on the switch. The unique VCI value is used to retain the identity of the source (e.g., ingress ISR). However, because the VCI values may share a common VPI, a unique range of VCI values must be communicated to the ingress ISR devices to prevent VCI collisions.

VPCI/VCI (VIRTUAL PATH CONNECTION IDENTIFIER/VIRTUAL CHANNEL IDENTIFIER). A combination of two numbers, one that identifies the VP and one that identifies the VCI. It is a label for IP switching solutions that operate over ATM data links.

VPI (VIRTUAL PATH IDENTIFIER). An 8-bit value in the cell header that identifies the VP and, accordingly, the VC the cell belongs to. VC can be used as a label in the IP switching technologies.

VPN (VIRTUAL PRIVATE NETWORK). A logical network overlayed on top of the public Internet. Intra-VPNs are secure channels that are formed using the IPsec protocols.

WAN (WIDE AREA NETWORK). A network that covers long-haul areas and usually utilizes public telephone circuits.

INDEX

ABOUT THE AUTHOR

Chris Metz is a recognized authority on IP and ATM switching technology, the co-author of McGraw-Hill's best-selling *ATM and Multiprotocol Networking* and of *Internetworking over ATM*, from the Prentice-Hall Redbook Series, and a regular lecturer at industry conferences. His principal areas of interest are integrated switching and routing, service differentiation, and multicast. He can be reached at metzy@ibm.net.